NONLINEAR DIFFERENTIAL EQUATIONS IN ORDERED SPACES

by

S. Carl and S. Heikkilä

CRC Press
Taylor & Francis Group
Boca Raton London New York

CRC Press is an imprint of the
Taylor & Francis Group, an **informa** business

CRC Press
Taylor & Francis Group
6000 Broken Sound Parkway NW, Suite 300
Boca Raton, FL 33487-2742

First issued in paperback 2019

© 2000 by Taylor & Francis Group, LLC
CRC Press is an imprint of Taylor & Francis Group, an informa business

No claim to original U.S. Government works

ISBN-13: 978-1-58488-068-4 (hbk)
ISBN-13: 978-0-367-39847-7 (pbk)

Library of Congress Cataloging-in-Publication Data

Catalog record is available from the Library of Congress

Visit the Taylor & Francis Web site at
http://www.taylorandfrancis.com

and the CRC Press Web site at
http://www.crcpress.com

Introduction

In order to give an idea of the methods and to prepare for the subject of this monograph, consider the following linear elliptic boundary value problem

$$-\Delta u + mu = h \quad \text{in } \Omega, \quad u = 0 \quad \text{on } \partial\Omega, \tag{1}$$

where m is a nonnegative constant. The solutions of (1) are order preserving, which means that if u_1 and u_2 are two solutions of (1) in some bounded domain $\Omega \subset \mathbb{R}^N$ corresponding to the data h_1 and h_2, respectively, then the following property holds:

$$h_1 \leq h_2 \quad \text{in } \Omega \quad \text{implies} \quad u_1 \leq u_2 \text{ in } \Omega.$$

This property which immediately implies a uniqueness result is usually called the *inverse monotonicity* of the operator $L = -\Delta + mI$, and follows easily from the maximum principle. It is a key property to apply the monotone iteration method to the nonlinear boundary value problem

$$-\Delta u = f(u) \quad \text{in } \Omega, \quad u = 0 \quad \text{on } \partial\Omega. \tag{2}$$

Assuming the existence of an upper solution $\bar{u}$ and a lower solution $\underline{u}$ of (2) such that $\underline{u} \leq \bar{u}$, the existence of extremal solutions, i.e., least and greatest solutions within the interval $[\underline{u}, \bar{u}]$ can be proved by the monotone iteration method, for instance, if the nonlinearity $f : \mathbb{R} \to \mathbb{R}$ is continuous and satisfies a one-sided Lipschitz condition

$$f(s_1) - f(s_2) \geq -m(s_1 - s_2), \quad s_2 \leq s_1. \tag{3}$$

The monotone iterative technique combined with the upper and lower solution method has widely been used as a powerful tool to get constructive existence results for different kind of problems for both ordinary and partial differential equations, cf. [165, 189]. By means of a generalized iteration principle this technique has been extended in the monograph [141] to deal with discontinuous differential equations, such as problem (2) with a nonlinearity f satisfying (3) but which need not be continuous. This

1

was possible, since due to the inverse monotonicity of the elliptic operator $-\Delta + mI$ problem (2) can be reformulated as a fixed point equation
$u = Gu$ with an increasing (not necessarily continuous) fixed point operator $G = (-\Delta + mI)^{-1} \circ (F + mI)$, where F denotes the Nemytskij operator
related with f. Many differential operators occurring in models of real phenomena are inverse monotone.

If the nonlinearity f in (2) is only supposed to be continuous, then the
BVP (2) cannot be reduced to a fixed point equation with an increasing
operator, in general, and thus the monotone iteration method cannot be
applied. However, assuming the existence of upper and lower solutions $\bar{u}$
and $\underline{u}$, such that $\underline{u} \leq \bar{u}$, one can use other methods to show the existence
of least and greatest solutions of (2) within the interval $[\underline{u}, \bar{u}]$ (cf. chapter
5). This existence result holds likewise for the BVP

$$Lu = h \quad \text{in } \Omega, \quad u = 0 \quad \text{on } \partial\Omega, \tag{4}$$

with a given right-hand side h, where $-Lu := \Delta u + f(u)$. Moreover, the
operator that assigns h to the least (greatest) solution of (4) can be proved
to be increasing. In this sense we may consider L to be inverse monotone
among its extremal solutions within some order interval of upper and lower
solutions. Once the operator L has this property, it can be used to treat
more general nonlinear problems

$$Lu = Nu \quad \text{in } \Omega, \quad u = 0 \quad \text{on } \partial\Omega, \tag{5}$$

where N may be a discontinuous operator depending on u and also on Lu,
so that (5) becomes a discontinuous implicit elliptic equation.

The main subject of the monograph will be ordinary and partial differential equations which can be represented as

$$Lu = Nu, \tag{6}$$

where L denotes some differential operator and N stands for the lower order
terms, and may depend also on Lu in the implicit case. The generality of
the problems under consideration lies in the fact that the nonlinearities of
the differential equations and corresponding initial and boundary conditions
are allowed to depend discontinuously on the solution of the problem. For
this purpose we first develop a theory for an abstract operator equation in
the form (6) which is based on fixed point results in ordered normed spaces,
and on the assumption that the operator L is inverse monotone among the
extremal solutions of the equation

$$Lu = h, \tag{7}$$

that is, these extremal solutions exist and are increasing with respect to h. In fact, the major part of this monograph is devoted to the study of extremal solutions of explicit differential equations which can be represented in the form (7).

By means of the method developed here we prove that the considered implicit and explicit problems have extremal solutions between the assumed lower and upper solutions, and study the dependence of these solutions on the data. We also present sufficient conditions for the existence of such lower and upper solutions that all the solutions lie between them, thus obtaining the least and the greatest of all the solutions of the problem in question. Uniqueness and well-posedness results, as well as conditions which ensure compactness of the solution set of some considered problems, are also presented. Theoretical and worked examples, as well as special cases are given to demonstrate the developed theory. As shown in many examples, the successive approximations combined with their numerical estimations can be used to infer the exact formula of the extremal solutions of discontinuous problems.

This monograph contains seven chapters and an appendix, which provides prerequisites used throughout the book to make it self-contained. Chapter 1 serving as a preparatory chapter is followed by the ODE part, which consists of chapters 2–4 dealing with first order and second order discontinuous implicit and explicit ordinary and functional differential equations. The PDE part, which consists of chapters 5–7 treats quasilinear elliptic and parabolic equations and inclusions of hemivariational type as well as implicit elliptic and parabolic problems. Both these parts can be read independently.

In chapter 1 we first present a fixed point theorem for an increasing mapping G in an ordered normed space by assuming that increasing (decreasing) sequences of the range of G possess weak limits and the common lower (upper) bound in the domain. The proof based on the generalized iteration method is given in the appendix. A consequence of this theorem is then applied to derive existence and comparison results for extremal solutions of equation (6). After considering cases where ordinary iteration methods are applicable we provide, as an introduction to forthcoming chapters, first applications for implicit ordinary and partial differential equations involving discontinuous nonlinearities.

In chapter 2 we prove existence and comparison results for extremal solutions of first order ordinary and functional differential equations. We shall first treat initial value problems of explicit ordinary differential equations. The obtained results are then extended to the case when the initial condition is replaced by a functional boundary condition. Applying also the results derived in chapter 1 for equation (6) we prove next extremality results for the explicit and implicit functional differential equations

$$\begin{cases} Lu(t) = h(t) \text{ and } Lu(t) = f(t, u(t), u, Lu(t)), \text{ where} \\ Lu(t) = \frac{d}{dt}\varphi(u(t)) - g(t, u(t), u), \end{cases}$$

equipped with functional boundary conditions. The function φ is assumed to be an increasing homeomorphism on $\mathbb{R}$, whereas g and f above, as well as the functions in the boundary condition, may depend discontinuously on the whole solution u. The function g need not even be sup-measurable.

Chapter 3 is devoted to comparison and uniqueness studies. We begin with first order explicit and implicit quasilinear initial and boundary value problems, proving also well-posedness results. Next we derive maximum principles for a differential operator combined by a generalized phi-Laplacian operator and a part containing lower order terms. These maximum principles are then applied to present uniqueness and comparison results for second order quasilinear differential equations of the form

$$-\frac{d}{dt}\varphi(t, u'(t)) = g(t, u(t), u'(t)),$$

equipped with separated, periodic, Neumann or Dirichlet boundary conditions. No continuity or measurability hypotheses are imposed on the functions φ and g.

In chapter 4 we prove existence and comparison results for second order ordinary and functional differential equations. We start with explicit functional Sturm-Liouville differential equations with separated boundary conditions. Next we consider the corresponding implicit functional equations with implicit boundary conditions, which both may contain discontinuous nonlinearities. Special cases where the existence of extremal solutions can be proved by the method of successive approximations are also considered. In the remaining part of chapter 4 we provide extremality and comparison results for explicit and implicit functional phi-Laplacian equations

$$\begin{cases} Lu(t) = h(t) \text{ and } Lu(t) = f(t, u, u', u'(t), Lu(t)), \text{ where} \\ Lu(t) = \frac{d}{dt}\varphi(u'(t)) - g(t, u, u', u'(t)), \end{cases}$$

with functional initial conditions. Both the differential equations and initial conditions may contain discontinuous functions. As in chapter 2 the function g above need not be sup-measurable.

In chapter 5 we prove the existence of extremal solutions within an interval of upper and lower solutions of the Dirichlet problem for general quasilinear elliptic equations of the form

$$-\sum_{i=1}^{N} \frac{\partial}{\partial x_i} a_i(x, u(x), \nabla u(x)) = f(x, u(x), \nabla u(x)),$$

as well as for the initial boundary value problem of its quasilinear parabolic counterpart. Moreover, we provide compactness results of the solution set enclosed by the given upper and lower solutions and treat discontinuous problems by applying an abstract fixed point result for increasing (not necessarily continuous) mappings in partially ordered sets. Even though only Dirichlet boundary conditions have been considered, other boundary conditions such as, e.g., nonlinear flux conditions, nonlocal and periodic boundary conditions can also be treated by the methods developed in this chapter.

Chapter 6 deals with differential inclusions of hemivariational type which can be considered as the multivalued versions of some hemivariational inequalities of the form

$$u \in X : \quad \langle Au - h, v - u \rangle + J^o(u; v - u) \geq 0 \quad \text{for all } v \in X,$$

where $J^o(u; v)$ denotes the generalized directional derivative in the sense of Clarke (cf. [96]) of a locally Lipschitz functional $J : X \to \mathbb{R}$, on a reflexive Banach space X and $A : X \to X^*$ is some pseudomonotone and coercive operator (see section D) satisfying certain continuity conditions. The field of hemivariational inequalities initiated with the pioneering work of Panagiotopoulos (cf., e.g., [187, 188]) has attracted increasing attention over the last years mainly due to its many applications in Mechanics and Engineering. This new type of variational inequalities arises, e.g., in mechanical problems governed by nonconvex, possibly nonsmooth energy functionals (so-called superpotentials), which appear if nonmonotone, multivalued constitutive laws are taken into account. The main goal of this chapter is to develop an appropriate method of upper and lower solutions for such kinds of differential inclusions, and to prove existence and enclosure results. In case the multivalued term is generated by a state-dependent subdifferential we are able to show the existence of *extremal* solutions within a sector of an ordered pair of appropriately defined upper and lower solutions. In recent papers of the first author, extremality results were obtained when the nonconvex superpotential J is given in form of a d.c.-functional, which can be represented as the difference of two convex functionals. Comments and notes in this direction can be found at the end of the chapter.

In chapter 7 we apply the results of chapters 1 and 5 to provide extremality results for discontinuous implicit elliptic problems, and for parabolic initial-boundary value problems of the form

$$\begin{cases} Lu(x,t) = h(x,t) \text{ and } Lu(x,t) = f(x,t,u(x,t),Lu(x,t)), \text{ where} \\ Lu(x,t) = \frac{\partial u(x,t)}{\partial t} - \sum_{i,j=1}^{N} \frac{\partial}{\partial x_i} \left(a_i(x,t)\frac{\partial u_i}{\partial x_j} \right) + g(x,t,u(x,t)). \end{cases}$$

The function f may be discontinuous in all its variables.

The appendix contains basics of the theory of partially ordered spaces, Sobolev spaces, pseudomonotone and quasilinear elliptic operators, and first order evolution equations, as well as fixed point results in ordered normed spaces used in the treatment of the considered problems. Furthermore, basic inequalities and facts from nonsmooth analysis are provided.

The contents of the monograph is mainly based on results obtained by the authors during the last few years. Due to the limited space many interesting problems could not be included such as, e.g., discontinuous ODE in ordered Banach spaces or discontinuous quasilinear elliptic equations and hemivariational inequalities in unbounded domains, which have been studied by the authors over the last years. To all these subjects we have provided references of recent works.

The monograph is intended to serve as a sourcebook for pure and applied mathematicians who model phenomena involving discontinuous changes, researchers of ODEs, FDEs, and PDEs, and as a textbook for upper level undergraduate and graduate students in these disciplines.

Acknowledgments

We are grateful to Professors Gary F. Roach and V. Lakshmikantham who encouraged us to write this monograph, to our colleagues who contributed to this work by giving helpful comments, and to the editorial staff of CRC Press, particularly to Dr. Sunil Nair, for efficient cooperation. We also acknowledge the support provided by the Deutsche Akademische Austauschdienst (DAAD), Deutsche Forschungsgemeinschaft (DFG), and by the Academy of Finland.

Contents

Chapter 1

Operator equations in ordered spaces and first applications

Our basic method in treating implicit problems is to represent them as an operator equation

$$Lu = Nu, \qquad (1.1.1)$$

where the dependence of the operator N on u is implicit, e.g., of the form $Nu = Q(u, Lu)$. For this purpose we study in this chapter the solvability of (1.1.1) in the case when L and N are mappings from a partially ordered set (poset) to an ordered normed space. No linearity or continuity hypotheses are imposed on the operators L and N, since no algebraic or topological structures are assumed for their domain. This general setting gives us tools to treat also implicit problems involving discontinuous nonlinearities.

In section 1.1 we present fixed point theorems in ordered spaces, and apply them to prove existence and comparison results for equation (1.1.1). Section 1.2, which serves as an introduction to the forthcoming chapters, contains elementary applications of the results of section 1.1 to implicit ordinary and partial differential equations governed by discontinuous nonlinearities. Concrete examples of such equations are also solved to describe the applicability of the given tools.

1.1 Operator equations in ordered spaces

In this section we first present a fixed point theorem for an increasing mapping G in an ordered normed space. The proof, based on a generalized iteration method, will be given in section A of the appendix. As a consequence we prove another fixed point result which is then applied to study the solvability of equation (1.1.1). Finally, we consider cases where ordinary iteration methods are available when domains and ranges of the operators L, N, and G are ordered topological spaces.

1.1.1 Fixed point results. Our first fixed point theorem, which forms a basis to other results of this section, will be needed also to study existence and comparison of weak solutions of partial differential equations.

Theorem 1.1.1. *Let P be a subset of an ordered normed space, $G \colon P \to P$ an increasing mapping, and $G[P] = \{Gx \mid x \in P\}$.*

a) If $G[P]$ has a lower bound in P and increasing sequences of $G[P]$ converge weakly in P, then G has the least fixed point x_*, and $x_* = \min\{x \mid Gx \leq x\}$.

b) If $G[P]$ has an upper bound and decreasing sequences of $G[P]$ converge weakly in P, then G has the greatest fixed point $x^* = \max\{x \mid x \leq Gx\}$.

The proof of Theorem 1.1.1, which is outlined in section 1.3, is given in section A of the appendix, where also basic concepts of ordered spaces are presented. For instance, an ordered normed space X is an *ordered topological space* with respect to both weak and strong topology. This means that the sets $[y) = \{x \in X \mid y \leq x\}$ and $(z] = \{x \in X \mid x \leq z\}$, and hence also the order intervals

$$[y, z] = \{x \in X \mid y \leq x \leq z\}, \quad y, z \in X,$$

are both weakly and strongly closed. Moreover, strongly convergent sequences of X converge weakly, whence the next result is an immediate consequence of Theorem 1.1.1.

Proposition 1.1.1. *Let $[\underline{x}, \overline{x}]$ be a nonempty order interval in an ordered normed space X, and let $G \colon [\underline{x}, \overline{x}] \to [\underline{x}, \overline{x}]$ be an increasing mapping. If monotone sequences of $G[\underline{x}, \overline{x}]$ converge weakly or strongly in X, then G has the least fixed point x_* and the greatest fixed point x^*. Moreover,*

$$x_* = \min\{x \mid Gx \leq x\} \quad and \quad x^* = \max\{x \mid x \leq Gx\}. \tag{1.1.2}$$

Proposition 1.1.1 will be applied to prove our main existence and comparison theorem for equation (1.1.1). The following consequence is used in the study of explicit ordinary and functional differential equations.

Corollary 1.1.1. *Let $[a, b]$ be a closed interval in $\mathbb{R}$, and $[\underline{x}, \overline{x}]$ an order interval in the pointwise ordered space $C([a, b])$ of continuous functions $x \colon [a, b] \to \mathbb{R}$. If $G \colon [\underline{x}, \overline{x}] \to [\underline{x}, \overline{x}]$ is an increasing mapping, and if monotone sequences of $G[\underline{x}, \overline{x}]$ have pointwise limits in $C([a, b])$, then G has least and greatest fixed points x_* and x^*, and (1.1.2) holds.*

Proof. The given hypotheses and Dini's theorem ensure that the monotone sequences of $G[\underline{x}, \overline{x}]$ converge uniformly on $[a, b]$. Thus the hypotheses of Proposition 1.1.1 hold when $X = C([a, b])$ is ordered pointwise, or equivalently, by the cone $C_+([a, b])$, and is normed by $\|x\| = \sup\limits_{a \leq x \leq b} |x(t)|$. □

1.1.2 On extremal solutions of equation (1.1.1). In this subsection we study the existence of *extremal solutions*, i.e., least and greatest solutions, of equation (1.1.1). In order to find hypotheses for the operators L and N which are relevant in the study of implicit equations, consider the following example.

Example 1.1.1. Let $[a, b]$ be a closed interval in $\mathbb{R}$. Show that the equation

$$Lu = Q(u, Lu) \tag{1.1.3}$$

has extremal solutions if $L : [a, b] \to \mathbb{R}$ is continuous, if $Q : [a, b] \times \mathbb{R} \to \mathbb{R}$ is increasing in its both arguments, and if $Q[[a, b] \times L[a, b]] \subseteq [La, Lb]$.

Denoting $V = [a, b]$ and $X = \mathbb{R}$, and defining

$$Nu := Q(u, Lu), \quad u \in V, \tag{1.1.4}$$

we obtain operators $L, N : V \to X$ such that equation (1.1.3) is of the form (1.1.1). The given hypotheses imply that the asserted results follow from the next theorem by setting $h_- = La$ and $h_+ = Lb$.

Theorem 1.1.2. *Given an ordered normed space X, a poset V, mappings $L, N : V \to X$, and elements $h_\pm$ of X, $h_- \le h_+$, assume that the following hypotheses are valid.*

(I) *If $u, v \in V$, $u \le v$ and $Lu \le Lv$, then $h_- \le Nu \le Nv \le h_+$.*

(II) *Equation $Lu = h$ has for each $h \in [h_-, h_+]$ extremal solutions, and they are increasing with respect to h.*

(III) *Monotone sequences of $N[V]$ converge weakly or strongly in X.*

Then equation (1.1.1) has extremal solutions, and they are increasing with respect to N.

Proof. The hypotheses (I) and (II) ensure that the relation

$$Gx := Nu, \quad \text{where } u \text{ is the least solution of equation } Lu = x, \tag{1.1.5}$$

defines an increasing mapping $G : [h_-, h_+] \to [h_-, h_+]$. Since monotone sequences of $G[h_-, h_+]$ are contained in $N[V]$ by (1.1.5), they have weak or strong limits in X by (III). Thus G satisfies the hypotheses of Proposition 1.1.1 when $[\underline{x}, \overline{x}] = [h_-, h_+]$, so that G has the least fixed point x_*. Denoting by u_* the least solution of $Lu = x_*$, then $Lu_* = x_* = Gx_* = Nu_*$, whence u_* is a solution of equation (1.1.1).

To prove the remaining assertions, assume that the hypotheses (I) and (III) hold also for a mapping $\hat{N} \colon V \to X$, and that

$$Nu \le \hat{N}u \quad \text{for all } u \in V. \tag{1.1.6}$$

The above proof shows the existence of $v \in V$ such that $Lv = \hat{N}v$. Denoting by $\hat{u}$ the least solution of $Lu = x := \hat{N}v$, we have $\hat{u} \le v$ and $L\hat{u} = Lv$. These relations and (I) imply that $\hat{N}\hat{u} \le \hat{N}v$. Applying this inequality, (1.1.5), and (1.1.6), we obtain $Gx = N\hat{u} \le \hat{N}\hat{u} \le \hat{N}v = x$. In view of (1.1.2) we then have $x_* \le x$, whence $u_* \le \hat{u}$ by (II). Since $\hat{u} \le v$, then $u_* \le v$, where v can be any solution of $Lu = \hat{N}u$, and hence any solution of $Lu = Nu$. Consequently, u_* is the least solution of (1.1.1), and $u_* \le \hat{u}_*$, where $\hat{u}_*$ is the least solution of $Lu = \hat{N}u$, which implies that the least solution of (1.1.1) is increasing with respect to N.

The proof that equation $Lu = Nu$ has the greatest solution which is increasing with respect to N is similar. $\qquad\qquad\qquad\qquad\qquad\qquad\square$

The following consequence of Theorem 1.1.2 will be used in the study of discontinuous implicit ordinary and partial differential equations.

Corollary 1.1.2. *Given a poset V, a Lebesgue measurable subset Ω of $\mathbb{R}^N$, and $p \ge 1$. Let $L^p(\Omega)$ be ordered almost everywhere (a.e.) pointwise, i.e., $u \le v$ iff $u(x) \le v(x)$ a.e. in Ω. If the operators L, $N \colon V \to L^p(\Omega)$ satisfy the hypotheses (I) and (II) of Theorem 1.1.2 for some $h_\pm \in L^p(\Omega)$, then equation (1.1.1) has extremal solutions, and they are increasing with respect to N.*

Proof. Since $N[V] \subseteq [h_-, h_+]$ by (I), it follows from the dominated convergence theorem that the monotone sequences of $N[V]$ converge with respect to the p-norm of $L^p(\Omega)$. Thus the hypotheses of Theorem 1.1.2 hold when $X = L^p(\Omega)$, ordered a.e. pointwise, or equivalently, by the cone $L^p_+(\Omega)$, and equipped with the L^p-norm. $\qquad\qquad\qquad\qquad\qquad\qquad\qquad\qquad\square$

Remark 1.1.1. Equations

$$Au = Fu \quad \text{and} \quad Au + Tu = Fu + Tu$$

have the same solutions whenever A, F, $T \colon V \to X$. If T can be chosen so that $L = A + T$ and $N = F + T$ satisfy the assumptions of Theorem 1.1.2, then equation $Au = Fu$ has also extremal solutions.

1.1.3 Applicability of ordinary iteration methods. Consider first the case where a method of successive approximations can be applied to find extremal solutions of equation (1.1.1) when the domain and the range of the operators L and N are ordered topological spaces.

Proposition 1.1.2. *Let V and X be ordered topological spaces, and assume that $h_\pm \in X$, and that L, $N\colon V \to X$ have properties (I) and (II). Let v_0 be the least solution of $Lu = h_-$, and w_0 the greatest solution of $Lu = h_+$, and define*

$$\begin{cases} v_{n+1} := \text{ the least solution of } Lu = Nv_n, \ n \in \mathbb{N}, \\ w_{n+1} := \text{ the greatest solution of } Lu = Nw_n, \ n \in \mathbb{N}. \end{cases} \tag{1.1.7}$$

a) If $u_* = \lim\limits_{n \to \infty} v_n$ exists, and if $Lv_n \to Lu_*$ and $Nv_n \to Nu_*$, then u_* is the least solution of (1.1.1).
b) If $u^* = \lim\limits_{n \to \infty} w_n$ exists, and if $Lw_n \to Lu^*$ and $Nw_n \to Nu^*$, then u^* is the greatest solution of (1.1.1).

Proof. We prove the case b). Property (II), the assumptions of b) and the definition (1.1.7) imply that

$$Lu^* = \lim_{n \to \infty} Lw_n = \lim_{n \to \infty} Lw_{n+1} = \lim_{n \to \infty} Nw_n = Nu^*.$$

Thus u^* is a solution of equation (1.1.1).

Assume next that u is another solution of equation (1.1.1). Let $w = \hat{u}$ denote the greatest solution of $Lw = Nu$. Since $u \le \hat{u}$ and $Lu = L\hat{u}$, then $Nu \le N\hat{u}$ by (I). Inequalities $\hat{u} \le w_n$ and $L\hat{u} \le Lw_n$ hold when $n = 0$. If they hold for some $n \in \mathbb{N}$, they imply by (1.1.7) and (I) that $L\hat{u} \le N\hat{u} \le Nw_n = Lw_{n+1}$, whence $\hat{u} \le w_{n+1}$ by (II). Thus $\hat{u} \le w_n$ for each $n \in \mathbb{N}$. Because $u^* = \lim\limits_{n \to \infty} w_n$, we get $\hat{u} \le u^*$. Since $u \le \hat{u}$, we then have $u \le u^*$, so that u^* is the greatest solution of (1.1.1).

The assertions of case a) can be proved similarly. $\qquad\Box$

The following result is also needed in the sequel.

Proposition 1.1.3. *Let $[\underline{x}, \overline{x}]$ be a nonempty order interval in an ordered topological space, and let $G\colon [\underline{x}, \overline{x}] \to [\underline{x}, \overline{x}]$ be an increasing mapping.*
a) If $G^n\underline{x} \to x_*$, and if $x_* = Gx_*$, then x_* is the least fixed point of G.
b) If $G^n\overline{x} \to x^*$, and if $x^* = Gx^*$, then x^* is the greatest fixed point of G.

Proof. Let x be a fixed point of G. Since G is increasing, it is easy to show by induction that $G^n\underline{x} \le G^n x = x \le G^n\overline{x}$ for each $n \in \mathbb{N}$. When $n \to \infty$, we get the asserted results. $\qquad\Box$

1.2 Applications to differential equations

The results derived in section 1.1 for the operator equation (1.1.1) will now be applied to implicit initial and boundary value problems of ordinary and partial differential equations.

1.2.1 Applications to ordinary differential equations. Consider first an explicit initial value problem (IVP)

$$u'(t) = g(t, u(t)) + h(t) \text{ a.e. in } J = [t_0, t_1], \quad u(t_0) = x_0. \qquad (1.2.1)$$

We assume that the function $h \colon J \to \mathbb{R}$ belongs to the set $L^1(J)$ of all Lebesgue integrable functions, and that the function $g \colon J \times \mathbb{R} \to \mathbb{R}$ is an L^1-*bounded Carathéodory function*, i.e., $g(t, x)$ is measurable in t for all $x \in \mathbb{R}$, continuous in x for almost every (a.e.) $t \in J$, and there exists a function $m \in L^1(J)$ such that $|g(t, x)| \le m(t)$ for all $x \in \mathbb{R}$ and a.e. $t \in J$. By a well known existence result due to Carathéodory (cf. [44]), problem (1.2.1) has a solution in the set $AC(J)$ of all absolutely continuous functions $u \colon J \to \mathbb{R}$. In the following we assume that a partial ordering $\le$ is defined in $AC(J)$ pointwise and in $L^1(J)$ a.e. pointwise. It is well known that a function $u \in AC(J)$ is a.e. differentiable and $u' \in L^1(J)$. Define $u'(t) = 0$ at those points of J where u is not differentiable.

The next lemma is a special case of Proposition 2.1.1, and it is also an easy consequence of [97, Theorem 1.2] and its proof.

Lemma 1.2.1. *If $g \colon J \times \mathbb{R} \to \mathbb{R}$ is an L^1-bounded Carathéodory function, then the IVP (1.2.1) has for each $h \in L^1(J)$ the least and the greatest solution in $AC(J)$, and they are increasing with respect to h.*

As an application of Lemma 1.2.1 and Corollary 1.1.2 we prove an existence and comparison result for extremal solutions, i.e., least and greatest solutions, of the following implicit IVP

$$u'(t) = g(t, u(t)) + f(t, u(t), u'(t) - g(t, u(t))) \text{ a.e. in } J, \quad u(t_0) = x_0, \quad (1.2.2)$$

where $f \colon J \times \mathbb{R} \times \mathbb{R} \to \mathbb{R}$ and $g \colon J \times \mathbb{R} \to \mathbb{R}$.

Proposition 1.2.1. *The IVP (1.2.2) has extremal solutions in $AC(J)$, and they are increasing with respect to f, if g is an L^1-bounded Carathéodory function, if f is L^1-bounded and sup-measurable, and if $f(t, x, y)$ is increasing in x and y for a.e. $t \in J$.*

Proof. The given hypotheses imply that defining

$$\begin{cases} V := \{u \in AC(J) \mid u(t_0) = x_0\}, \\ Lu(t) := u'(t) - g(t, u(t)), \ Nu(t) := f(t, u(t), Lu(t)), \quad t \in J, \end{cases} \qquad (1.2.3)$$

we obtain mappings $L, N \colon V \to L^1(J)$. Obviously, $u \in V$ is a solution of the IVP (1.2.2) if and only if u is a solution of equation (1.1.1). Let M be an L^1-bound of f, and let $u, v \in V$, satisfy $u \leq v$ and $Lu \leq Lv$. Denoting $h_\pm = \pm M$, and noticing that $f(t, x, y)$ is increasing in x and in y for a.e. $t \in J$, we obtain

$$h_-(t) = -M(t) \leq f(t, u(t), Lu(t)) \leq f(t, v(t), Lv(t)) \leq M(t) = h_+(t)$$

for a.e. $t \in J$. These inequalities, (1.2.3) and Lemma 1.2.1 imply that the following properties hold.

(I) If $u, v \in V$, $u \leq v$, and $Lu \leq Lv$, then $h_- \leq Nu \leq Nv \leq h_+$.

(II) Equation $Lu = h$ has for each $h \in [h_-, h_+]$ extremal solutions, and they are increasing with respect to h.

It then follows from Corollary 1.1.2 that equation (1.1.1), where L and N are defined by (1.2.3), has least and greatest solutions u_* and u^* in the set V, and they are increasing with respect to N. This implies by (1.2.3) that u_* and u^* are extremal solutions of (1.2.2), and they are increasing with respect to f.   □

Consider next the IVP (1.2.2) in the case when also f is a Carathéodory function.

Proposition 1.2.2. *Assume that g is an L^1-bounded Carathéodory function, and that f is L^1-bounded and $f(t, x, y)$ is increasing and jointly continuous in x and y for a.e. $t \in J$. Let V and $L, N: V \to L^1(J)$ be defined by (1.2.3). Then the sequences (v_n) and (w_n), given by (1.1.7), converge uniformly on J to least and greatest solutions of the IVP (1.2.2), respectively.*

Proof. It follows from the proof of Proposition 1.2.1 that the operators L and N have properties (I) and (II). They imply by induction that $(Lw_n)_{n=0}^\infty$ is a decreasing sequence in the order interval $[h_-, h_+]$ of $L^1(J)$, and that $(w_n)_{n=0}^\infty$ is a decreasing sequence in the order interval $[\underline{u}, \overline{u}]$, so that $(w_n(t))$ is a decreasing sequence in $[\underline{u}(t), \overline{u}(t)]$ for each $t \in J$. Thus the limits

$$u^*(t) := \lim_{n \to \infty} w_n(t), \ t \in J, \quad \text{and} \quad h(t) := \lim_{n \to \infty} Lw_n(t) \text{ a.e. in } J \quad (1.2.4)$$

exist. It follows from (1.1.7) and (1.2.3) that

$$w_n(t) = x_0 + \int_{t_0}^t (g(s, w_n(s)) + Lw_n(s))ds, \quad t \in J, \ n \in \mathbb{N}.$$

When $n \to \infty$ we get, by applying (1.2.4), continuity of $g(t, \cdot)$ and the dominated convergence theorem,

$$u^*(t) = x_0 + \int_{t_0}^t (g(s, u^*(s)) + h(s))ds, \quad t \in J.$$

Thus $u^* \in V = \{u \in AC(J) \mid u(t_0) = x_0\}$, and $Lu^* = h$. In view of this, (1.2.3), (1.2.4), and joint continuity of f in its last two variables we obtain

$$\begin{cases} Lw_n(t) \to Lu^*(t) \quad \text{and} \\ Nw_n(t) = f(t, w_n(t), Lw_n(t)) \to f(t, u^*(t), Lu^*(t)) = Nu^*(t) \text{ a.e. in } J. \end{cases}$$

Since (Lw_n) and (Nw_n) are decreasing and L^1-bounded, then $Lw_n \to Lu^*$ and $Nw_n \to Nu^*$ in the L^1-norm. The above proof shows that the hypotheses of Proposition 1.1.2.b hold, whence u^* is the greatest solution of (1.1.1) in V, and hence by (1.2.3) the greatest solution of the IVP (1.2.2) in $AC(J)$. Moreover, it follows from Dini's theorem that (w_n) converges uniformly on J to u^*. The proof that the sequence (v_n) converges uniformly on J to the least solution of (1.2.2) is similar. $\qquad\Box$

Remarks 1.2.1. Generalizations to Proposition 1.2.1 will be presented in chapter 2.

If in Proposition 1.2.2 the IVP (1.2.1) has for each $h \in L^1(J)$ a unique solution, it follows from (1.1.7) and (1.2.3) that the sequence (u_n), given by

$$\begin{cases} u'_{n+1}(t) = g(t, u_{n+1}(t)) + f(t, u_n(t), Lu_n(t)) \text{ a.e. in } J, \\ u_{n+1}(t_0) = x_0, \end{cases} \qquad (1.2.5)$$

is equal to (v_n) if $u_0 = \underline{u}$ and to (w_n) if $u_0 = \overline{u}$. If M denotes an L^1-bound of f, we can choose $\underline{u} = u_-$ and $\overline{u} = u^+$, where the functions $u_\pm$ are the solutions of the IVPs

$$u'_\pm(t) = g(t, u_\pm(t)) \pm M(t) \text{ a.e. in } J, \quad u_\pm(t_0) = x_0.$$

The above sequences may converge to the extremal solutions also when $f(t, \cdot, \cdot)$ is not continuous. This holds, e.g., when $u_{n+1} = u_n$ for some $n \in \mathbb{N}$. In particular, the proof given in Proposition 1.2.2 for the existence of the maximal solution of (1.2.2) holds also when the continuity of $f(t, \cdot, \cdot)$ is replaced by the following property: $f(t, x_n, y_n) \to f(t, x, y)$ for a.e. $t \in J$ whenever the sequences (x_n) and (y_n) are decreasing, $x_n \to x$ and $y_n \to y$.

Example 1.2.1. Let H be the Heaviside function: $H(x) = \begin{cases} 1 \text{ if } x \geq 0, \\ 0 \text{ if } x < 0, \end{cases}$ and let $[x]$ denote the greatest integer $\leq x$. Consider the IVP

$$\begin{cases} u'(t) = H(1 - 2t) + \frac{[4 - 3t + u(t)]}{1 + |[4 - 3t + u(t)]|} + \frac{[u'(t) - H(1 - 2t)]}{2(1 + |[u'(t) - H(1 - 2t)]|)} \\ u(0) = 0, \end{cases} \qquad (1.2.6)$$

on the interval $J = [0, 1]$. The IVP (1.2.6) is of the form (1.2.2) with

$$\begin{cases} f(t, x, y) = \frac{[4-3t+x]}{1+|[4-3t+x]|} + \frac{[y]}{2(1+|[y]|)}, \\ g(t, x) = H(1 - 2t), \quad t \in [0, 1], \quad x, y \in \mathbb{R}. \end{cases}$$

It is easy to see that the hypotheses of Proposition 1.2.1 hold. Thus the IVP (1.2.7) has extremal solutions u_* and u^*. To determine these solutions, calculate the successive approximations

$$\begin{cases} u'_{n+1}(t) = H(1 - 2t) + \frac{[4-3t+u_n(t)]}{1+|[4-3t+u_n(t)]|} + \frac{[u'_n(t)-H(1-2t)]}{2(1+|[u'_n(t)-H(1-2t)]|)}, \\ u_{n+1}(0) = 0, \end{cases} \quad (1.2.7)$$

where $u_0 = u_\pm$ are solutions of

$$u'_\pm(t) = H(1 - 2t) \pm 3/2 \quad \text{a.e. in } J, \quad u_\pm(0) = 0,$$

i.e.,

$$u_+(t) = \begin{cases} \frac{5}{2}t, & 0 \le t \le \frac{1}{2}, \\ \frac{1}{2} + \frac{3}{2}t, & \frac{1}{2} < t \le 1, \end{cases} \quad u_-(t) = \begin{cases} -\frac{1}{2}t, & 0 \le t \le \frac{1}{2}, \\ \frac{1}{2} - \frac{3}{2}t, & \frac{1}{2} < t \le 1. \end{cases}$$

Since the functions $[\cdot]$ and H are right-continuous, Proposition 1.2.2 and Remarks 1.2.1 ensure that u^* is a limit of the successive approximations u_n when $u_0 = u_+$.

Calculate first u_1. When $n = 1$ and $u_0 = u_+$, it follows from (1.2.7) that

$$u'_1(t) = \begin{cases} 1 + \frac{[4-\frac{5}{2}]}{1+|[4-\frac{5}{2}]|} + \frac{[\frac{5}{2}-1]}{2(1+|[\frac{5}{2}-1]|)} = 1 + \frac{3}{4} + \frac{1}{4} = 2, & 0 \le t \le \frac{1}{2}, \\ 0 + \frac{[\frac{9}{2}-\frac{3}{2}t]}{1+|[\frac{9}{2}-\frac{3}{2}t]|} + \frac{[\frac{3}{2}]}{2(1+|[\frac{3}{2}]|)} = \frac{3}{4} + \frac{1}{4} = 1, & \frac{1}{2} \le t \le 1, \end{cases}$$

$u_1(0) = 0.$

Thus

$$u_1(t) = \begin{cases} 2t, & 0 \le t \le \frac{1}{2}, \\ \frac{1}{2} + t, & \frac{1}{2} \le t \le 1. \end{cases}$$

Similar calculations yield the following representations of u_2 and u_3.

$$u_2(t) = \begin{cases} 2t, & 0 \le t \le \frac{1}{2}, \\ \frac{1}{2} + t, & \frac{1}{2} \le t \le \frac{3}{4}, \\ \frac{9}{11} + \frac{11}{12}t, & \frac{3}{4} \le t \le 1, \end{cases} \quad u_3(t) = \begin{cases} 2t, & 0 \le t < \frac{1}{2}, \\ \frac{1}{2} + t, & \frac{1}{2} \le t < \frac{3}{4}, \\ \frac{3}{4} + \frac{2}{3}t, & \frac{3}{4} \le t \le 1. \end{cases}$$

It turns out that $u_4 = u_3$, whence $u^* = u_3$ is the greatest solution of (1.2.6) by Proposition 1.1.2.

When $u_0 = u_-$, we get, after a finite number of steps, the least solution u_* of (1.2.6). Its exact representation is

$$u_*(t) = \begin{cases} \frac{7}{4}t, & 0 \le t < \frac{1}{2}, \\ \frac{1}{2} + \frac{3}{4}t, & \frac{1}{2} \le t < \frac{2}{3}, \\ \frac{5}{9} + \frac{2}{3}t, & \frac{2}{3} \le t \le 1. \end{cases}$$

Moreover, between u_* and u^* there is a continuum of chaotically behaving solutions of (1.2.6). For, denoting $\Omega = \{(t,x) \mid u_*(t) \le x \le u^*(t),\ t \in J\}$, the differential equation of (1.2.6) can be reduced to an inclusion equation

$$u'(t) \in \mathcal{H}(t, u(t)) = \begin{cases} \{\frac{7}{4}, 2\}, & (t, u(t)) \in \Omega,\ t \le \frac{1}{2}, \\ \{\frac{3}{4}, 1\}, & (t, u(t)) \in \Omega,\ t > \frac{1}{2},\ u(t) \ge 3t - 1, \\ \{\frac{2}{3}\}, & (t, u(t)) \in \Omega,\ t > \frac{1}{2},\ u(t) < 3t - 1. \end{cases}$$

Thus each point in the set $\{(t,x) \in \Omega \mid x \ge 3t - 1\}$ is a bifurcation point for the solutions of (1.2.6).

1.2.2 Applications to implicit elliptic boundary value problems.

Let us first consider the following semilinear elliptic boundary value problem (BVP)

$$Au(x) = g(x, u(x)) + h(x) \text{ in } \Omega, \quad u = 0 \text{ on } \partial\Omega, \tag{1.2.8}$$

in a bounded domain $\Omega \subset \mathbb{R}^N$ with Lipschitz boundary $\partial\Omega$. We assume that A is a second order strongly elliptic operator in divergence form

$$Au(x) = -\sum_{i=1}^{N} \frac{\partial}{\partial x_i}\left(a_{ij}(x)\frac{\partial u(x)}{\partial x_j}\right),$$

where the coefficients $a_{ij} \in L^\infty(\Omega)$ satisfy for some positive constant μ the following ellipticity condition:

$$\sum_{i=1}^{N} a_{ij}(x)\xi_i\xi_j \ge \mu\,|\xi|^2 \text{ for a.e. } x \in \Omega, \text{ and for all } \xi \in \mathbb{R}^N.$$

Let $W^{1,2}(\Omega)$ denote the usual Sobolev space of square integrable functions having square integrable generalized derivatives of first order, and denote by $W_0^{1,2}(\Omega)$ its subspace whose elements have zero boundary values in the sense of traces; see C.1.

The following result can be easily proved by means of Theorem 5.1.1 of chapter 5.

Lemma 1.2.2. *Let $g : \Omega \times \mathbb{R} \to \mathbb{R}$ be an L^2-bounded Carathéodory function. Then for each $h \in L^2(\Omega)$ problem (1.2.8) has the least and the greatest solution in $W_0^{1,2}(\Omega)$, and they are increasing with respect to h.*

Proof. In view of the L^2-roundedness of $g : \Omega \times \mathbb{R} \to \mathbb{R}$ there exists a function $m \in L^2(\Omega)$ such that $|g(t, s)| \leq m(x)$ for a.e. $x \in \Omega$ and for all $s \in \mathbb{R}$. Let $w \in W_0^{1,2}(\Omega)$ be the unique solution of the linear BVP

$$Au = m + |h| \text{ in } \Omega, \quad u = 0 \text{ on } \partial\Omega.$$

Then by the maximum principle one easily verifies that $w \geq 0$, and that w is an upper solution and $-w$ is a lower solution of the BVP (1.2.8). Moreover, any solution of (1.2.8) must belong to the order interval $[-w, w]$. As a simple consequence of Theorem 5.1.1 provided in chapter 5 the BVP (1.2.8) has extremal solutions within $[-w, w]$, which proves the lemma.   □

By means of Lemma 1.2.2 and Corollary 1.1.2 we are now able to prove an existence result for the implicit BVP

$$Au = g(x, u) + f(x, u, Au - g(x, u)) \text{ in } \Omega, \quad u = 0 \text{ on } \partial\Omega, \qquad (1.2.9)$$

without imposing any continuity hypotheses on f.

Proposition 1.2.3. *Let $g : \Omega \times \mathbb{R} \to \mathbb{R}$ be an L^2-bounded Carathéodory function, let $f : \Omega \times \mathbb{R} \times \mathbb{R} \to \mathbb{R}$ be sup-measurable and L^2-bounded, and assume that the function $f(x, r, s)$ is increasing in r and in s for a.e. $x \in \Omega$. Then the BVP (1.2.9) has extremal, i.e., greatest and least solutions in the set $V = \{u \in W_0^{1,2}(\Omega) \mid Au - g(\cdot, u) \in L^2(\Omega)\}$. Moreover, these extremal solutions are increasing with respect to f.*

Proof. We reduce first the BVP (1.2.9) to an operator equation $Lu = Nu$. Define $L, N : V \to L^2(\Omega)$ by

$$Lu := Au - g(\cdot, u), \quad \text{and} \quad Nu := f(\cdot, u, Lu). \qquad (1.2.10)$$

It is easy to see that $u \in V$ is a solution of (1.2.9) if and only if u is a solution of equation $Lu = Nu$. If M is an L^2-bound of f, and $u \in V$, then

$$|Nu(x)| = |f(x, u(x), Lu(x))| \leq M(x) \text{ a.e. in } \Omega.$$

Thus $N[V] \subseteq [h_-, h_+]$, where $h_\pm = \pm M$. This property, the above definition of N, the monotonicity of f in its last two arguments and Lemma 1.2.2 imply that the following properties hold.

(I) If $u, v \in V$, $u \leq v$, and $Lu \leq Lv$, then $h_- \leq Nu \leq Nv \leq h_+$.

(II) Equation $Lu = h$ has for each $h \in [h_-, h_+]$ extremal solutions, and they are increasing with respect to h.

Hence the equation $Lu = Nu$ has by Corollary 1.1.2 least and greatest solutions u_* and u^* in V. This result and the above definitions of V, L, and N imply that u_* and u^* are extremal solutions of (1.2.9). Since u_* and u^* are increasing with respect to N, they are also increasing with respect to f, because $Nu := f(\cdot, u, Lu)$.   □

A generalized version of Proposition 1.2.3 will be proved in chapter 7.

Example 1.2.2. Consider the BVP

$$\begin{cases} -u''(x) = 2\frac{[u(x)+2x+1]}{1+|[u(x)+2x+1]|} + \frac{[-u''(x)]}{1+|[-u''(x)]|} & \text{a.e. in } \Omega = (0,1), \\ u(0) = u(1) = 0, \end{cases} \quad (1.2.11)$$

where $[z]$ means the greatest integer $\leq z$. Problem (1.2.11) is of the form (1.2.9), where

$$g(x,s) \equiv 0, \quad \text{and} \quad f(x,s,r) = 2\,\frac{[s+2x+1]}{1+|[s+2x+1]|} + \frac{[r]}{1+|[r]|}. \quad (1.2.12)$$

The hypotheses of Proposition 1.2.3 are satisfied, so that problem (1.2.11) has extremal solutions. To determine them, notice that $Au = Lu = -u''$, and that $V = \{u: J \to \mathbb{R} \mid u' \in AC(J),\ u'' \in L^2(\Omega),\ u(0) = u(1) = 0\}$, where $J = [0,1]$. By elementary calculations one can show that for each $h \in L^2(\Omega)$ the function

$$u(x) = (1-x)\int_0^x th(t)dt + x\int_x^1 (1-t)h(t)dt, \quad x \in J \quad (1.2.13)$$

is a solution of equation $Lu = h$ in V. Uniqueness follows from well-known comparison results (cf. [197]), and is also a consequence of Corollary 3.4.4.

The above results, (1.2.10), and (1.2.12) imply that equation $Lu = Nu$ is reduced to (1.2.13), where

$$h(x) = 2\frac{[u(x)+2x+1]}{1+|[u(x)+2x+1]|} + \frac{[-u''(x)]}{1+|[-u''(x)]|} \quad \text{a.e. in } (0,1).$$

The function $M(t) \equiv 3$ is an L^2-bound of f, and the solutions of (1.2.13) with $h(x) = h_\pm(x) \equiv \pm 3$ are $u_\pm(x) = \pm\frac{3}{2}x(1-x)$, $x \in J$. In this case the successive approximations $Lu_{n+1} = Nu_n$ can be rewritten as

$$\begin{cases} u_{n+1}(x) = (1-x)\int_0^x th_n(t)dt + x\int_x^1 (1-t)h_n(t)dt, \quad x \in J, \\ h_n(x) = 2\frac{[u_n(x)+2x+1]}{1+|[u_n(x)+2x+1]|} + \frac{[-u_n''(x)]}{1+|[-u_n''(x)]|} \quad \text{a.e. in } (0,1). \end{cases} \quad (1.2.14)$$

Since the function $z \mapsto [z]$ is right-continuous, the greatest solution of the BVP (1.2.11) can be obtained by Proposition 1.1.2 as the uniform limit of the successive approximations u_n when $u_0 = u_+$. Calculating these approximations numerically by Simpson rule one obtains the following estimate for the greatest solution of (1.2.11).

$$u^*(x) \approx \begin{cases} -.75x^2 + .839x + 0, & 0 \leq x < .392, \\ -x^2 + 1.04x - .4, & .392 < x \leq 1. \end{cases}$$

In view of this one can infer that the greatest solution of (1.2.11) is of the form

$$u^*(x) = \begin{cases} -\frac{3}{4}x^2 + ax, & 0 \leq x < d, \\ -x^2 + bx + 1 - b, & d \leq x \leq 1. \end{cases}$$

It remains to determine a, b, and d. Because u and u' are continuous at d, we get two equations. Solving from these equations a and b in terms of d, and substituting these values into the above formula of u^*, we get

$$u^*(x) = \begin{cases} -\frac{3}{4}x^2 + (\frac{1}{4}d^2 - \frac{1}{2}d + 1)x, & 0 \leq x < d, \\ -x^2 + (\frac{1}{4}d^2 + 1)x - \frac{1}{4}d^2, & d \leq x \leq 1. \end{cases}$$

To determine d, notice first that $[u(x) + 2x + 1] = 1$, when $0 \leq x < d$, since $u(0) = 0$ and $u''(x) = -\frac{3}{4}$ when $0 \leq x < d$. At a point $x = d$ the second derivative of u has a jump, and our approximation of u^* shows that it is caused by the jump of $[u(x) + 2x + 1]$ to the value 2 at that point, i.e., $u(d) + 2d + 1 = 2$. This and the latest formula of u^* imply that d is the positive solution of equation $x^3 - 5x^2 + 12x - 4 = 0$, i.e.,

$$d = \frac{1}{3}[5 + (6\sqrt{267} - 91)^{\frac{1}{3}} - (6\sqrt{267} + 91)^{\frac{1}{3}}] \approx .3924782332621779.$$

In this case also the least solution u_* of (1.2.11) can be obtained as a limit of successive approximations (1.2.14) when $u_0 = u_-$. Similar reasoning as above yields

$$u_*(x) = \begin{cases} -\frac{3}{4}x^2 + (\frac{1}{6}c^2 - \frac{1}{3}c + \frac{11}{12})x, & 0 \leq x < c, \\ -\frac{11}{12}x^2 + (\frac{1}{6}c^2 + \frac{11}{12})x - \frac{1}{6}c^2, & c \leq x \leq 1, \end{cases}$$

where c is the positive solution of equation $2x^3 - 13x^2 + 35x - 12 = 0$, i.e.,

$$c = \frac{1}{6}(13 + (9\sqrt{20141} - 1250)^{\frac{1}{3}} - (9\sqrt{20141} + 1250)^{\frac{1}{3}}) \approx .3981235265036852.$$

1.3 Notes and comments

Theorem 1.1.1 is proved in section A of the appendix by using the following generalized iteration method (see also [141]): Given an element $\underline{x}$ of a poset P and a mapping $G \colon P \to P$, there is by Lemma A.2.2 a unique well-ordered chain C in P which has properties

(C) $\underline{x} = \min C$, and if $\underline{x} < x \in P$, then $x \in C$ iff $x = \sup G\{y \in C | y < x\}$.

Assuming that G is increasing, that $\underline{x}$ is a lower bound of $G[P]$, and that $x_* = \sup G[C]$ exists, then $x_* = \max C$, and x_* is the least fixed point of G by Proposition A.2.1.

To obtain the result of Theorem 1.1.1.a), notice first that its hypotheses and the fact that $G[C]$ is well-ordered as a subset of C (cf. Lemma A.2.3) imply that increasing sequences of $G[C]$ have weak limits in P. This ensures by Lemma A.3.1 the existence of $x_* = \sup G[C]$, whence the above results imply that $x_* = \max C$, and that x_* is the least fixed point of G. Moreover, applying [164, 24.1,(7)] one can show (cf. Lemma A.3.1) that $\sup G[C]$ is a weak limit of an increasing sequence of $G[C]$. This result and [141, Lemma 1.1.4] can be used to prove that $G[C]$ and C are countable.

The results of Lemma A.2.3 imply that the first elements of C are the iterations $G^n \underline{x}$, $n \in \mathbb{N}$, if G is increasing and $\underline{x}$ is a lower bound of $G[P]$. In the case when G is defined by (1.1.5), the successive approximations v_n given by (1.1.7) are the first elements of C when $\underline{x} = v_0$. Thus the classical iteration methods applied in Propositions 1.1.2 and 1.1.3 are special cases of the generalized iteration method described above. In [134] this method is compared to other chain methods used in the fixed point theory (cf., e.g., [107, 166, 217]).

The existence and comparison results of Theorem 1.1.2 for the equation $Lu = Nu$ are proved in [81] when the convergence in the hypothesis (III) is strong (see also [76]). In [80] a generalized iteration method is presented for equation $Lu = Nu$. This method is then used to prove fixed point results and existence and comparison results for equation (1.1.1) under weaker hypotheses than those of Theorems 1.1.1 and 1.1.2. The special case where the operator L is linear is considered, e.g., in [168, 175, 184].

Section 1.2, whose results are adapted from [80], is intended to serve as an introduction to the theory of discontinuous explicit and implicit differential equations which will be presented in chapters 2, 4, and 7.

Chapter 2
Extremality results for first order differential equations

Recently, the existence of extremal solutions of the differential equation $u'(t) = g(t, u(t))$ with a given initial condition has been proved in [125] without assuming the sup-measurability of the function g, i.e., the measurability of $g(\cdot, u(\cdot))$ for all measurable u. This result is applied in [194] to derive extremality results for the above differential equation equipped with discontinuous functional boundary conditions.

In sections 2.1 and 2.2 this research is continued as follows.
- The above cited results are extended to equation $\frac{d}{dt}\varphi(u(t)) = g(t, u(t))$, where $\varphi \colon \mathbb{R} \to \mathbb{R}$ is an increasing homeomorphism.
- L^1-boundedness of g, assumed in [125], is replaced by a weaker growth condition.
- When g is nonnegative-valued, local conditions are introduced which allow a new type of discontinuity for g.
- Dependence of extremal solutions on the function g and on the given initial and boundary conditions is studied.
- The obtained results are applied to initial and boundary value problems of the differential equation $u'(t) = q(u(t))g(t, u(t))$.

Applying the results of section 2.2 and fixed point theorems of section 1.1 we prove in section 2.3 existence and comparison results for extremal solutions of the functional differential equation $\frac{d}{dt}\varphi(u(t)) = g(t, u(t), u)$, equipped with a functional boundary condition.

In section 2.4 we present extremality results for the implicit problem

$$\begin{cases} \frac{d}{dt}\varphi(u(t)) = g(t, u(t), u) + f(t, u(t), u, \frac{d}{dt}\varphi(u(t)) - g(t, u(t), u)) \\ \text{a.e. in } [t_0, t_1], \quad u(t) = B_0(u(t_0), u) + B_1(t, u(t), u), \quad t \in [t_0 - r, t_0]. \end{cases}$$

All the functions g, f, B_0, and B_1 may depend discontinuously on the whole solution u defined on $[t_0 - r, t_1]$, and thus may be strongly interrelated. In the proofs we reduce the problem to an operator equation $Lu = Nu$ in some ordered function spaces, and then apply results of sections 1.1 and 2.3.

2.1 Explicit initial value problems

In this section we derive existence and comparison results for extremal solutions of first order explicit scalar initial value problems.

2.1.1 Hypotheses and main results. Consider the initial value problem

$$\frac{d}{dt}\varphi(u(t)) = g(t, u(t)) \quad \text{for a.e. } t \in J = [t_0, t_1], \quad u(t_0) = x_0. \qquad (2.1.1)$$

Definition 2.1.1. A function $u\colon J \to \mathbb{R}$ is said to be a *lower solution* of (2.1.1) if u belongs to the set

$$Y = \{u \in C(J) \mid \varphi \circ u \in AC(J)\},$$

and if

$$\frac{d}{dt}\varphi(u(t)) \leq g(t, u(t)) \quad \text{for a.e. } t \in J, \quad u(t_0) \leq x_0.$$

If the reversed inequalities hold, then $u \in Y$ is called an *upper solution* of (2.1.1), and a *solution* of (2.1.1) if equalities hold. If u_* and u^* are such solutions of (2.1.1) that $u_*(t) \leq u(t) \leq u^*(t)$ on J for every solution u of (2.1.1), we say that u_* is the *least solution* and u^* is the *greatest solution* of (2.1.1), and that u_* and u^* are the *extremal solutions* of (2.1.1).

Assuming that the functions $\varphi\colon \mathbb{R} \to \mathbb{R}$ and $g\colon J \times \mathbb{R} \to \mathbb{R}$ have the following properties:

(φ0) φ is an increasing homeomorphism,
(g0) for each $x \in \mathbb{R}$ the function $g(\cdot, x)$ is measurable, and
$$\limsup_{y \to x-} g(t, y) \leq g(t, x) \leq \liminf_{y \to x+} g(t, y) \quad \text{for a.e. } t \in J,$$
(A) (2.1.1) has a lower solution $\underline{u}$ and an upper solution $\overline{u}$ such that $\underline{u} \leq \overline{u}$, and g is L^1-bounded in $\Omega = \{(t, x) \mid t \in J,\ \underline{u}(t) \leq x \leq \overline{u}(t)\}$,

we prove that there exists the least and the greatest among those solutions u of (2.1.1) for which $\underline{u}(t) \leq u(t) \leq \overline{u}(t)$ on J. Replacing (A) by assumption

(g2) $|g(t, x)| \leq p_1(t)\psi(|\varphi(x)|)$ for all $x \in \mathbb{R}$ and a.e. $t \in J$, where p_1 belongs to $L^1_+(J)$, $\psi\colon \mathbb{R}_+ \to (0, \infty)$ is increasing and $\int_0^\infty \frac{dx}{\psi(x)} = \infty$,

we prove that (2.1.1) has extremal solutions, and that they are increasing with respect to g and x_0. When g is nonnegative-valued we give a localized version to property (g0) which allows also downward jumps for $g(t, \cdot)$. Conditions which ensure that the dependence of extremal solutions of (2.1.1) on g and x_0 is continuous from the right or from the left are also given.

The so-obtained results are then shown to hold for the IVP

$$u'(t) = q(u(t))g(t, u(t)) \quad \text{a.e. in} \quad J, \quad u(t_0) = x_0,$$

if $(\varphi 0)$ is replaced by the following assumption.

(q0) $q \colon \mathbb{R} \to (0, \infty)$ and $\frac{1}{q}$ belong to $L_{loc}^\infty(\mathbb{R})$, and $\int_0^{\pm\infty} \frac{dz}{q(z)} = \pm\infty$.

Finally, examples and counterexamples are given to illustrate the obtained results and the need of given hypotheses.

Remarks 2.1.1. Properties (g0) and (q0) allow the functions g and q to be discontinuous. (g0) holds, e.g., if g is a Carathéodory function, and also when $g(t, x)$ is measurable in t for all $x \in \mathbb{R}$ and increasing in x for a.e. $t \in J$. In the latter case g need not be sup-measurable, or equivalently, a standard function in the sense of Shragin (cf. [12]).

If φ is locally absolutely continuous, then $(\varphi 0)$ holds if and only if φ' is a.e. positive-valued and $\int_0^{\pm\infty} \varphi'(x) \, dx = \pm\infty$.

One can replace $\psi(|\varphi(x)|)$ by $\psi(|x|)$ in (g2) if φ is Lipschitz-continuous. For if $|\varphi(x) - \varphi(y)| \le K|x - y|$, $x, y \in \mathbb{R}$, for some $K > 0$, then $|\varphi(x)| \le K|x| + |\varphi(0)|$, $x \in \mathbb{R}$, and the function $z \mapsto \psi(Kz + |\varphi(0)|)$ has the properties given for ψ in (g2). This holds, e.g., if φ is locally absolutely continuous and φ' is essentially bounded.

2.1.2 Preliminaries. We begin with a result of Hassan and Rzymowski (cf. [125]), which is used in the proofs of our main existence theorems for the IVP (2.1.1).

Theorem 2.1.1. *The IVP*

$$u'(t) = f(t, u(t)) \quad \text{for a.e. } t \in J, \quad u(t_0) = x_0, \qquad (2.1.2)$$

has extremal solutions in $AC(J)$ for each $x_0 \in \mathbb{R}$, if $f \colon J \times \mathbb{R} \to \mathbb{R}$ is an L^1-bounded function and has the following property.

(HR) *For each $x \in \mathbb{R}$ the function $f(\cdot, x)$ is measurable, and*
$$\limsup_{y \to x-} f(t, y) \le f(t, x) \le \liminf_{y \to x+} f(t, y) \quad \text{for a.e. } t \in J.$$

Hints to the proof. It can be shown that the functions f_* and f^*, given by

$$f_*(t, x) = \limsup_{y \to x-} f(t, y), \quad f^*(t, x) = \liminf_{y \to x+} f(t, y), \quad (t, x) \in J \times \mathbb{R},$$

satisfy for almost all $t \in J$ and all $x \in \mathbb{R}$

$$\begin{cases} \limsup_{y \to x-} f^*(t, y) \le f^*(t, x) = \liminf_{y \to x+} f^*(t, y), \\ \limsup_{y \to x-} f_*(t, y) = f_*(t, x) \le \liminf_{y \to x+} f_*(t, y). \end{cases}$$

Applying these properties, denoting by M an L^1-bound of f, and defining

$$\begin{cases} \mathcal{X}_0 = \{u \colon J \to \mathbb{R} \mid u(t_0) = x_0, \ |u(t) - u(s)| \le \left| \int_s^t M(\tau) d\tau \right|, \ s, t \in J\}, \\ \mathcal{X} = \{v \in \mathcal{X}_0 \mid v' \le f(\cdot, v(\cdot))\}, \quad u^*(t) = \sup_{v \in \mathcal{X}} v(t), \quad t \in J, \\ \mathcal{Y} = \{w \in \mathcal{X}_0 \mid w(t) \ge u^*(t), \ t \in J, \ \text{and} \ w' \ge f^*(\cdot, w(\cdot))\}, \end{cases}$$

one can show that the function

$$u_+(t) = \inf_{w \in \mathcal{Y}} w(t), \qquad t \in J,$$

is the greatest solution of the IVP (2.1.2). The proof concerning the least solution is similar. As for details, see [125, Theorem 3.1]. □

Next we shall show that problem (2.1.1) can be converted to the IVP

$$v'(t) = g(t, \varphi^{-1}(v(t))) \ \text{a.e. in} \ J, \quad v(t_0) = \varphi(x_0). \tag{2.1.3}$$

Lemma 2.1.1. *If (φ0) holds, then $u \in Y$ is a lower solution, an upper solution or a solution of (2.1.1) if and only if $v = \varphi \circ u$ belongs to $AC(J)$ and is a lower solution, an upper solution, or a solution of (2.1.3), respectively.*

Proof. If $u \in Y$ is a lower solution of (2.1.1), then $v = \varphi \circ u \in AC(J)$, and

$$v'(t) = \frac{d}{dt} \varphi(u(t)) \le g(t, u(t)) = g(t, \varphi^{-1}(v(t))) \ \text{a.e. in} \ J.$$

Since $u(t_0) \le x_0$ and φ is increasing, then $v(t_0) = \varphi(u(t_0)) \le \varphi(x_0)$, whence v is a lower solution of the IVP (2.1.3).

Conversely, let $v \in AC(J)$ be a lower solution of (2.1.3). Then $u = \varphi^{-1} \circ v$ is continuous by (φ0), and $\varphi \circ u = v \in AC(J)$, whence $u \in Y$, and

$$\frac{d}{dt} \varphi(u(t)) = v'(t) \le g(t, \varphi^{-1}(v(t))) = g(t, u(t)) \ \text{a.e. in} \ J.$$

Since $v(t_0) \le \varphi(x_0)$ and φ^{-1} is increasing, then $u(t_0) = \varphi^{-1}(v(t_0)) \le x_0$. Thus u is a lower solution of the IVP (2.1.1).

Similar reasoning shows that $u \in Y$ is a solution or an upper solution of (2.1.1) if and only if $v = \varphi \circ u$ belongs to $AC(J)$ and is a solution or an upper solution of (2.1.3), respectively. □

2.1.3 Existence and comparison results. Denote by $\le$ the pointwise ordering of $C(J)$. If $\underline{u}, \overline{u} \in C(J)$, denote $[\underline{u}, \overline{u}] = \{u \in C(J) \mid \underline{u} \le u \le \overline{u}\}$.

Now we prove our first existence and comparison result for (2.1.1).

Theorem 2.1.2. *Assume that (φ0), (g0) and (A) hold. Then the IVP (2.1.1) has the least solution u_* and the greatest solution u^* in $[\underline{u}, \overline{u}]$, and*

$$\begin{cases} u_* = \min\{u_+ \mid u_+ \text{ is an upper solution of (2.1.1) in } [\underline{u}, \overline{u}]\}, \\ u^* = \max\{u_- \mid u_- \text{ is a lower solution of (2.1.1) in } [\underline{u}, \overline{u}]\}. \end{cases} \quad (2.1.4)$$

Proof. The assumption (A) and Lemma 2.1.1 imply that the functions $\underline{v} = \varphi \circ \underline{u}$ and $\overline{v} = \varphi \circ \overline{u}$ are lower and upper solutions of the IVP (2.1.3), and that $\underline{v} \leq \overline{v}$. If $t \in J$ and $x \in [\underline{v}(t), \overline{v}(t)]$, then $\varphi^{-1}(x) \in [\underline{u}(t), \overline{u}(t)]$. This result and (A) ensure the existence of such an $M \in L^1(J)$ that

$$|g(t, \varphi^{-1}(x))| \leq M(t) \quad \text{for a.e. } t \in J \text{ and for all } x \in [\underline{v}(t), \overline{v}(t)].$$

Thus the function $f \colon J \times \mathbb{R} \to \mathbb{R}$, defined by

$$f(t, x) := \begin{cases} g(t, \varphi^{-1}(\underline{v}(t))), & x < \underline{v}(t), \\ g(t, \varphi^{-1}(x)), & \underline{v}(t) \leq x \leq \overline{v}(t), \quad t \in J, \\ g(t, \varphi^{-1}(\overline{v}(t))), & x > \overline{v}(t), \end{cases} \quad (2.1.5)$$

is L^1-bounded. Applying properties (g0) and (φ0) it is also easy to see that f has the properties given in the assumption (HR). It then follows from Theorem 2.1.1 that the IVP

$$v'(t) = f(t, v(t)) \quad \text{for a.e. } t \in J, \quad v(t_0) = \varphi(x_0), \quad (2.1.6)$$

has the least solution v_* and the greatest solution v^*. To prove that v_* and v^* are extremal solutions of (2.1.3) in $[\underline{v}, \overline{v}]$ we show that every solution v of (2.1.6) belongs to the order interval $[\underline{v}, \overline{v}]$. For if $\underline{v} \not\leq v$, there exist $a, b \in J$, $a < b$, such that

$$v(a) = \underline{v}(a) \quad \text{and} \quad v(t) < \underline{v}(t) \text{ on } (a, b]. \quad (2.1.7)$$

Since v is a solution of (2.1.6) and $\underline{v}$ is a lower solution of (2.1.3), we obtain

$$\underline{v}'(t) - v'(t) \leq g(t, \varphi^{-1}(\underline{v}(t))) - g(t, \varphi^{-1}(\underline{v}(t))) = 0 \quad \text{for a.e. } t \in (a, b).$$

Thus

$$\underline{v}(t) - v(t) = \underline{v}(a) - v(a) + \int_a^t (\underline{v}'(s) - v'(s)) \, ds \leq 0, \quad t \in (a, b],$$

which contradicts (2.1.7), and hence implies that $\underline{v} \leq v$. Similarly, it can be shown that if v is a solution of (2.1.6), then $v \leq \overline{v}$. These results and (2.1.5) imply that v is a solution of (2.1.6) if and only if v is a solution of (2.1.3) in $[\underline{v}, \overline{v}]$. This proves that v_* and v^* are extremal solutions of (2.1.3) in $[\underline{v}, \overline{v}]$. Since φ^{-1} is strictly increasing, it then follows by Lemma 2.1.1 that $u_* = \varphi^{-1} \circ v_*$ and $u^* = \varphi^{-1} \circ v^*$ are extremal solutions of (2.1.1) in $[\underline{u}, \overline{u}]$.

To prove (2.1.4), let u_+ be an upper solution of (2.1.1) in $[\underline{u}, \overline{u}]$. Replacing $\overline{u}$ by u_+ in the above proof it follows that the IVP (2.1.1) has a solution $u \in [\underline{u}, u_+] \subseteq [\underline{u}, \overline{u}]$. But u_* is the least of all the solutions of (2.1.1) in $[\underline{u}, \overline{u}]$, so that $u_* \leq u_+$. Similarly, it can be shown that if u_- is a lower solution of (2.1.1) in $[\underline{u}, \overline{u}]$, then $u_- \leq u^*$. Noticing also that u_* is an upper solution and u^* a lower solution of (2.1.1), we obtain (2.1.4).   □

Remark 2.1.2. By the above proof it suffices that x is restricted in (g0) to $\underline{u}(t) \leq x \leq \overline{u}(t)$, $t \in J$. The so-restricted hypothesis (g0) holds, e.g., when

$$g(t, x) = \begin{cases} 0, & x > t, \\ 1, & x \leq t, \end{cases} \quad t \in J = [0, 1], \ x \in \mathbb{R}, \ \varphi(x) \equiv x \ \text{ and } x_0 = 0$$

(cf. [125, Ex. 1.2]). For these g, φ, and x_0 the functions $\underline{u}(t) \equiv 0$ and $\overline{u}(t) = t$ are such lower and upper solutions the IVP (2.1.1), and it has a unique solution $u(t) = t$, $t \in J$.

As an application of Lemma B.7.1, Lemma 2.1.1, and Theorem 2.1.2 we shall now prove the following generalization to Theorem 2.1.1.

Theorem 2.1.3. *Assume that the functions* $\varphi \colon \mathbb{R} \to \mathbb{R}$ *and* $g \colon J \times \mathbb{R} \to \mathbb{R}$ *have properties* $(\varphi 0)$, $(g0)$, *and* $(g2)$. *Then the IVP (2.1.1) has for each* $x_0 \in \mathbb{R}$ *the least solution* u_* *and the greatest solution* u^*. *Moreover,*

$$\begin{cases} u_*(t) = \min\{u_+(t) \mid u_+ \text{ is an upper solution of (2.1.1)}\}, \\ u^*(t) = \max\{u_-(t) \mid u_- \text{ is a lower solution of (2.1.1)}\}. \end{cases} \quad (2.1.8)$$

Proof. Let $x_0 \in \mathbb{R}$ be given. Choose $w_0 \in \mathbb{R}$ so that $|\varphi(x_0)| \leq w_0$. If $v \in AC(J)$ is a solution of (2.1.3), it follows from (g2) that

$$|v'(t)| = |g(t, \varphi^{-1}(v(t)))| \leq p_1(t)\psi(|v(t)|) \ \text{ a.e. in } \ J.$$

Thus

$$|v(t)| \leq |\varphi(x_0)| + \int_{t_0}^t |v'(s)| \, ds \leq w_0 + \int_{t_0}^t p_1(s)\psi(|v(s)|) \, ds$$

for all $t \in J$. This implies by Lemma B.7.1 that $|v(t)| \leq w(t)$ on J, where w is the solution of the IVP

$$w'(t) = p_1(t)\psi(w(t)) \quad \text{a.e. in } J, \quad w(t_0) = w_0. \tag{2.1.9}$$

Moreover, applying (g2) and (2.1.9) we obtain

$$|g(t, \varphi^{-1}(x))| \leq p_1(t)\psi(w(t)) = w'(t)$$

for a.e. $t \in J$ and all $x \in [-w(t), w(t)]$. This implies that $-w$ and w are lower and upper solutions of (2.1.3).

The above proof and Lemma 2.1.1 imply that $\underline{u} = \varphi^{-1} \circ (-w)$ and $\overline{u} = \varphi^{-1} \circ w$ are lower and upper solutions of the IVP (2.1.1), and that all the solutions of (2.1.1) belong to the order interval $[\underline{u}, \overline{u}]$. Moreover, if $t \in J$ and $x \in [\underline{u}(t), \overline{u}(t)]$, then $\varphi(x) \in [-w(t), w(t)]$, so that for a.e. $t \in J$ and for all $x \in [\underline{u}(t), \overline{u}(t)]$,

$$|g(t, x)| = |g(t, \varphi^{-1}(\varphi(x)))| \leq w'(t).$$

Thus the hypothesis (A) holds, whence the IVP (2.1.1) has by Theorem 2.1.2 extremal solutions u_* and u^* in $[\underline{u}, \overline{u}]$. Because all the solutions of (2.1.1) belong to $[\underline{u}, \overline{u}]$, then u_* and u^* are the extremal solutions of (2.1.1).

To prove the last assertion, let u_+ be an upper solution of (2.1.1). Choose w_0 above so that $-w_0 \leq \varphi(u_+(t))$ on J. Then $-w \leq \varphi \circ u_+$, whence $\underline{u} \leq u_+$, and the IVP (2.1.1) has by Theorem 2.1.2 a solution $u \in [\underline{u}, u_+]$. But u_* is the least of all the solutions of (2.1.1), so that $u_* \leq u_+$. Similarly, it can be shown that if u_- is a lower solution of (2.1.1), then $u_- \leq u^*$. Since u_* is an upper solution and u^* a lower solution of (2.1.1), we obtain (2.1.8). □

The hypothesis (g0) can be weakened if g is nonnegative-valued.

Theorem 2.1.4. *Assume that a function $\varphi \colon \mathbb{R} \to \mathbb{R}$ has property (φ0), and that a function $g \colon J \times \mathbb{R} \to \mathbb{R}$ satisfies the following hypotheses.*

(g01) *$g(\cdot, x)$ is measurable and nonnegative-valued for each $x \in \mathbb{R}$.*
(g02) *For each $(s, z) \in [t_0, t_1) \times \mathbb{R}$ there exist $\delta > 0$ and $\epsilon > 0$ such that*
$$\limsup_{y \to x-} g(t, y) \leq g(t, x) \quad \text{for a.e. } t \in [s, s+\delta] \text{ and all } x \in (z, z+\epsilon],$$
and $g(t, x) \leq \liminf_{y \to x+} g(t, y)$ for a.e. $t \in [s, s+\delta]$ and all $x \in [z, z+\epsilon)$.

a) *If the hypothesis (A) holds, then the IVP (2.1.1) has the least solution u_* and the greatest solution u^* in the order interval $[\underline{u}, \overline{u}]$, and (2.1.4) holds.*
b) *If the hypothesis (g2) holds, then (2.1.1) has for each $x_0 \in \mathbb{R}$ the least solution u_* and the greatest solution u^*, and (2.1.8) holds.*

Proof. Consider first the IVP

$$u'(t) = g(t, u(t)) \quad \text{for a.e. } t \in J, \quad u(t_0) = x_0, \tag{2.1.10}$$

where g is bounded by $p_1 \in L_+^1(J)$ and has properties (g01) and (g02), and where $x_0 \in \mathbb{R}$ is given. Choose $\delta > 0$ and $\epsilon > 0$ such that (g02) holds when $(s, z) = (t_0, x_0)$, and denote $J_0 = [t_0, t_0 + \delta]$. We may also assume that $\int_{J_0} p_1(t)dt < \epsilon$. Thus the function $f \colon J_0 \times \mathbb{R} \to \mathbb{R}$, defined by

$$f(t, x) := g(t, \max\{x_0, \min\{x, x_0 + \epsilon\}\}), \ t \in J_0, \ x \in \mathbb{R},$$

satisfies the hypotheses of Theorem 2.1.1 when J is replaced by J_0, whence the IVP (2.1.2) has extremal solutions u_* and u^* on J_0. Because g is nonnegative-valued, the above choices of δ and ϵ and the definitions of J_0 and f ensure that (2.1.2) and (2.1.10) have the same solutions on J_0. In particular, u_* is the least solution of (2.1.10) on J_0. Denote

$$t_2 = \sup\{t_3 \in J \mid (2.1.10) \text{ has the least solution } u_* \text{ on } [t_0, t_3]\}.$$

Obviously, $t_2 = t_1$, for otherwise we can repeat the above reasoning when $(s, z) = (t_2, u_*(t_2))$, and obtain a continuation of u_* to $J_1 = [t_0, t_2 + \delta]$, which contradicts the choice of t_2. This proves that (2.1.10) has the least solution on J, the proof for the existence of the greatest solution being similar. The proofs of a) and b) are then similar to the proofs of Theorems 2.1.2 and 2.1.3. $\square$

Remark 2.1.3. Theorem 2.1.4 differs from Theorem 2.1.3 in the sense that the hypothesis (g0) does not allow the function $g(t, \cdot)$ to have jumps downwards if t belongs to a complement of a fixed null set of J, whereas such discontinuities are possible under the hypotheses of Theorem 2.1.4.

2.1.4   Dependence on data. As a consequence of Theorem 2.1.3 we obtain the following result for the IVP

$$\frac{d}{dt}\varphi(u(t)) = g(t, u(t)) + h(t) \quad \text{for a.e. } t \in J, \quad u(t_0) = x_0. \tag{2.1.11}$$

Proposition 2.1.1. *If the hypotheses ($\varphi0$), (g0), and (g2) hold, then the IVP (2.1.11) has for all $h \in L^1(J)$ and $x_0 \in \mathbb{R}$ extremal solutions, and they are increasing with respect to x_0, h, and g.*

Proof. Given $x_0, \hat{x}_0 \in \mathbb{R}$, $h, \hat{h} \in L^1(J)$, and $g, \hat{g} \colon J \times \mathbb{R} \to \mathbb{R}$, assume that g and $\hat{g}$ have properties (g0) and (g2), and that

$$x_0 \leq \hat{x}_0, \ h \leq \hat{h} \ \text{and} \ g(\cdot, x) \leq \hat{g}(\cdot, x) \ \text{for all} \ x \in \mathbb{R}.$$

The functions $(t, x) \mapsto g(t, x) + h(t)$ and $(t, x) \mapsto \hat{g}(t, x) + \hat{h}(t)$ satisfy (g0), and also (g2) when p_1 and ψ are replaced by $t \mapsto p_1(t) + |h(t)| + |\hat{h}(t)|$ and $z \mapsto \psi(z) + 1$, respectively. Denoting by $\hat{u}$ the least solution of the IVP

$$\frac{d}{dt}\varphi(u(t)) = \hat{g}(t, u(t)) + \hat{h}(t) \quad \text{for a.e. } t \in J, \quad u(t_0) = \hat{x}_0, \qquad (2.1.12)$$

it follows from the above hypotheses that $\hat{u}$ is an upper solution of (2.1.11). This and (2.1.8) imply that $u_* \leq \hat{u}$. Similarly, it can be shown that if $\hat{u}$ is the greatest solution of (2.1.12), then $u^* \leq \hat{u}$, which concludes the proof. $\square$

Next we shall prove a result concerning right-continuity of the greatest solution of (2.1.1) with respect to x_0 and g.

Proposition 2.1.2. *Given a function $g: J \times \mathbb{R} \to \mathbb{R}$ and a decreasing sequence of functions $g_n: J \times \mathbb{R} \to \mathbb{R}$ which satisfy (g2) and the following hypotheses.*

(g0') *For each $x \in \mathbb{R}$ the function $g(\cdot, x)$ is measurable, and*
$$\limsup_{y \to x-} g(t, y) \leq g(t, x) = \lim_{y \to x+} g(t, y) \text{ for a.e. } t \in J.$$

(gn) $\lim\limits_{n \to \infty} \sup\limits_{x \in [a,b]} (g_n(s, x) - g(s, x)) = 0$ *for a.e. $s \in J$ and whenever $a \leq b$.*

Let $(x_n)_{n=1}^{\infty} \subset \mathbb{R}$ be a decreasing sequence which converges to $x_0 \in \mathbb{R}$, and let $\varphi: \mathbb{R} \to \mathbb{R}$ satisfy (φ0). Then the IVPs

$$\frac{d}{dt}\varphi(u(t)) = g_n(t, u(t)) \quad \text{a.e. in } J, \quad u(t_0) = x_n, \qquad (2.1.13)$$

have greatest solutions u_n, which converge uniformly on J to the greatest solution u of the IVP (2.1.1).

Proof. Hypothesis (g0') implies that (g0) holds for g and g_n, and that $g(\cdot, u(\cdot))$ and $g_n(\cdot, u(\cdot))$ are measurable when $u \in C(J)$. The proof of Theorem 2.1.3 ensures that the IVP

$$v'(t) = g_n(t, \varphi^{-1}(v(t))) \text{ a.e. in } J, \quad v(t_0) = \varphi(x_n), \qquad (2.1.14)$$

has for each $n = 1, 2, \ldots$ the greatest solution v_n. If $n < m$, then $x_m \leq x_n$ and $g_m(t, \varphi^{-1}(x)) \leq g_n(t, \varphi^{-1}(x))$ in $J \times \mathbb{R}$, so that v_m is a lower solution of (2.1.14). Thus $v_m \leq v_n$ by Theorem 2.1.3 and Lemma 2.1.1, whence the sequence $(v_n)_{n=1}^{\infty}$ is decreasing. Denote $w_0 = \sup\{|\varphi(x_n)| \mid n = 1, 2, \ldots\}$, and let w be the solution of the IVP

$$w'(t) = p_1(t)\psi(w(t)) \text{ a.e. in } J, \quad w(t_0) = w_0. \qquad (2.1.15)$$

Applying (g2) we obtain for each $n = 1, 2, \ldots$ and for all $t \in J$,

$$|v_n(t)| \le |\varphi(x_n)| + \int_{t_0}^t |g_n(s, \varphi^{-1}(v_n(s)))| ds \le w_0 + \int_{t_0}^t p_1(s)\psi(|(v_n(s)|) ds.$$

This implies by Lemma B.7.1 that $|v_n(t)| \le w(t)$ for each $t \in J$ and for each $n = 1, 2, \ldots$. In particular,

$$v_n(t) \in [-w(t_1), w(t_1)] \quad \text{for all } n = 1, 2, \ldots \text{ and } t \in J. \tag{2.1.16}$$

If $t_0 \le a \le b \le t_1$, we have for each $n = 1, 2, \ldots$,

$$\begin{aligned}
|v_n(b) - v_n(a)| &\le \int_a^b |g_n(s, \varphi^{-1}(v_n(s)))| ds \\
&\le \int_a^b p(s)\psi(w(s) ds = w(b) - w(a).
\end{aligned} \tag{2.1.17}$$

Thus the sequence $(v_n)_{n=1}^\infty$ is decreasing, uniformly bounded by (2.1.16) and equicontinuous by (2.1.17), whence it converges uniformly on J to a function v which has the property

$$|v(b) - v(a)| \le |w(b) - w(a)|, \quad a, b \in J.$$

In particular, $v \in AC(J)$. It follows from (2.1.16) when $n \to \infty$ that

$$v(t) \in [-w(t_1), w(t_1)] \quad \text{for all } t \in J.$$

Since $g(t, \cdot)$ is right-continuous for a.e. $t \in J$ by (g0'), since the sequence $(\varphi^{-1} \circ v_n)$ is decreasing and converges uniformly to $\varphi^{-1} \circ v$, and since (gn) holds, then

$$g_n(s, \varphi^{-1}(v_n(s))) \to g(s, \varphi^{-1}(v(s))) \quad \text{as } n \to \infty \text{ for a.e. } s \in J. \tag{2.1.18}$$

Each v_n satisfies the integral equation

$$v_n(t) = \varphi(x_n) + \int_{t_0}^t g_n(s, \varphi^{-1}(v_n(s))) \, ds, \quad t \in J. \tag{2.1.19}$$

Because $x_n \to x_0$ and $v_n(t) \to v(t)$, and since both φ and φ^{-1} are continuous, it follows from (2.1.19) when $n \to \infty$, applying also (2.1.18) and the dominated convergence theorem, that

$$v(t) = \varphi(x_0) + \int_{t_0}^t g(s, \varphi^{-1}(v(s))) \, ds, \quad t \in J. \tag{2.1.20}$$

This implies that v is a solution of the IVP (2.1.3).

Denote by $\hat{v}$ the greatest solution of (2.1.3). Since $\hat{v}$ is a lower solution of (2.1.14) for each $n \in \mathbb{N}$, then $\hat{v}(t) \leq v_n(t)$, $t \in J$, $n \in \mathbb{N}$. This implies when $n \to \infty$ that $\hat{v}(t) \leq v(t)$ on J. The reverse inequality holds since v is a solution of (2.1.3) and $\hat{v}$ is its greatest solution. Thus $v = \hat{v}$, i.e., v is the greatest solution of (2.1.3).

The above results and Lemma 2.1.1 imply that the function $u_n = \varphi^{-1} \circ v_n$ is for each $n = 1, 2, \ldots$, a greatest solution of the IVP (2.1.13), and that (u_n) is a decreasing sequence which converges uniformly to the greatest solution of the IVP (2.1.1). This concludes the proof. $\quad\square$

Remarks 2.1.4. The hypothesis that the sequence $(x_n)_{n=1}^{\infty}$ is decreasing is essential in Proposition 2.1.2. For instance, if H is the Heaviside function, then the IVP

$$u'(t) = H(u(t)) \quad \text{a.e. in } J, \quad u(0) = 0,$$

has $u(t) = t$ as its only solution, and the IVP

$$u_n'(t) = H(u_n(t)) \quad \text{a.e. in } J, \quad u_n(0) = -\frac{1}{n},$$

has for each $n = 1, 2, \ldots$ a unique solution $u_n(t) \equiv -\frac{1}{n}$ on J, so that the sequence $(u_n)_{n=1}^{\infty}$ does not converge even pointwise to u on J. This holds also for the solutions of the IVPs

$$u_n'(t) = H(u_n(t) - \frac{1}{n}) \quad \text{a.e. in } J, \quad u_n(0) = \frac{1}{2n},$$

so that the result of Proposition 2.1.2 does not necessarily hold if, instead of (gn), we assume that the sequence $(g_n)_{n=1}^{\infty}$ is increasing and converges pointwise to g.

Proposition 2.1.2 has an obvious dual for left-continuity of the least solution of (2.1.1) with respect to x_0 and g.

2.1.5 A special case. The IVP

$$u'(t) = q(u(t))g(t, u(t)) \quad \text{a.e. in } J, \quad u(t_0) = x_0, \qquad (2.1.21)$$

can be reduced to the IVP of the form

$$\frac{d}{dt}\varphi(u(t)) = g(t, u(t)) \quad \text{a.e. in } J, \quad u(t_0) = x_0, \qquad (2.1.1)$$

if $q \colon \mathbb{R} \to (0, \infty)$ satisfies one of the following hypotheses.

(q0) q and $\frac{1}{q}$ belong to $L_{loc}^{\infty}(\mathbb{R})$, and $\int_0^{\pm\infty} \frac{dz}{q(z)} = \pm\infty$.

(q1) $q \in L_{loc}^{\infty}(\mathbb{R})$, $\frac{1}{q} \in L_{loc}^1(\mathbb{R})$ and $\int_0^{\pm\infty} \frac{dz}{q(z)} = \pm\infty$.

This is shown in the next two Lemmas, the first one being an obvious consequence of the properties assumed for q in (q0) and (q1).

Lemma 2.1.2. *If (q0) or (q1) holds, then the function $\varphi \colon \mathbb{R} \to \mathbb{R}$, defined by*

$$\varphi(x) = \int_0^x \frac{dz}{q(z)}, \qquad x \in \mathbb{R}, \tag{2.1.22}$$

is an increasing homeomorphism, its inverse φ^{-1} is locally Lipschitz continuous if (q1) holds, and both φ and φ^{-1} are locally Lipschitz continuous if (q0) is valid. In particular, φ has property $(\varphi 0)$.

Lemma 2.1.3. *If (q0) holds, then $u \in AC(J)$ is a lower solution, an upper solution, or a solution of (2.1.21) if and only if u is a lower solution, an upper solution, or a solution of the IVP (2.1.1), where $\varphi \colon \mathbb{R} \to \mathbb{R}$ is defined by (2.1.22).*

Proof. Let u be a solution of (2.1.21). Then $u \in AC(J)$, and (q0) ensures that $\frac{1}{q}$ is measurable and locally essentially bounded. Thus an application of Lemma C.3.2 yields

$$\varphi(u(t)) - \varphi(u(t_0)) = \int_{u(t_0)}^{u(t)} \frac{dz}{q(z)} = \int_{t_0}^t \frac{u'(s)ds}{q(u(s))}, \qquad t \in J.$$

This implies that $\varphi \circ u \in AC(J)$, and that

$$\frac{d}{dt}\varphi(u(t)) = \frac{d}{dt} \int_{t_0}^t \frac{u'(s)ds}{q(u(s))} = \frac{u'(t)}{q(u(t))} = g(t, u(t)) \quad \text{a.e. in} \ \ J.$$

Thus u is a solution of the IVP (2.1.1).

Conversely, let $u \in Y$ be a solution of (2.1.1). Then $\varphi \circ u \in AC(J)$, and since φ^{-1} is locally Lipschitz continuous, then $u \in AC(J)$. Consequently,

$$\varphi(u(t)) - \varphi(u(t_0)) = \int_{u(t_0)}^{u(t)} \varphi'(z)dz = \int_{u(t_0)}^{u(t)} \frac{dz}{q(z)} = \int_{t_0}^t \frac{u'(s)ds}{q(u(s))}, \qquad t \in J,$$

so that

$$g(t, u(t)) = \frac{d}{dt}\varphi(u(t)) = \frac{d}{dt} \int_{t_0}^t \frac{u'(s)ds}{q(u(s))} = \frac{u'(t)}{q(u(t))} \quad \text{a.e. in} \ \ J.$$

Thus u is a solution of the IVP (2.1.21).

The above proof shows that every solution of (2.1.21) is a solution of (2.1.1) and vice versa. Obvious modifications to the above proof show that problems (2.1.21) and (2.1.1) have the same upper and lower solutions. □

According to Lemma 2.1.2 and Lemma 2.1.3 the results derived for the IVPs (2.1.1) and (2.1.11) have the following consequences.

Proposition 2.1.3. *The results of Theorems 2.1.2 and 2.1.3 hold for the IVP (2.1.21), and the results of Proposition 2.1.1 hold for the IVP*

$$u'(t) = q(u(t))(g(t, u(t)) + h(t)) \quad for \ a.e. \ t \in J, \quad u(t_0) = x_0, \quad (2.1.23)$$

if we replace the hypothesis ($\varphi 0$) by ($q 0$).

The next result is a consequence of Theorem 2.1.4.

Proposition 2.1.4. *Let $q : \mathbb{R} \to (0, \infty)$ and $g : J \times \mathbb{R} \to \mathbb{R}$ have properties ($q 1$), ($g 01$), ($g 02$) and ($g 2$). Then the IVP (2.1.21) has for each $x_0 \in \mathbb{R}$ extremal solutions, which are increasing with respect to x_0 and g.*

Proof. Because g is nonnegative-valued by ($g 01$), then each solution of (2.1.21) is increasing. Hence, applying the second part of Lemma C.3.2 one can show as in the proof of Lemma 2.1.3 that $u \in AC(J)$ is a solution of (2.1.21) if and only if u is a solution of the IVP (2.1.1), where $\varphi \colon \mathbb{R} \to \mathbb{R}$ is defined by (2.1.22). The given assumptions and Lemma 2.1.2 ensure that in such a case the hypotheses of Theorem 2.1.4 are valid, which concludes the proof.   □

Remarks 2.1.5. The function $\varphi \colon \mathbb{R} \to \mathbb{R}$, defined by

$$\varphi(x) = |x|^{p-2} x, \qquad x \in \mathbb{R}, \quad (2.1.24)$$

has property ($\varphi 0$) for each $p > 1$. But φ is not locally Lipschitz-continuous if $p \in (1, 2)$, and φ^{-1} is not locally Lipschitz-continuous if $p > 2$. It then follows from Lemma 2.1.2 that the function φ defined by (2.1.24) is of the form (2.1.22), where q has property ($q 0$) (resp. ($q 1$)), only when $p = 2$ (resp. $p \in (1, 2]$). Thus problem (2.1.1) is more general than problem (2.1.21).

By Remarks 2.1.1 we can replace $\psi(|\varphi(x)|)$ by $\psi(|x|)$ in ($g 2$) if φ is Lipschitz-continuous. This holds for the function φ, defined by (2.1.22) if $\frac{1}{q}$ is essentially bounded.

2.1.6 Examples and counterexamples.

Example 2.1.1. Choose $J = [0, 1]$, and define a function $q \colon \mathbb{R} \to (0, \infty)$ by

$$q(z) = \sum_{m=1}^{\infty} \sum_{k=1}^{\infty} \frac{(2 + [k^{\frac{1}{m}} z] - k^{\frac{1}{m}} z)}{(km)^2} \left(2 + \sin \frac{1}{1 + [k^{\frac{1}{m}} z] - k^{\frac{1}{m}} z} \right), \quad (2.1.25)$$

where $[x]$ denotes the greatest integer $\leq x$. It is easy to see that q is discontinuous at $\frac{n}{k^{\frac{1}{m}}}$ for all $n \in \mathbb{Z}$, k, $m = 1, 2, \dots$. Moreover, $1 \leq q(z) \leq \frac{\pi^4}{6}$ for each $z \in \mathbb{R}$, so that q has property ($q 0$).

The function

$$g(t,x) = \sum_{m=-\infty}^{\infty}\sum_{n=1}^{\infty}\frac{f(t,x-\frac{m}{n})}{2^{|m|+n}}, \text{ where } f(t,x) = \begin{cases} \cos\frac{1}{x-t}+2, & x > t, \\ \chi_U(t), \ U \subset J, & x = t, \\ \cos\frac{1}{x-t}-2, & x < t, \end{cases}$$

(2.1.26)

has properties (g0), and (g2). The set of all the discontinuity points of g is $\{(t,t+q) \mid t \in J, q \in \mathbb{Q}\}$. If $u(t) \equiv t$, then $f(\cdot,u(\cdot))$ is equal to the characteristic function χ_U of U, which is not measurable if U is nonmeasurable.

Proposition 2.1.3 implies that the IVP (2.1.21) has extremal solutions when q is given by (2.1.25) and g by (2.1.26).

The function $g \colon J \times \mathbb{R} \to \mathbb{R}$, defined by

$$g(t,x) = \sum_{m=-\infty}^{\infty}\sum_{n=1}^{\infty}\frac{f(t,x-\frac{m}{n})}{2^{|m|+n}}, \text{ where } f(t,x) = \begin{cases} (x-t)\cos\frac{1}{x-t}, & x > t, \\ 0, & x = t, \\ \cos\frac{1}{x-t}-2, & x < t, \end{cases}$$

has properties (g2) and (g0'), assumed in Proposition 2.1.2. The discontinuity points of g are the same as those of the function given by (2.1.26).

Example 2.1.2. The points

$$c(n_0,\ldots,n_m) = 1 - 2^{-m-1} - \sum_{k=0}^{m} 2^{-k-m-2}\prod_{j=0}^{k}2^{-n_j} - 2^{-2m-2}\prod_{j=0}^{m}2^{-n_j},$$

where $m, n_0, \ldots, n_m \in \mathbb{N}$, form a well-ordered set C of rational numbers with $\min C = 0$ and $\sup C = 1$ (cf. [141, Ex. 1.1.1]). Define

$$f(z) = \frac{z - c(n_0,\ldots,n_m)}{c(n_0,\ldots,n_m+1) - c(n_0,\ldots,n_m)},$$
$$c(n_0,\ldots,n_m) \le z < c(n_0,\ldots,n_m+1), \quad m, n_0, \ldots, n_m \in \mathbb{N},$$

and

$$g(t,x) = f(t+x-[t+x]), \quad t \in J = [0,1], \ x \in \mathbb{R}.$$

It is easy to see that g has properties (g01), (g02) and (g2), so that the IVP

$$\frac{d}{dt}(|u(t)|^{p-2}u(t)) = f(t+u(t) - [t+u(t)]) \quad \text{a.e. in } J, \quad u(0) = x_0,$$

has by Theorem 2.1.4 and Remarks 2.1.4 extremal solutions when $x_0 \in \mathbb{R}$ and $p > 1$.

The IVP
$$u'(t) = g(t, u(t)) \quad \text{a.e. in } J, \quad u(0) = 0$$

has no solution on $J = [0, T]$ for any $T > 0$ if g is one of the functions

$$g(t, x) = \begin{cases} 2, & x < t, \\ \frac{1}{2}, & x \geq t, \end{cases} \qquad g(t, x) = \begin{cases} 1, & x \leq 0, \\ 0, & x > 0, \end{cases} \qquad t \in J.$$

Thus the property $\limsup\limits_{y \to x-} g(t, y) \leq g(t, x) \leq \liminf\limits_{y \to x+} g(t, y)$ is needed at least between assumed lower or upper solutions (cf. Remark 2.1.2) or locally (cf. Theorem 2.1.4).

The functions $\varphi \colon \mathbb{R} \to \mathbb{R}$ and $g \colon J \times \mathbb{R} \to \mathbb{R}$, defined by

$$\varphi(x) = x^{\frac{1}{3}}, \quad g(t, x) = \frac{2}{3} t x^{\frac{4}{3}}, \quad t \in J, x \in \mathbb{R},$$

satisfy conditions (φ0) and (g0) but not condition (g2). When $t_0 = 0$ and $x_0 = 1$, the IVP (2.1.1) can be rewritten in this case as

$$u'(t) = 2t u(t)^2 \quad \text{a.e. in } J = [0, t_1], \quad u(0) = 1.$$

This IVP does not have any solution in $AC(J)$ if $t_1 \geq 1$, since the only possible solution is $u(t) = \frac{1}{1-t^2}$. Thus condition (g2) cannot be omitted in general.

The IVP

$$\frac{d}{dt} u^3(t) = 3t^2 \cos^3(\frac{\pi}{t}) + t \cos^2(\frac{\pi}{t}) \sin(\frac{\pi}{t}) \quad \text{a.e. in } J = [0, T], \quad u(0) = 0,$$

is of the form (2.1.1), where $\varphi(x) = x^3$, and

$$g(t, x) = \begin{cases} 3t^2 \cos^3(\frac{\pi}{t}) + t \cos^2(\frac{\pi}{t}) \sin(\frac{\pi}{t}), & t \in (0, T], x \in \mathbb{R}, \\ 0, & t = 0. \end{cases}$$

These functions φ and g have properties (φ0), (g0) and (g2). It is easy to see that the only possible solution is

$$u(t) = \begin{cases} t \cos(\frac{\pi}{t}), & t \in (0, T], \\ 0, & t = 0. \end{cases}$$

Since $u \in C(J)$ and $\varphi \circ u = u^3 \in AC(J)$, then $u \in Y$, whence u is a solution in the sense of Definition 2.1.1. On the other hand, u is not of bounded variation, and hence not absolutely continuous on $J = [0, T]$ for any $T > 0$. This justifies the choice of the solution set Y in Definition 2.1.1 to be a subset of $C(J)$.

Constant multiples of a Cantor function (cf. [155], p. 338) are solutions to problem

$$u'(t) = 0 \text{ a.e. in } J = [0, 1], \quad u(0) = 0,$$

which then has a continuum of continuous and monotone solutions having no extremal solutions. This justifies condition $\varphi \circ u \in AC(J)$ in Definition 2.1.1.

2.2 Explicit boundary value problems

In this section we present existence and comparison results for first order discontinuous differential equations equipped with discontinuous, implicit, and functional boundary conditions. Some results of subsection 2.1.3 are used in the proofs.

2.2.1 Hypotheses and preliminaries. Consider the boundary value problem (BVP)

$$\frac{d}{dt}\varphi(u(t)) = g(t, u(t)) \text{ a.e. in } J = [t_0, t_1], \quad B(u(t_0), u) = 0, \quad (2.2.1)$$

where $g: J \times \mathbb{R} \to \mathbb{R}$, $\varphi: \mathbb{R} \to \mathbb{R}$ and $B: \mathbb{R} \times C(J) \to \mathbb{R}$. We assume that $C(J)$ is equipped with the pointwise ordering $\leq$.

Definition 2.2.1. We say that a function $u \in C(J)$ is a *lower solution* of (2.2.1) if $\varphi \circ u \in AC(J)$ and

$$\frac{d}{dt}\varphi(u(t)) \leq g(t, u(t)) \text{ a.e. in } J, \quad B(u(t_0), u) \leq 0,$$

and an *upper solution* of (2.2.1) if the reversed inequalities hold. If equalities hold, we say that u is a *solution* of (2.2.1).

The following hypotheses are imposed on the functions φ, g, and B.

(φ0) φ is an increasing homeomorphism.

(g0) For each $x \in \mathbb{R}$ the function $g(\cdot, x)$ is measurable, and
$$\limsup_{y \to x-} g(t, y) \leq g(t, x) \leq \liminf_{y \to x+} g(t, y) \text{ for a.e. } t \in J.$$

(B0) For each $x \in \mathbb{R}$ the function $B(x, \cdot)$ is decreasing, and
$$\liminf_{y \to x-} B(y, u) \geq B(x, u) \geq \limsup_{y \to x+} B(y, u) \text{ for all } u \in C(J).$$

Moreover, if

(A) (2.2.1) has lower and upper solutions $\underline{u}$ and $\overline{u}$ such that $\underline{u} \leq \overline{u}$, and g is L^1-bounded in the set $\Omega = \{(t, x) \in J \times \mathbb{R} \mid \underline{u}(t) \leq x \leq \overline{u}(t)\}$,

we prove that there exists the least and the greatest among those solutions of (2.2.1) which belong to the order interval $[\underline{u}, \overline{u}] = \{u \in C(J) \mid \underline{u} \leq u \leq \overline{u}\}$.

If (A) is replaced by the following hypotheses:

(g2) $|g(t, x)| \leq p_1(t)\psi(|\varphi(x)|)$ for all $x \in \mathbb{R}$ and for a.e. $t \in J$, where p_1 belongs to $L^1_+(J)$, $\psi \colon \mathbb{R}_+ \to (0, \infty)$ is increasing and $\int_0^\infty \frac{dx}{\psi(x)} = \infty$,

(B1) $|x - B(x, v)| \leq c|x| + d$ for all $(x, v) \in \mathbb{R} \times C(J)$, where $c \in [0, 1)$ and $d \geq 0$,

we prove that (2.2.1) has extremal solutions, i.e., the least and the greatest of all its solutions, and that they are increasing with respect to g and decreasing with respect to B. These results are then applied to the BVP

$$u'(t) = q(u(t))g(t, u(t)) \quad \text{a.e. in } J, \quad B(u(t_0), u) = 0. \tag{2.2.2}$$

Existence of the extremal solutions of (2.2.2) is also proved under growth conditions which are different from (g2) and (B1). Examples and counterexamples are given to illustrate the obtained results.

The following two Lemmas are used in the proof of our first existence result.

Lemma 2.2.1. *Assume that the hypothesis (B0) holds. If $u_1, \ldots, u_n$ are solutions of the BVP (2.2.1), then $\max\{u_1, \ldots, u_n\}$ is a lower solution of (2.2.1), and $\min\{u_1, \ldots, u_n\}$ is an upper solution of (2.2.1).*

Proof. Assume that $u_1, \ldots, u_n$ are solutions of the BVP (2.2.1). Since u_i belongs to $C(J)$ and $\varphi \circ u_i$ belongs to $AC(J)$ for each $i = 1, \ldots, n$, and since φ is strictly increasing, then $u = \max\{u_1, \ldots, u_n\}$ belongs to $C(J)$ and $\varphi \circ u = \max\{\varphi \circ u_1, \ldots, \varphi \circ u_n\}$ belongs to $AC(J)$. Moreover, it is easy to show that

$$\frac{d}{dt}\varphi(u(t)) = g(t, u(t)) \quad \text{a.e. in } J.$$

(B0) implies that if $u_i(t_0) = \max\{u_1(t_0), \ldots, u_n(t_0)\}$, then

$$B(u(t_0), u) = B(u_i(t_0), u) \leq B(u_i(t_0), u_i) = 0.$$

Thus $u = \max\{u_1, \ldots, u_n\}$ is a lower solution of the BVP (2.2.1). The proof that $\min\{u_1, \ldots, u_n\}$ is an upper solution of (2.2.1) is similar. $\square$

The next Lemma is an immediate consequence of Theorems 2.1.2 and 2.1.4.

Lemma 2.2.2. *If the hypotheses (g0), (φ0), and (A) are valid, and if v and w are lower and upper solutions of (2.2.1) such that $\underline{u} \le v \le w \le \overline{u}$, then the IVP*

$$\frac{d}{dt}\varphi(u(t)) = g(t, u(t)) \quad a.e. \ in \ J, \quad u(t_0) = x_0, \qquad (2.1.1)$$

has for each $x_0 \in [v(t_0), w(t_0)]$ extremal solutions in the order interval $[v, w]$. This result holds also when (g0) is replaced by the hypotheses (g01) and (g02) given in Theorem 2.1.4.

2.2.2 Existence of extremal solutions of (2.2.1) in $[\underline{u}, \overline{u}]$. We shall first prove that the BVP (2.2.1) has at least one solution between the assumed lower and upper solutions $\underline{u}$ and $\overline{u}$.

Proposition 2.2.1. *If the hypotheses (φ0), (B0), and (A), and either (g0), or (g01) and (g02) are valid, then the BVP (2.2.1) has a solution in the order interval $[\underline{u}, \overline{u}]$.*

Proof. Denote by X the set of those $x_0 \in [\underline{u}(t_0), \overline{u}(t_0)]$ with the property that (2.1.1) has a solution $u \in [\underline{u}, \overline{u}]$ for which $B(u(t_0), u) \le 0$. The set X is nonempty, because (2.1.1) has for $x_0 = \underline{u}(t_0)$ a solution u in $[\underline{u}, \overline{u}]$ by Lemma 2.2.2, and (B0) implies that

$$B(u(t_0), u) = B(\underline{u}(t_0), u) \le B(\underline{u}(t_0), \underline{u}) \le 0.$$

Denote $x^* = \sup X$, and let $(x_n)_{n=0}^{\infty}$ be an increasing sequence in X which converges to x^*. The definition of X allows us to choose for each $n \in \mathbb{N}$ a function $v_n \in [\underline{u}, \overline{u}]$ such that $\varphi \circ v_n \in AC(J)$, and that

$$\begin{cases} \frac{d}{dt}\varphi(v_n(t)) = g(t, v_n(t)) \quad a.e. \ in \ J, \quad v_n(t_0) = x_n, \\ B(v_n(t_0), v_n) \le 0. \end{cases} \qquad (2.2.3)$$

We may assume that $(v_n)_{n=0}^{\infty}$ is increasing, for otherwise we obtain an increasing sequence $(u_n)_{n=0}^{\infty}$ in $[\underline{u}, \overline{u}]$ by defining $u_n = \max\{v_0, \ldots, v_n\}$, and showing as in the proof of Lemma 2.2.1 that (2.2.3) holds when v_n is replaced by u_n. The hypothesis (A) implies the existence of a function $M \in L^1(J)$ such that

$$|\frac{d}{dt}\varphi(v_n(t))| = |g(t, v_n(t))| \le M(t) \quad a.e. \ in \ J,$$

whence

$$|\varphi(v_n(t_3)) - \varphi(v_n(t_2))| \le \int_{t_2}^{t_3} M(t)dt \quad \text{for all} \ t_2, t_3 \in J, \ t_2 \le t_3.$$

Thus $(\varphi \circ v_n)_{n=0}^{\infty}$ is an absolutely continuous and increasing sequence in $[\varphi \circ \underline{u}, \varphi \circ \overline{u}]$. This and $(\varphi 0)$ imply that $(v_n) \subset [\underline{u}, \overline{u}]$ is an increasing sequence, so that it converges to a function $v \colon J \to \mathbb{R}$ on J. When $n \to \infty$ in the above inequality we obtain

$$|\varphi(v(t_3)) - \varphi(v(t_2))| \leq \int_{t_2}^{t_3} M(t)dt \quad \text{for all } t_2, t_3 \in J, \ t_2 \leq t_3.$$

This implies that $\varphi \circ v \in AC(J)$, whence $v \in C(J)$ by $(\varphi 0)$. It follows from (2.2.3) that

$$\varphi(v_n(t_3)) - \varphi(v_n(t_2)) = \int_{t_2}^{t_3} g(t, v_n(t))dt$$

for all $n \in \mathbb{N}$ and $t_2, t_3 \in J, \ t_2 \leq t_3$. Allowing n to tend to infinity and applying Fatou's Lemma we obtain

$$\int_{t_2}^{t_3} \frac{d}{dt} \varphi(v(t))dt = \varphi(v(t_3)) - \varphi(v(t_2)) \leq \int_{t_2}^{t_3} \limsup_{n \to \infty} g(t, v_n(t))dt$$

for all $t_2, t_3 \in J, \ t_2 \leq t_3$. In view of this inequality and $(g0)$ we get

$$\frac{d}{dt} \varphi(v(t)) \leq \limsup_{n \to \infty} g(t, v_n(t)) \leq \limsup_{y \to v(t)-} g(t, y) \leq g(t, v(t)) \qquad (2.2.4)$$

a.e. in J. The last inequality of (2.2.3) and the hypothesis (B0) ensure that

$$B(v_n(t_0), v) \leq B(v_n(t_0), v_n) \leq 0 \quad \text{for all } n \in \mathbb{N}.$$

These inequalities and another application of (B0) yields

$$B(v(t_0), v) \leq \liminf_{y \to v(t_0)-} B(y, v) \leq \liminf_{n \to \infty} B(v_n(t_0), v) \leq 0,$$

which, together with (2.2.4), implies that v is a lower solution of (2.2.1).

To prove that the BVP (2.2.1) has a solution in the order interval $[v, \overline{u}]$, assume first that $x^* = v(t_0) = \overline{u}(t_0)$. Then the IVP (2.1.1), with $x_0 = x^*$, has by Lemma 2.2.2 a solution u in $[v, \overline{u}]$. Since $u(t_0) = v(t_0) = \overline{u}(t_0)$ and $v \leq u \leq \overline{u}$, the hypothesis (B0) implies that

$$0 \geq B(v(t_0), v) \geq B(u(t_0), u) \geq B(\overline{u}(t_0), \overline{u}) \geq 0.$$

Thus $B(u(t_0), u) = 0$, and since u is a solution of (2.1.1) when $x_0 = x^*$, then u is also a solution of (2.2.1) in $[v, \overline{u}] \subseteq [\underline{u}, \overline{u}]$.

Assume next that $x^* = v(t_0) < \overline{u}(t_0)$, and choose a decreasing sequence (y_n) from $[x^*, \overline{u}(t_0)]$ which converges to x^*. Dual arguments to those used in the construction of the sequence (v_n) above show an existence of a decreasing sequence (w_n) in the order interval $[v, \overline{u}]$ such that each $\varphi \circ w_n$ belongs to $AC(J)$ and

$$\frac{d}{dt}\varphi(w_n(t)) = g(t, w_n(t)) \quad \text{a.e. in} \quad J, \quad w_n(t_0) = x_n \text{ and } B(w_n(t_0), w_n) \geq 0,$$

and which converges on J to an upper solution w of the BVP (2.2.1). Because $v(t_0) = w(t_0)$ and $v \leq w$, then replacing $\overline{u}$ by w in the above reasoning when $v(t_0) = \overline{u}(t_0)$ one can prove that (2.2.1) has a solution in $[v, w] \subset [\underline{u}, \overline{u}]$. This concludes the proof.   □

Now we are ready to prove the main result of this subsection.

Theorem 2.2.1. *If the hypotheses ($\varphi 0$), (B0), and (A), and either (g0), or (g01) and (g02), are valid, then the BVP (2.2.1) has extremal solutions u_* and u^* in the order interval $[\underline{u}, \overline{u}]$. Moreover,*

$$\begin{cases} u_* = \min\{u_+ \mid u_+ \text{ is an upper solution of (2.2.1) in} [\underline{u}, \overline{u}]\}, \\ u^* = \max\{u_- \mid u_- \text{ is a lower solution of (2.2.1) in} [\underline{u}, \overline{u}]\}. \end{cases} \quad (2.2.5)$$

Proof. Denote

$$S = \{u \in C(J) \mid u \text{ is a solution of the BVP (2.2.1) in } [\underline{u}, \overline{u}]\},$$

and define a mapping $w \colon J \to \mathbb{R}$ by

$$w(t) := \sup_{u \in S} u(t), \qquad t \in J.$$

Let $D = \{t_j\}_{j \in \mathbb{N}}$ be a dense subset of J, and choose for each $j \in \mathbb{N}$ such a sequence $(v_k^j)_{k=0}^\infty$ from the solution set S that

$$\lim_{k \to \infty} v_k^j(t_j) = w(t_j), \qquad j \in \mathbb{N}.$$

It follows from Lemma 2.2.1 that the functions $v_n \colon J \to \mathbb{R}$, $n \in \mathbb{N}$, defined by

$$v_n(t) = \max\{v_k^j(t) \mid j, k \in \{1, \ldots, n\}\}, \qquad t \in J,$$

are lower solutions of (2.2.1). Moreover, $(v_n)_{n=0}^\infty$ is an increasing sequence in $[\underline{u}, \overline{u}]$. It can be shown as in the proof of Proposition 2.2.1 that the sequence $(v_n)_{n=0}^{\infty}$ converges on J to a lower solution v of (2.2.1). The above construction implies also that $v \in [\underline{u}, \overline{u}]$, and that

$$v(t_j) = w(t_j) = \sup_{u \in S} u(t_j) \quad \text{for each} \quad j \in \mathbb{N}. \qquad (2.2.6)$$

In particular, the hypotheses of Proposition 2.2.1 hold when $\underline{u}$ is replaced by v, whence the BVP (2.2.1) has a solution u^* in $[v, \overline{u}]$. Thus $u^* \in S$ and $v \leq u^*$. These relations imply by (2.2.6) that if u is any solution of (2.2.1) in $[\underline{u}, \overline{u}]$, then

$$u(t_j) \leq w(t_j) = u^*(t_j), \qquad j \in \mathbb{N}.$$

Since $D = \{t_j\}_{j \in \mathbb{N}}$ is a dense subset of J, it then follows that $u(t) \leq u^*(t)$ on J. Thus u^* is the greatest solution of the BVP (2.2.1) in $[\underline{u}, \overline{u}]$.

The proof that the BVP (2.2.1) has the least solution u_* in $[\underline{u}, \overline{u}]$ is similar. To prove (2.2.5), let u_+ be an upper solution of (2.2.1) in $[\underline{u}, \overline{u}]$. Replacing $\overline{u}$ by u_+ in the above proof it follows that the BVP (2.2.1) has a solution $u \in [\underline{u}, u_+] \subseteq [\underline{u}, \overline{u}]$. But u_* is the least of all the solutions of (2.2.1) in $[\underline{u}, \overline{u}]$, so that $u_* \leq u_+$. Similarly, it can be shown that if u_- is a lower solution of (2.2.1) in $[\underline{u}, \overline{u}]$, then $u_- \leq u^*$. Noticing also that u_* is an upper solution and u^* a lower solution of (2.2.1), we obtain (2.2.5). $\quad \square$

2.2.3 Existence of extremal solutions of (2.2.1). Next we are going to prove a result which ensures the existence of extremal solutions of problem (2.2.1) in the whole solution space.

Theorem 2.2.2. *Assume that the functions $\varphi \colon \mathbb{R} \to \mathbb{R}$, $g \colon J \times \mathbb{R} \to \mathbb{R}$ and $B \colon \mathbb{R} \times C(J) \to \mathbb{R}$ have properties $(\varphi 0)$, $(g0)$, $(g2)$, $(B0)$, and $(B1)$. Then problem (2.2.1) has least and greatest solutions u_* and u^*. Moreover,*

$$\begin{cases} u_*(t) = \min\{u_+(t) \mid u_+ \text{ is an upper solution of } (2.2.1)\}, \\ u^*(t) = \max\{u_-(t) \mid u_- \text{ is a lower solution of } (2.2.1)\}. \end{cases} \qquad (2.2.7)$$

These results hold also when $(g0)$ is replaced by $(g01)$ and $(g02)$.

Proof. Assume first that u is a solution of (2.2.1). Applying (2.2.1) and the hypotheses (B1) and (g2) we obtain

$$|u(t_0)| = |u(t_0) - B(u(t_0), u)| \leq c|u(t_0)| + d, \quad \text{i.e. } |u(t_0)| \leq \frac{d}{1-c},$$

$$\left| \frac{d}{dt} \varphi(u(t)) \right| = |g(t, u(t))| \leq p_1(t) \psi(|\varphi(u(t))|) \quad \text{a.e. in} \quad J.$$

By choosing $w_0 \in \mathbb{R}$ so that

$$-w_0 \leq \varphi(-\frac{d}{1-c}), \varphi(\frac{d}{1-c}) \leq w_0,$$

we get

$$|\varphi(u(t))| \leq |\varphi(u(t_0))| + \int_{t_0}^t p_1(s)\psi(|\varphi(u(s))|)ds \leq w_0 + \int_{t_0}^t p_1(s)\psi(|\varphi(u(s))|)ds$$

for all $t \in J$. This implies by Lemma B.7.1 that $|\varphi(u(t))| \leq w(t)$ on J, where w is the solution of the IVP

$$w'(t) = p_1(t)\psi(w(t)), \quad \text{a.e. in } J, \quad w(t_0) = w_0. \qquad (2.2.8)$$

Defining

$$\underline{u}(t) = \varphi^{-1}(-w(t)), \quad t \in J, \quad \text{and} \quad \overline{u}(t) = \varphi^{-1}(w(t)), \quad t \in J, \qquad (2.2.9)$$

the above considerations, the special choice of w_0 and the hypothesis $(\varphi 0)$ imply that $u \in [\underline{u}, \overline{u}]$. Next we shall show that $\underline{u}$ and $\overline{u}$ are lower and upper solutions of (2.2.1). Since w, as a solution of (2.2.8), belongs to $AC(J)$, it follows from (2.2.9) that $\varphi \circ \underline{u} = -w$ and $\varphi \circ \overline{u} = w$ belong to $AC(J)$. Thus $\underline{u}$ and $\overline{u}$ belong to $C(J)$ by $(\varphi 0)$. Applying (g2), (2.2.8) and (2.2.9) we obtain

$$\begin{cases} \frac{d}{dt}\varphi(\underline{u}(t)) = -w'(t) = -p_1(t)\psi(w(t)) = -p_1(t)\psi(|\varphi(\underline{u}(t))|) \leq g(t, \underline{u}(t)), \\ \frac{d}{dt}\varphi(\overline{u}(t)) = w'(t) = p_1(t)\psi(w(t)) = p_1(t)\psi(|\varphi(\overline{u}(t))|) \geq g(t, \overline{u}(t)), \end{cases}$$

for a.e. $t \in J$. The above choice of w_0 and the monotonicity of φ^{-1} imply

$$\underline{u}(t_0) = \varphi^{-1}(-w_0) \leq -\frac{d}{1-c}, \frac{d}{1-c} \leq \varphi^{-1}(w_0) = \overline{u}(t_0).$$

Hence,

$$-B(\overline{u}(t_0), \overline{u}) = \overline{u}(t_0) - B(\overline{u}(t_0), \overline{u}) - \overline{u}(t_0)$$
$$\leq c|\overline{u}(t_0)| + d - \overline{u}(t_0) = c\overline{u}(t_0) + d - \overline{u}(t_0) \leq 0,$$

$$B(\underline{u}(t_0), \underline{u}) = \underline{u}(t_0) + B(\underline{u}(t_0), \underline{u}) - \underline{u}(t_0)$$
$$\leq c|\underline{u}(t_0)| + d + \underline{u}(t_0) = -c\underline{u}(t_0) + d + \underline{u}(t_0) \leq 0.$$

Moreover, applying (g2), (2.2.8) and (2.2.9) we see that

$$|g(t,x)| \leq p_1(t)\psi(|\varphi(x)|) \leq p_1(t)\psi(w(t)) = w'(t)$$

for a.e. $t \in J$, and for all $x \in [\underline{u}(t), \overline{u}(t)]$, whence the hypothesis (A) holds.

The above proof shows that the hypotheses of Theorem 2.2.1 are satisfied, whence problem (2.2.1) has extremal solutions u_* and u^* in $[\underline{u}, \overline{u}]$. Because all the solutions of (2.2.1) belong to $[\underline{u}, \overline{u}]$, then u_* and u^* are the extremal solutions of (2.2.1). To prove (2.2.7), let u_+ be an upper solution of (2.2.1). We may choose d in (B1) so that $-\frac{d}{1-c} \leq \varphi(u_+(t))$ on J. Then $-w \leq \varphi \circ u_+$, whence $\underline{u} \leq u_+$, so that problem (2.2.1) has by Theorem 2.2.1 a solution $u \in [\underline{u}, u_+]$. But u_* is the least of all the solutions of (2.2.1), whence $u_* \leq u_+$. Similarly, it can be shown that if u_- is a lower solution of (2.2.1), then $u_- \leq u^*$. Noticing also that u_* is an upper solution and u^* is a lower solution of (2.2.1), we obtain (2.2.7). $\qquad\square$

As a consequence of Theorem 2.2.2 we obtain the following result.

Proposition 2.2.2. *If $(\varphi 0)$, $(B0)$, $(g0)$, and $(g2)$ are valid, then problem*

$$\frac{d}{dt}\varphi(u(t)) = g(t, u(t)) + h(t) \quad \text{for a.e.} \quad t \in J, \quad B(u(t_0), u) = 0, \quad (2.2.10)$$

has for all $h \in L^1(J)$ extremal solutions and they are increasing with respect to h and decreasing with respect to B.

Proof. Given $h, \hat{h} \in L^1(J)$ and $B, \hat{B} \colon J \times C(J) \to \mathbb{R}$, assume that B and $\hat{B}$ have properties (B0) and (B1), and that

$$h \leq \hat{h} \quad \text{and} \quad B(x, u) \geq \hat{B}(x, u) \text{ for all } (x, u) \in \mathbb{R} \times C(J).$$

The functions $(t, x) \mapsto g(t, x) + h(t)$ and $(t, x) \mapsto \hat{g}(t, x) + \hat{h}(t)$ satisfy condition (g0), and also condition (g2) when p_1 and ψ are replaced by $t \mapsto p_1(t) + |h(t)| + |\hat{h}(t)|$ and $z \mapsto \psi(z) + 1$, respectively. Denoting by $\hat{u}$ the least solution of problem

$$\frac{d}{dt}\varphi(u(t)) = \hat{g}(t, u(t)) + \hat{h}(t) \quad \text{for a.e.} \quad t \in J, \quad \hat{B}(u(t_0), u) = 0,$$

then $\hat{u}$ is an upper solution of (2.2.10). This and (2.2.7) imply that $u_* \leq \hat{u}$. Similarly, it can be shown that if $\hat{u}$ is the greatest solution of (2.2.10), then $u^* \leq \hat{u}$, which concludes the proof. $\qquad\square$

Example 2.2.1. Choose $J = [0,1]$ and consider the problem

$$\begin{cases} u'(t) = H(u(t) - 4t) + \frac{[u(t)]}{1+|[u(t)]|} & \text{a.e. in } J, \\ u(0) = \frac{[u(0)]}{1+|[u(0)]|} + \frac{2[\int_0^1 u(s)ds]}{1+|[\int_0^1 u(s)ds]|}, \end{cases} \qquad (2.2.11)$$

where H is the Heaviside function and $[x]$ denotes the greatest integer $\le x$. Problem (2.2.11) is of the form (2.2.1) with

$$\begin{cases} g(t,x) = H(x - 4t) + \frac{[x]}{1+|[x]|}, & t \in J, \ x \in \mathbb{R}, \\ B(x,u) = x - \frac{[x]}{1+|[x]|} - \frac{2[\int_0^1 u(t)dt]}{1+|[\int_0^1 u(t)dt]|}, & x \in \mathbb{R}, \ u \in C(J). \end{cases}$$

It is easy to see that the hypotheses of Theorem 2.2.2 hold, whence problem (2.2.11) has the least solution u_* and the greatest solution u^*. These extremal solutions can be determined by using numerical integration methods and inference. Denoting by χ_W the characteristic function of a subset W of $\mathbb{R}$, we get the following representations for u_* and u^*.

$$\begin{cases} u_*(t) = (-\frac{9}{4} - \frac{3}{4}t)\chi_{[0,1]}(t), \ t \in J, \\ u^*(t) = (\frac{13}{6} + \frac{5}{3}t)\chi_{[0,\frac{1}{2}]}(t) + (\frac{17}{8} + \frac{7}{4}t)\chi_{[\frac{1}{2},\frac{17}{18}]}(t) + (\frac{221}{72} + \frac{3}{4}t)\chi_{[\frac{17}{18},1]}(t). \end{cases}$$

The function $u(t) \equiv 0$ is also a solution of (2.2.11).

2.2.4 Special cases. In this subsection we shall consider solvability of the BVP

$$u'(t) = q(u(t))g(t, u(t)) \quad \text{a.e. in } J = [t_0, t_1], \quad B(u(t_0), u) = 0, \qquad (2.2.2)$$

where $g: J \times \mathbb{R} \to \mathbb{R}$, $q: \mathbb{R} \to (0, \infty)$ and $B: \mathbb{R} \times C(J) \to \mathbb{R}$. A function $u \in AC(J)$ is said to be a *lower solution* of the BVP (2.2.2) if

$$u'(t) \le q(u(t))g(t, u(t)) \quad \text{for a.e. } t \in J, \text{ and } B(u(t_0), u) \le 0, \qquad (2.2.12)$$

and an *upper solution* if the reversed inequalities hold. If equalities hold in (2.2.12), we say that u is a *solution* of the BVP (2.2.2).

Lemma 2.1.2 and the proof of Lemma 2.1.3 imply the following result.

Lemma 2.2.3. *Assume that $q: \mathbb{R} \to (0, \infty)$ has the following property.*

(q0) *q and $\frac{1}{q}$ belong to $L^\infty_{loc}(\mathbb{R})$, and $\int_0^{\pm\infty} \frac{dz}{q(z)} = \pm\infty$.*

Then $u \in AC(J)$ is a lower solution, an upper solution or a solution of (2.2.2) if and only if u is a lower solution, an upper solution or a solution of the BVP (2.2.1), respectively, where $\varphi: \mathbb{R} \to \mathbb{R}$ is defined by $\varphi(x) = \int_0^x \frac{dz}{q(z)}$, $x \in \mathbb{R}$. Moreover, φ has property (φ0).

In view of Lemma 2.2.3 we obtain the following result.

Proposition 2.2.3. *The results of Theorems 2.2.1 and 2.2.2 hold for problem (2.2.2) if the hypothesis (φ0) is replaced by property (q0).*

The next result gives another sufficient conditions for the existence of extremal solutions of the BVP (2.2.2).

Theorem 2.2.3. *Assume that functions g, B, and q have properties (g0), (B0), and*

(g3) $|g(t, x)| \leq M(t) + p_1(t)|x|$ *for a.e.* $t \in J$ *and all* $x \in \mathbb{R}$, *where* $p_1, M \in L^1_+(J)$;

(q2) $q \colon \mathbb{R} \to (0, \infty)$ *is measurable and essentially bounded, and* $\frac{1}{q}$ *is locally essentially bounded.*

Assume also the existence of such constants $a > 0$ *and* $b, c \geq 0$ *and such a bounded and nonnegative linear functional* I *on* $C(J)$ *that*

(B2) $|B(x, u) - ax + bI(u)| \leq c$ *for all* $x \in \mathbb{R}$ *and* $u \in C(J)$;

(A1) $a > bI(e^{P(\cdot)})$, *where* $P(t) = \int_{t_0}^{t} \|q\|_\infty p_1(s)\, ds$, $t \in J$.

Then the BVP (2.2.2) has extremal solutions, and all the solutions of (2.2.2) lie within the order interval $[-w, w]$, *where*

$$
\begin{aligned}
w(t) = {} & e^{P(t)} \frac{c + bI(\tau \mapsto \int_{t_0}^{\tau} e^{P(\tau) - P(s)} \|q\|_\infty M(s) ds)}{a - bI(e^{P(\cdot)})} \\
& + \int_{t_0}^{t} e^{P(t) - P(s)} \|q\|_\infty M(s)\, ds, \quad t \in J.
\end{aligned}
\tag{2.2.13}
$$

Proof. It follows from Lemma 2.2.3 that we can assume that q is bounded, i.e., $q(x) \leq \|q\|_\infty$ for all $x \in \mathbb{R}$. It is elementary to verify that w, given by (2.2.13), is a unique solution of the BVP

$$
w'(t) = \|q\|_\infty (M(t) + p_1(t)w(t)) \text{ a.e. in } J, \quad a\, w(t_0) - bI(w) = c. \quad (2.2.14)
$$

Applying the hypotheses (q2) and (g3) we obtain

$$
\begin{cases}
w'(t) = \|q\|_\infty (M(t) + p_1(t)w(t)) \geq q(w(t))g(t, w(t)), \\
-w'(t) = \|q\|_\infty (-M(t) - p_1(t)w(t)) \leq q(-w(t))g(t, -w(t)),
\end{cases}
$$

for a.e. $t \in J$. The boundary condition of (2.2.14) and the assumption (B2) imply that

$$
\begin{cases}
B(w(t_0), w) = B(w(t_0), w) - aw(t_0) + bI(w) + c \geq 0, \\
B(-w(t_0), -w) = B(-w(t_0), -w) - a(-w(t_0)) + bI(-w) - c \leq 0.
\end{cases}
$$

Thus $\underline{u} = -w$ is a lower solution and $\bar{u} = w$ is an upper solution of (2.2.2). Moreover

$$|g(t, u)| \leq M(t) + p_1(t)w(t) \text{ for a.e. } t \in J \text{ and for all } u \in [-w(t), w(t)],$$

so that g is L^1-bounded in Ω. Thus the hypothesis (A) holds when $\underline{u} = -w$ and $\bar{u} = w$. Since (g0) and (B0) are assumed to hold, and since (q2) implies the validity of (q0), then the BVP (2.2.2) has by Proposition 2.2.3 extremal solutions u_* and u^* in $[-w, w]$.

If u is a solution of (2.2.2), it follows from (2.2.2) and (g3) that

$$u'(s) - \|q\|_\infty p_1(s)u(s) \leq \|q\|_\infty M(s) \text{ a.e. in } J. \tag{2.2.15}$$

Multiplying both sides of (2.2.15) by $e^{-P(s)}$ and integrating from t_0 to t, we obtain

$$u(t) \leq e^{P(t)}u(t_0) + \int_{t_0}^{t} e^{P(t)-P(s)}\|q\|_\infty M(s)ds, \quad t \in J. \tag{2.2.16}$$

In view of (B2) and the boundary condition of (2.2.2) we have

$$au(t_0) - b I(u) = au(t_0) - b I(u) - B(u(t_0), u) \leq c.$$

This and (2.2.16) imply that

$$c \geq au(t_0) - b I(u)$$
$$\geq au(t_0) - b I(u(t_0)e^{P(\cdot)}) - b I(\tau \mapsto \int_{t_0}^{\tau} e^{P(\tau)-P(s)}\|q\|_\infty M(s)ds),$$

so that

$$u(t_0) \leq \frac{c + b I(\tau \mapsto \int_{t_0}^{\tau} e^{P(\tau)-P(s)}\|q\|_\infty M(s)ds)}{a - b I(e^{P(\cdot)})}. \tag{2.2.17}$$

It then follows from (2.2.13), (2.2.16), and (2.2.17) that $u(t) \leq w(t)$ for each $t \in J$, i.e., $u \leq w$. Similarly one can show that $-w \leq u$, so that $u \in [-w, w]$.

The above proof shows that all the solutions of (2.2.2) belong to the order interval $[-w, w]$, whence u_* and u^* are least and greatest of all the solutions of (2.2.2). $\qquad\square$

In the case when $B(x, u) = ax - bu(t_1) - c$ and $I(u) = u(t_1)$ we get the following consequence of Theorem 2.2.3.

Proposition 2.2.4. *If the functions g and q have properties (g0), (g3), and (q2), and if positive constants a and b satisfy $\int_{t_0}^{t_1} p_1(s)\, ds < \frac{1}{\|q\|_\infty} \ln \frac{a}{b}$, then the BVP*

$$u'(t) = q(u(t))g(t, u(t)) \quad \text{a.e. in} \quad J = [t_0, t_1], \quad au(t_0) - bu(t_1) = c,$$
$$(2.2.18)$$

has extremal solutions.

Example 2.2.2. Let the functions $q\colon \mathbb{R} \to \mathbb{R}$ and $g\colon J \times \mathbb{R} \to \mathbb{R}$, $J = [0, 1]$, be defined by (2.1.25) and (2.1.26). It follows from Example 2.1.1 that q and g have properties (q2), (g0), and (g3) when $p_1(t) \equiv 0$ and $M(t) \equiv 9$.

The function

$$B(x, u) = 2x - \int_J u(t)\, dt + \sum_{m=-\infty}^{\infty} \sum_{n=1}^{\infty} \frac{\mu(x - \frac{m}{n})}{2^{|m|+n}} \frac{1 - [\int_J u(t)\, dt]}{1 + |[\int_J u(t)\, dt]|},$$

$$\mu(x) = (\cos \frac{1}{x} + 2)\chi_{(-\infty, 0)}(x) + 0\chi_{\{0\}}(x) + (\cos \frac{1}{x} - 2)\chi_{(0, \infty)}(x),$$

has properties (B0) and (B2) when $a = 2$, $b = 1$, $c = 9$, and $I(u) = \int_J u(t)\, dt$. Also the hypothesis (A1) holds. Thus the BVP (2.2.2) has for these functions q, g, and B extremal solutions by Theorem 2.2.3.

2.3 Explicit functional problems

In this section we shall prove extremality and comparison results for the problem

$$\begin{cases} \frac{d}{dt}\varphi(u(t)) = g(t, u(t), u) & \text{for a.e.} \ \ t \in J = [t_0, t_1], \\ u(t) = B_0(u(t_0), u) + h_0(t), & t \in J_0 = [t_0 - r, t_0], \end{cases} \quad (2.3.1)$$

where $\varphi\colon \mathbb{R} \to \mathbb{R}$, $g\colon J \times \mathbb{R} \times \mathcal{F} \to \mathbb{R}$, $B_0\colon \mathbb{R} \times \mathcal{F} \to \mathbb{R}$ with $\mathcal{F} = C([t_0 - r, t_1])$, $r \geq 0$, and $h_0 \in C(J_0)$. In the following we assume that $\mathcal{F}$ is ordered pointwise.

Definition 2.3.1. A function $u \in \mathcal{F}$ is said to be a *lower solution* of (2.3.1) if $\varphi \circ u_{|J}$ belongs to $AC(J)$, and if

$$\begin{cases} \frac{d}{dt}\varphi(u(t)) \leq g(t, u(t), u) & \text{for a.e.} \ \ t \in J, \\ u(t) \leq B_0(u(t_0), u) + h_0(t), & t \in J_0, \end{cases}$$

and an *upper solution* of (2.3.1) if the reversed inequalities are satisfied. If equalities hold, we say that u is a *solution* of (2.3.1).

2.3.1 Hypotheses and main results. We are going to prove that if the functions φ and g satisfy the following hypotheses:

(φ0) φ is an increasing homeomorphism,

(g0)   for all $x \in \mathbb{R}$ and $v \in \mathcal{F}$ the function $g(\cdot, x, v)$ is measurable, and
$$\limsup_{y \to x-} g(t, y, v) \leq g(t, x, v) \leq \liminf_{y \to x+} g(t, y, v) \text{ for a.e. } t \in J,$$

(g1) $g(t, x, \cdot)$ is increasing for a.e. $t \in J$ and all $x \in \mathbb{R}$,

(B0)   for each $x \in \mathbb{R}$ the function $B_0(x, \cdot)$ is increasing, and
$$\limsup_{y \to x-} B_0(y, v) \leq B_0(x, v) \leq \liminf_{y \to x+} B_0(y, v) \text{ for all } v \in \mathcal{F},$$

(A)   problem (2.3.1) has such lower and upper solutions $\underline{u}$ and $\overline{u}$ that $\underline{u} \leq \overline{u}$, and that the function g is L^1-bounded in the set
$$\Omega = \{(t, x, v) \in J \times \mathbb{R} \times \mathcal{F} \mid \underline{u}(t) \leq x \leq \overline{u}(t), \underline{u} \leq v \leq \overline{u}\},$$

there exists the least and the greatest among those solutions of (2.3.1) which belong to the order interval $[\underline{u}, \overline{u}] = \{u \in \mathcal{F} \mid \underline{u} \leq u \leq \overline{u}\}$. Assuming also that B_0 is bounded, and replacing (A) by the following hypothesis:

(g2) $|g(t, x, v)| \leq p_1(t)\psi(|\varphi(x)|)$ for all $x \in \mathbb{R}$ and $v \in \mathcal{F}$ and a.e. $t \in J$, where $p_1 \in L_+^1(J)$, $\psi \colon \mathbb{R}_+ \to (0, \infty)$ is increasing and $\int_0^\infty \frac{dx}{\psi(x)} = \infty$,

we prove that problem (2.3.1) has extremal solutions, i.e., the least and the greatest of all its solutions, and that they are increasing with respect to the functions g, B_0, and h_0.

Finally we show that the above results are valid for the problem

$$u'(t) = q(u(t))g(t, u(t), u) \text{ a.e. in } J, \quad u(t) = B_0(u(t_0), u) + h_0(t), \ t \in J_0,$$

if (φ0) is replaced by the following hypothesis.

(q0) $q \colon \mathbb{R} \to (0, \infty)$ and $\frac{1}{q}$ belong to $L_{loc}^\infty(\mathbb{R})$, and $\int_0^{\pm\infty} \frac{dz}{q(z)} = \pm\infty$.

Examples are given to illustrate the obtained results.

Remarks 2.3.1. The hypotheses given above allow the functions g, B_0, and q to be discontinuous. $g(\cdot, u(\cdot), u)$ is not necessarily measurable for all $u \in C(J)$. By Remarks 2.1.1 we can replace $\psi(|\varphi(x)|)$ by $\psi(|x|)$ in the hypothesis (g2) if φ is Lipschitz-continuous.

2.3.2 Preliminaries. As an application of Theorem 2.2.1 we obtain

Lemma 2.3.1. *Assume that the hypotheses (φ0), (g0), (g1), (B0), and (A) hold. Then for each $v \in [\underline{u}, \overline{u}]$ problem*

$$\begin{cases} \frac{d}{dt}\varphi(u(t)) = g(t, u(t), v) & a.e. \text{ in } J, \\ u(t) = B_0(u(t_0), v) + h_0(t), & t \in J_0, \end{cases} \tag{2.3.2}$$

has extremal solutions in $[\underline{u}, \overline{u}]$, and they are increasing with respect to v.

Proof. Let $v \in [\underline{u}, \overline{u}]$ be given, and consider the IVP

$$\frac{d}{dt}\varphi(u(t)) = g(t, u(t), v) \text{ a.e. in } J, \quad u(t_0) = B_0(u(t_0), v) + h_0(t_0). \quad (2.3.3)$$

The hypotheses (g1) and (A) imply that

$$\frac{d}{dt}\varphi(\underline{u}(t)) \leq g(t, \underline{u}(t), v) \text{ a.e. in } J, \quad \underline{u}(t_0) \leq B_0(\underline{u}(t_0), v) + h_0(t_0),$$
$$\frac{d}{dt}\varphi(\overline{u}(t)) \geq g(t, \overline{u}(t), v) \text{ a.e. in } J, \quad \overline{u}(t_0) \geq B_0(\overline{u}(t_0), v) + h_0(t_0).$$

Thus $\underline{u}_{|J}$ and $\overline{u}_{|J}$ are lower and upper solutions of (2.3.3) and $\underline{u}_{|J} \leq \overline{u}_{|J}$. The hypotheses ($\varphi$0), (g0), and (B0) imply that the hypotheses of Theorem 2.2.1 hold when $(t, x) \mapsto g(t, x, v)$ and $(x, u) \mapsto x - B_0(x, v) - h_0(t_0)$ stand for g and B. Thus the IVP (2.3.3) has extremal solutions u_- and u_+ in $[\underline{u}_{|J}, \overline{u}_{|J}]$. These solutions can be extended by $u_{\pm}(t) = B_0(u_{\pm}(t_0), v) + h_0(t)$, $t \in J_0$, to extremal solutions of problem (2.3.2) in $[\underline{u}, \overline{u}]$.

Let $\hat{v}$ be another function from $[\underline{u}, \overline{u}]$, and assume that $v \leq \hat{v}$. Denoting by $\hat{u}_-$ the least solution of the IVP

$$\frac{d}{dt}\varphi(u(t)) = g(t, u(t), \hat{v}) \text{ a.e. in } J, \quad u(t_0) = B_0(u(t_0), \hat{v}) + h_0(t_0) \quad (2.3.4)$$

in $[\underline{u}_{|J}, \overline{u}_{|J}]$, it follows from (g1) that $\hat{u}_-$ is an upper solution of (2.3.3) in $[\underline{u}_{|J}, \overline{u}_{|J}]$. Thus the first relation of (2.2.5) implies that $u_-(t) \leq \hat{u}_-(t)$ on J. Applying the second relation of (2.2.5) it can be shown similarly that $u_+(t) \leq \hat{u}_+(t)$ on J, where $\hat{u}_+$ denotes the greatest solution of the IVP (2.3.4). Defining $\hat{u}_{\pm}(t) = B_0(\hat{u}_{\pm}(t_0), \hat{v}) + h_0(t)$, $t \in J_0$, the functions $\hat{u}_{\pm}$ are extended to extremal solutions of problem

$$\begin{cases} \frac{d}{dt}\varphi(u(t)) = g(t, u(t), \hat{v}) \text{ a.e. in } J, \\ u(t) = B_0(u(t_0), \hat{v}) + h_0(t), \ t \in J_0. \end{cases} \quad (2.3.5)$$

Moreover, $u_-(t) \leq \hat{u}_-(t)$ and $u_+(t) \leq \hat{u}_+(t)$ on $[t_0 - r, t_1]$. This proves the second assertion. $\qquad \square$

2.3.3 Existence and comparison results. We shall first prove the existence of extremal solutions of (2.3.1) between its lower and upper solutions.

Theorem 2.3.1. *Assume that the hypotheses ($\varphi 0$), (g0), (g1), (B0), and (A) hold. Then problem (2.3.1) has the least solution u_* and the greatest solution u^* in $[\underline{u}, \overline{u}]$. Moreover,*

$$\begin{cases} u_* = \min\{u_+ \mid u_+ \text{ is an upper solution of (2.3.1) in } [\underline{u}, \overline{u}]\}, \\ u^* = \max\{u_- \mid u_- \text{ is a lower solution of (2.3.1) in } [\underline{u}, \overline{u}]\}. \end{cases} \qquad (2.3.6)$$

Proof. The results of Lemma 2.3.1 imply that relation

$$u = Gv \text{ is the least solution of (2.3.2) in } [\underline{u}, \overline{u}], \ v \in [\underline{u}, \overline{u}], \qquad (2.3.7)$$

defines an increasing mapping $G: [\underline{u}, \overline{u}] \to [\underline{u}, \overline{u}]$. If $u, v \in [\underline{u}, \overline{u}]$, the hypothesis (A) implies the existence of such a function $M \in L^1(J)$ that

$$|g(t, u(t), v)| \leq M(t) \text{ for a.e. } t \in J.$$

In view of this inequality, the definition (2.3.7) of G, and (2.3.2) we see that if $u \in G[\underline{u}, \overline{u}]$, then

$$\begin{cases} |\varphi(u(t_3)) - \varphi(u(t_2))| \leq |\int_{t_2}^{t_3} M(t)\, dt|, & t_2, t_3 \in J, \\ |u(t_3) - u(t_2)| = |h(t_3) - h(t_2)|, & t_2, t_3 \in J_0. \end{cases} \qquad (2.3.8)$$

These relations and property ($\varphi 0$) imply that $G[\underline{u}, \overline{u}]$ is an equicontinuous subset of $[\underline{u}, \overline{u}]$, whence monotone sequences of $G[\underline{u}, \overline{u}]$ converge uniformly on $[t_0 - r, t_1]$.

The above proof shows that the mapping G, defined by (2.3.7), satisfies the hypotheses of Corollary 1.1.1 when $[a, b] = [t_0 - r, t_1]$, so that G has the least fixed point u_*. The definition (2.3.7) of G implies that u_* is a solution of problem (2.3.1) in $[\underline{u}, \overline{u}]$. If u is any solution of (2.3.1) in $[\underline{u}, \overline{u}]$, then u is a solution of (2.3.2) when $v = u$. Since Gu is the least solution of (2.3.2) in $[\underline{u}, \overline{u}]$ when $v = u$, then $Gu \leq u$. Since $u_* = \min\{u \mid Gu \leq u\}$ by Corollary 1.1.2, then $u_* \leq u$. This verifies that u_* is the least solution of (2.3.1) in $[\underline{u}, \overline{u}]$. The proof that problem (2.3.1) has the greatest solution in $[\underline{u}, \overline{u}]$ is similar.

To prove (2.3.6), let u_+ be an upper solution of (2.3.1) in $[\underline{u}, \overline{u}]$. Replacing $\overline{u}$ by u_+ in the above proof it follows that problem (2.3.1) has a solution $u \in [\underline{u}, u_+] \subseteq [\underline{u}, \overline{u}]$. But u_* is the least of all the solutions of (2.3.1) in $[\underline{u}, \overline{u}]$, so that $u_* \leq u \leq u_+$. Similarly, it can be shown that if u_- is a lower solution of (2.3.1) in $[\underline{u}, \overline{u}]$, then $u_- \leq u^*$. Noticing also that u_* is an upper solution and u^* a lower solution of (2.3.1), we obtain (2.3.6). $\square$

According to Proposition 1.1.3 the least fixed point of an increasing mapping $G\colon [\underline{u},\overline{u}] \to [\underline{u},\overline{u}]$ which satisfies the hypotheses of Lemma 2.3.2 can sometimes be obtained by calculating the successive approximations $u_{n+1} = Gu_n$, $n \in \mathbb{N}$, $u_0 = \underline{u}$. In particular, if G is defined by (2.3.7), $u_{n+1} = Gu_n$ is the least solution of problem

$$\begin{cases} \frac{d}{dt}\varphi(u(t)) = g(t, u(t), u_n) & \text{a.e. in } J, \\ u(t) = B_0(u(t_0), u_n) + h_0(t), & t \in J_0, \end{cases} \tag{2.3.9}$$

in $[\underline{u},\overline{u}]$. If there is $n \in \mathbb{N}$ such that $u_{n+1} = u_n$, then $u_* = u_n$ is the least solution of problem (2.3.1) in $[\underline{u},\overline{u}]$. $\lim_{n\to\infty} u_n$ is the next possible candidate for u_*. Dual results hold for the greatest solution if we choose $u_0 = \overline{u}$ and u_{n+1} to be the greatest solution of problem (2.3.9) in $[\underline{u},\overline{u}]$.

Example 2.3.1. Consider the problem

$$\begin{cases} u'(t) = \frac{[1-t+u(1-2t)+u(t)]}{1+|[1-t+u(1-2t)+u(t)]|} & \text{a.e. in } J = [0,1], \\ u(t) = -t, & t \in J_0 = [-1,0], \end{cases} \tag{2.3.10}$$

where $[x]$ denotes the greatest integer less than or equal to x. Problem (2.3.10) is of the form (2.3.1) with

$$\begin{cases} g(t,x,u) = \frac{[1-t+u(1-2t)+x]}{1+|[1-t+u(1-2t)+x]|}, & t \in J, \ x \in \mathbb{R}, \ u \in \mathcal{F} = C[-1,1], \\ \varphi(x) = x, \ B_0(x,u) = 0, \ x \in \mathbb{R}, \ u \in \mathcal{F}, \quad h_0(t) = -t, \quad t \in [-1,0]. \end{cases}$$

It is easy to see that the hypotheses of Theorem 2.3.1 hold when

$$\overline{u}(t) = |t|, \quad t \in [-1,1], \quad \text{and} \quad \underline{u}(t) = -t, \quad t \in [-1,1].$$

Obviously, every solution of (2.3.10) belongs to the order interval $[\underline{u},\overline{u}]$. Thus problem (2.3.10) has the least solution u_* and the greatest solution u^*. Calculating the successive approximations u_n by (2.3.9) with $u_0 = \overline{u}$ we see that $u_1(t) = \frac{t}{2}$, $t \in J$, and that

$$u_n(t) = \frac{t}{2}\chi_{[0,\frac{1}{3}]}(t) + \frac{1}{6}\chi_{[\frac{1}{3},a_n]}(t) + (\frac{1}{6} + \frac{1}{2}(t - a_n))\chi_{[a_n,1]}(t), \quad t \in [0,1],$$

where χ_U denotes the characteristic function of U, and $a_n \uparrow \frac{5}{6}$ as $n \to \infty$. The limit function

$$u(t) = \frac{t}{2}\chi_{[0,\frac{1}{3}]}(t) + \frac{1}{6}\chi_{[\frac{1}{3},\frac{5}{6}]}(t) + (\frac{t}{2} - \frac{3}{12})\chi_{[\frac{5}{6},1]}(t), \quad t \in [0,1],$$

satisfies the differential equation (2.3.10), whence the greatest solution of (2.3.10) is

$$u^*(t) = -t\chi_{[-1,0]}(t) + \frac{t}{2}\chi_{[0,\frac{1}{3}]}(t) + \frac{1}{6}\chi_{[\frac{1}{3},\frac{5}{6}]}(t) + (\frac{t}{2} - \frac{3}{12})\chi_{[\frac{5}{6},1]}(t), \ t \in [-1,1].$$

The successive approximations u_n with $u_0 = \underline{u}$ are equal to zero-function when $n \geq 2$. Thus the least solution of (2.3.10) is

$$u_*(t) = -t\chi_{[-1,0]}(t) + 0\chi_{[0,1]}(t), \quad t \in [-1,1].$$

It is also easy to see that the functions

$$u_a(t) = -t\chi_{[-1,0]}(t) + \frac{t}{2}\chi_{[0,a]}(t) + \frac{a}{2}\chi_{[a,1-\frac{a}{2}]}(t) + (\frac{a}{2} + \frac{1}{2}(t-1+\frac{a}{2}))\chi_{[1-\frac{a}{2},1]}(t),$$

where $a \in [0,\frac{1}{3}]$ form the solution set of (2.3.10). Thus the solutions of (2.3.10) form a continuum, and each point $(a, \frac{a}{2})$, $0 \leq a < \frac{1}{3}$, is a bifurcation point of the solutions of (2.3.10).

As an application of Lemma B.7.1 and Theorem 2.3.1 we shall now prove the following generalization to Theorem 2.1.3.

Theorem 2.3.2. *Assume that the functions $\varphi \colon \mathbb{R} \to \mathbb{R}$, $g \colon J \times \mathbb{R} \times \mathcal{F} \to \mathbb{R}$ and $B_0 \colon \mathbb{R} \to \mathbb{R}$ have properties $(\varphi 0)$, $(g0)$, $(g1)$, $(g2)$, and $(B0)$, and that B_0 is bounded. Then problem (2.3.1) has the least solution u_* and the greatest solution u^*. Moreover,*

$$\begin{cases} u_*(t) = \min\{u_+(t) \mid u_+ \text{ is an upper solution of (2.3.1)}\}, \\ u^*(t) = \max\{u_-(t) \mid u_- \text{ is a lower solution of (2.3.1)}\}. \end{cases} \tag{2.3.11}$$

Proof. Assume first that $u \in \mathcal{F}$ is a solution of (2.3.1). Applying (2.3.1) and (g2) we obtain

$$|\frac{d}{dt}\varphi(u(t))| = |g(t, u(t), u)| \leq p_1(t)\psi(|\varphi(u(t))|) \quad \text{a.e. in } J.$$

Denoting $b_0 = \max\{|B_0(x,u)| + |h_0(t)| \mid t \in J_0, x \in \mathbb{R}\}$ and choosing $w_0 \in \mathbb{R}$ so that

$$-w_0 \leq \varphi(-b_0), \varphi(b_0) \leq w_0,$$

we see that

$$|\varphi(u(t))| \leq |\varphi(u(t_0))| + \int_{t_0}^{t} p_1(s)\psi(|\varphi(u(s))|)ds \leq w_0 + \int_{t_0}^{t} p_1(s)\psi(|\varphi(u(s)|)ds$$

for all $t \in J$. This implies by Lemma B.7.1 that $|\varphi(u(t))| \leq w(t)$ on J, where w is the solution of the IVP

$$w'(t) = p_1(t)\psi(w(t)), \quad \text{a.e. in } J, \quad w(t_0) = w_0. \tag{2.3.12}$$

Defining

$$\underline{u}(t) = \begin{cases} \varphi^{-1}(-w_0), & t \in J_0, \\ \varphi^{-1}(-w(t)), & t \in J, \end{cases} \quad \overline{u}(t) = \begin{cases} \varphi^{-1}(w_0), & t \in J_0, \\ \varphi^{-1}(w(t)), & t \in J, \end{cases} \tag{2.3.13}$$

the above considerations, the choice of w_0 and property ($\varphi 0$) imply that $u \in [\underline{u}, \overline{u}]$. Next we shall show that $\underline{u}$ and $\overline{u}$ are lower and upper solutions of (2.3.1). Since w, as a solution of (2.3.12), belongs to $AC(J)$, it follows from (2.3.13) that $\varphi \circ \underline{u}_{|J} = -w$ and $\varphi \circ \overline{u}_{|J} = w$ belong to $AC(J)$. Thus $\underline{u}_{|J}$ and $\overline{u}_{|J}$ belong to $C(J)$ by ($\varphi 0$). Applying (g2), (2.3.12), and (2.3.13) we obtain

$$\frac{d}{dt}\varphi(\underline{u}(t)) = -w'(t) = -p_1(t)\psi(w(t)) = -p_1(t)\psi(|\varphi(\underline{u}(t))|) \leq g(t, \underline{u}(t), \underline{u}),$$

$$\frac{d}{dt}\varphi(\overline{u}(t)) = w'(t) = p_1(t)\psi(w(t)) = p_1(t)\psi(|\varphi(\overline{u}(t))|) \geq g(t, \overline{u}(t), \overline{u}),$$

for a.e. $t \in J$. The choice of w_0, b_0 and the monotonicity of φ^{-1} imply that

$$\underline{u}(t) = \varphi^{-1}(-w_0) \leq -b_0 \leq B_0(\underline{u}(t_0), \underline{u}) + h_0(t), \quad t \in J_0,$$

$$B_0(\overline{u}(t_0), \overline{u}) + h_0(t) \leq b_0 \leq \varphi^{-1}(w_0) = \overline{u}(t), \quad t \in J_0.$$

The inequalities derived above imply that $\underline{u}$ and $\overline{u}$ are lower and upper solutions of (2.3.1). Moreover, applying (g2), (2.3.12), and (2.3.13) we see that

$$|g(t, x, u)| \leq p_1(t)\psi(|\varphi(x)|) \leq p_1(t)\psi(w(t)) = w'(t)$$

for a.e. $t \in J$, and for all $x \in [\underline{u}(t), \overline{u}(t)]$ and $u \in [\underline{u}, \overline{u}]$, whence the hypothesis (A) holds.

The above proof shows that the hypotheses of Theorem 2.3.1 are satisfied, so that problem (2.3.1) has extremal solutions u_* and u^* in $[\underline{u}, \overline{u}]$. Because all the solutions of (2.3.1) belong to $[\underline{u}, \overline{u}]$, then u_* and u^* are the extremal solutions of (2.3.1). To prove (2.3.11), let u_+ be an upper solution of (2.3.1). Choose b_0 in the above proof so that $-b_0 \leq u_+(t)$, $t \in [t_0 - r, t_1]$. Then $-w \leq \varphi \circ u_+$, whence $\underline{u} \leq u_+$, so that problem (2.3.1) has by Theorem 2.3.1 a solution $u \in [\underline{u}, u_+]$. But u_* is the least of all the solutions of (2.3.1), whence $u_* \leq u_+$. Similarly, it can be shown that if u_- is a lower solution of (2.3.1), then $u_- \leq u^*$. Since u_* is an upper solution and u^* a lower solution of (2.3.1), we obtain (2.3.11). □

The following result is a consequence of Theorem 2.3.2.

Proposition 2.3.1. *If the hypotheses of Theorem 2.3.2 hold, then problem*

$$\begin{cases} \frac{d}{dt}\varphi(u(t)) = g(t, u(t), u) + h_1(t) & \text{for a.e. } t \in J, \\ u(t) = B_0(u(t_0), u) + h_0(t), & t \in J_0, \end{cases} \quad (2.3.14)$$

has for all $h_1 \in L^1(J)$ and $h_0 \in C(J_0)$ extremal solutions and they are increasing with respect to g, B_0, h_1, and h_0.

Proof. Assume that the hypotheses of Theorem 2.3.2 hold for the functions g, $\hat{g}$, B_0, and $\hat{B}_0$, that h_1, $\hat{h}_1 \in L^1(J)$ and h_0, $\hat{h}_0 \in C(J_0)$, and that

$$\begin{cases} g(t, x, u) \le \hat{g}(t, x, u) & \text{for all } (t, x, u) \in J_0 \times \mathbb{R} \times \mathcal{F}, \\ B_0(x, u) \le \hat{B}_0(x, u) & \text{for all } (x, u_\in \mathbb{R} \times \mathcal{F}, \\ h_1(t) \le \hat{h}_1(t) & \text{for a.e. } t \in J \text{ and } h_0(t) \le \hat{h}_0(t) \text{ for all } t \in J_0. \end{cases}$$

The functions $(t, x, u) \mapsto g(t, x, u) + h_1(t)$ and $(t, x, u) \mapsto \hat{g}(t, x, u) + \hat{h}_1(t)$ satisfy (g0) and (g1), and also (g2) when p_1 and ψ are replaced by $t \mapsto p_1(t) + |h_1(t)| + |\hat{h}_1(t)|$ and $z \mapsto \psi(z) + 1$, respectively. Denoting by $\hat{u}$ the least solution of problem

$$\begin{cases} \frac{d}{dt}\varphi(u(t)) = \hat{g}(t, u(t), u) + \hat{h}_1(t) & \text{for a.e. } t \in J, \\ u(t) = \hat{B}_0(u(t_0), u) + \hat{h}_0(t), & t \in J_0, \end{cases} \quad (2.3.15)$$

then $\hat{u}$ is an upper solution of (2.3.14). This and (2.3.11) imply that $u_* \le \hat{u}$. Similarly, it can be shown that if $\hat{u}$ is the greatest solution of (2.3.15), then $u^* \le \hat{u}$, which concludes the proof. $\qquad \square$

Example 2.3.2. Choose $J = [0, 1]$, $J_0 = [-1, 0]$ and $\mathcal{F} = C([-1, 1])$, and define a function $g \colon J \times \mathbb{R} \times \mathcal{F} \to \mathbb{R}$ by

$$g(t, x, u) = \sum_{m=-\infty}^{\infty} \sum_{n=1}^{\infty} \frac{g_1(t, x - \frac{m}{n}) + g_2(x + \max_{t \in [-1,1]} u(t) - \frac{m}{n})}{2^{|m|+n}}, \quad \text{where}$$

$$g_1(t, x) = \begin{cases} \cos\frac{1}{x-t} - 2, & x < t, \\ \chi_U(t), \ U \subset J, & x = t, \\ \cos\frac{1}{x-t} + 2, & x > t, \end{cases} \quad \text{and} \quad g_2(z) = \begin{cases} -1, & z < 0, \\ 0, & z = 0, \\ 1, & z < 0. \end{cases}$$

It is easy to see that g has properties (g0), (g1), and (g2).
The function $B_0 \colon \mathbb{R} \times \mathcal{F} \to \mathbb{R}$, defined by

$$B_0(x, u) = \sum_{m=-\infty}^{\infty} \sum_{n=1}^{\infty} \frac{g_1(\frac{m}{n}, x)}{2^{|m|+n}} + \arctan([\int_{-1}^{1} u(t)\, dt]), \quad (2.3.16)$$

has property (B0). The function $\varphi \colon \mathbb{R} \to \mathbb{R}$, defined by $\varphi(x) = |x|^{p-2}x$, has property (φ0) for each $p > 1$.

Thus problem (2.3.14) has with these functions g, B_0, and φ for all $h_1 \in L^1(J)$ and $h_0 \in C(J_0)$ extremal solutions and they are increasing with respect to h_1 and h_0.

2.3.4 A special case. In this subsection we consider the following problem:

$$\begin{cases} u'(t) = q(u(t))g(t, u(t), u) \quad \text{a.e. in } J, \\ u(t) = B_0(u(t_0), u) + h_0(t), \quad t \in J_0. \end{cases} \qquad (2.3.17)$$

It follows from the proof of Lemma 2.1.3 that if

(q0) $q \colon \mathbb{R} \to (0, \infty)$ and $\frac{1}{q}$ belong to $L^\infty_{loc}(\mathbb{R})$, and $\int_0^{\pm\infty} \frac{dz}{q(z)} = \pm\infty$,

then problem (2.3.17) has the same solutions, lower solutions, and upper solutions as problem (2.3.1), where $\varphi \colon \mathbb{R} \to \mathbb{R}$ is defined by $\varphi(x) = \int_0^x \frac{dz}{q(z)}$, $x \in \mathbb{R}$. Moreover, (q0) implies that φ has property (φ0), whence we obtain

Proposition 2.3.2. *The results of Theorems 2.3.1 and 2.3.2 hold for problem (2.3.17), and the results of Proposition 2.3.1 hold for problem*

$$\begin{cases} u'(t) = q(u(t))(g(t, u(t), u) + h_1(t)) \quad \text{a.e. in } J, \\ u(t) = B_0(u(t_0), u) + h_0(t), \quad t \in J_0, \end{cases} \qquad (2.3.18)$$

when (φ0) is replaced by (q0).

2.4 Implicit functional problems

In this section we consider an implicit problem of the form

$$\begin{cases} \frac{d}{dt}\varphi(u(t)) = g(t, u(t), u) + f(t, u(t), u, \frac{d}{dt}\varphi(u(t)) - g(t, u(t), u)) \\ \qquad \text{for a.e. } t \in J = [t_0, t_1], \\ u(t) = B_0(u(t_0), u) + B_1(t, u(t), u), \quad t \in J_0 = [t_0 - r, t_0], \end{cases} \qquad (2.4.1)$$

where $\varphi \colon \mathbb{R} \to \mathbb{R}$, $g \colon J \times \mathbb{R} \times \mathcal{F} \to \mathbb{R}$, $f \colon J \times \mathbb{R} \times \mathcal{F} \times \mathbb{R} \to \mathbb{R}$, $B_0 \colon \mathbb{R} \times \mathcal{F} \to \mathbb{R}$ and $B_1 \colon J_0 \times \mathbb{R} \times \mathcal{F} \to \mathbb{R}$, with $\mathcal{F} = C[t_0 - r, t_1]$. Results derived for the explicit problem

$$\begin{cases} \frac{d}{dt}\varphi(u(t)) = g(t, u(t), u) + h_1(t) \quad \text{for a.e. } t \in J, \\ u(t) = B_0(u(t_0), u) + h_0(t), \quad t \in J_0, \end{cases} \qquad (2.4.2)$$

in section 2.3 will be used in the sequel.

Assuming that $\mathcal{F}$ is equipped with pointwise ordering, we are going to prove that problem (2.4.1) has extremal solutions if φ and g have properties ($\varphi 0$), (g0), (g1), and (g2) given in subsection 2.3.1, and if f, B_0, and B_1 satisfy the following hypotheses:

- (f0) The function $t \mapsto f(t, u(t), u, v(t))$ is measurable when $u \in \mathcal{F}$ and $v \in L^1(J)$, and $f(t, x, u, y)$ is increasing in x, u and y for a.e. $t \in J$.
- (f1) $|f(t, x, u, y)| \leq p_2(t)\psi(|\varphi(x)|) + \lambda(t)|y|$ for a.e. $t \in J$ and all $x, y \in \mathbb{R}$ and $u \in \mathcal{F}$, where $\lambda \colon J \to [0, 1)$, $\frac{p_2}{1-\lambda} \in L^1_+(J)$, and ψ is as in (g2).
- (B0) $B_0(x, \cdot)$ is increasing for all $x \in \mathbb{R}$, and
  $$\limsup_{y \to x-} B_0(y, v) \leq B_0(x, v) \leq \liminf_{y \to x+} B_0(y, v) \text{ for all } x \in \mathbb{R}, \ v \in \mathcal{F},$$
- (B1) $B_1(t, x, u)$ is increasing in x and in u for each $t \in J_0$, and the set $\{B_1(\cdot, u_{|J_0}(\cdot), u) \mid u \in \mathcal{F}\}$ is an equicontinuous subset of $C(J_0)$.
- (B01) B_0 is bounded and $|B_1(t, x, u)| \leq c|x| + d$ for all $t \in J_0$, $x \in \mathbb{R}$ and $u \in \mathcal{F}$, where $c \in [0, 1)$ and $d \in \mathbb{R}_+$.

2.4.1 Preliminaries. In this subsection we reduce problem (2.4.1) to an operator equation of the form $Lu = Nu$, and show that the hypotheses of Theorem 1.1.2 are satisfied, by assuming that the hypotheses ($\varphi 0$), (g0), (g1), (g2), (f0), (f1), (B0), (B1), and (B01) hold. Denote

$$\begin{cases} X = L^1(J) \times C(J_0), \\ Y = \{u \in \mathcal{F} \mid \varphi \circ u_{|J} \in AC(J) \text{ and } g(\cdot, u(\cdot), u) \text{ is measurable}\}, \end{cases} \qquad (2.4.3)$$

and assume that X is ordered and normed as follows.

$$\begin{cases} (h_1, h_2) \leq (k_1, k_2) \text{ iff } h_1(t) \leq k_1(t) \text{ a.e. in } J \text{ and } h_2(t) \leq k_2(t) \text{ on } J_0, \\ \|(h_1, h_2)\| = \int_J |h_1(t)|dt + \sup_{t \in J_0} |h_2(t)|. \end{cases}$$
$$\qquad (2.4.4)$$

When $u \in Y$, define $(\varphi \circ u)'(t) := 0$ at those $t \in J$ where $\varphi \circ u$ is not differentiable.

Lemma 2.4.1. *Denoting for each $u \in Y$,*

$$\begin{cases} L_1 u(t) = (\varphi \circ u)'(t) - g(t, u(t), u), & t \in J, \\ N_1 u(t) = f(t, u(t), u, L_1 u(t)), & t \in J, \\ L_2 u(t) = u(t) - B_0(u(t_0), u), & t \in J_0, \\ N_2 u(t) = B_1(t, u(t), u), & t \in J_0, \end{cases} \qquad (2.4.5)$$

we obtain mappings $L = (L_1, L_2)$ and $N = (N_1, N_2)$ from Y to X. Moreover, $u \in Y$ is a solution of problem (2.4.1) if and only if $Lu = Nu$.

Proof. The assertions are direct consequences of (2.4.1) and (2.4.5) and the given hypotheses. $\qquad \square$

Next we define a subset V of Y which contains all the possible solutions of (2.4.1) in Y. Choose b_0, $w_0 \in \mathbb{R}$ such that

$$\begin{cases} |B_0(x,u)| \le b_0, & x \in \mathbb{R}, \ u \in \mathcal{F}, \\ \varphi^{-1}(-w_0) \le \pm \frac{2b_0+d}{1-c} \le \varphi^{-1}(w_0). \end{cases} \quad (2.4.6)$$

Let $z \in Y$ be the solution of the IVP

$$z'(t) = \left(p_1(t) + \frac{p_2(t)}{1-\lambda(t)} \right) \psi(z(t)) \quad \text{a.e. in } J, \quad z(t_0) = w_0, \quad (2.4.7)$$

and denote

$$V = \{u \in Y \mid h_- \le Lu \le h_+\}, \quad h_\pm = \left(\frac{\pm p_2(\cdot)\psi(z(\cdot))}{1-\lambda(\cdot)}, \pm \frac{b_0+d}{1-c} \right). \quad (2.4.8)$$

Lemma 2.4.2. *If $u \in Y$ is a solution of problem (2.4.1), then $u \in V$.*

Proof. The hypotheses (f1) and (g2) imply by Lemma B.7.1 that the IVP (2.4.7) has a unique solution $z \in AC(J)$. Assume that $u \in Y$ is a solution of (2.4.1). Applying (f1) we get for a.e. $t \in J$,

$$|(\varphi \circ u)'(t) - g(t, u(t), u)| = |f(t, u(t), u, (\varphi \circ u)'(t) - g(t, u(t), u))|$$
$$\le p_2(t)\psi(|\varphi(u(t))|) + \lambda(t)\,|(\varphi \circ u)'(t) - g(t, u(t), u)|,$$

so that

$$|(\varphi \circ u)'(t) - g(t, u(t), u)| \le \frac{p_2(t)}{1-\lambda(t)}\psi(|\varphi(u(t))|) \text{ for a.e. } t \in J. \quad (2.4.9)$$

This inequality and (g2) imply that

$$|(\varphi \circ u)'(t)| \le |(\varphi \circ u)'(t) - g(t, u(t), u)| + |g(t, u(t), u)|)$$
$$\le \frac{p_2(t)\psi(|\varphi(u(t))|)}{1-\lambda(t)} + p_1(t)\psi(|\varphi(u(t))|) = \left(p_1(t) + \frac{p_2(t)}{1-\lambda(t)} \right)\psi(|\varphi(u(t))|)$$

for a.e. $t \in J$. In view of (2.4.6) and (B01) we see that for each $t \in J_0$,

$$|u(t)| \le |B_0(u(t_0), u)| + |B_1(t, u(t), u)| \le b_0 + c|u(t)| + d,$$

whence

$$|u(t)| \le \frac{b_0+d}{1-c} \le \varphi^{-1}(w_0).$$

Thus we have

$$
|\varphi(u(t))| \le |\varphi(u(t_0))| + \int_{t_0}^{t} |(\varphi \circ u)'(s)| ds
$$

$$
\le w_0 + \int_{t_0}^{t} \left(p_1(s) + \frac{p_2(s)}{1 - \lambda(s)} \right) \psi(|\varphi(u(s))|) ds, \quad t \in J.
$$

Noticing that z is a solution of the IVP (2.4.7), this implies by Lemma B.7.1 that $|\varphi(u(t))| \le z(t)$ on J. In view of this property, (2.4.9), and the monotonicity of ψ we obtain

$$
|L_1 u(t)| = |(\varphi \circ u)'(t) - g(t, u(t), u)| \le \frac{p_2(t)\psi(z(t))}{1 - \lambda(t)} \quad \text{for a.e.} \ \ t \in J.
$$

The hypothesis (B01) and the initial condition of (2.4.1) imply that

$$
|L_2 u(t)| = |u(t) - B_0(u(t_0), u)| = |B_1(t, u(t), u)|
$$

$$
\le c|u(t)| + d \le c\frac{b_0 + d}{1 - c} + d \le \frac{b_0 + d}{1 - c}.
$$

These inequalities show by (2.4.8) that $h_- \le Lu \le h_+$, whence $u \in V$. $\square$

Lemma 2.4.3. *Let V be given by (2.4.8). If $u \in V$, then $h_- \le Nu \le h_+$.*

Proof. It follows from (2.4.5) and (2.4.8) that if $u \in V$, then

$$
|L_1 u(t)| = |(\varphi \circ u)'(t) - g(t, u(t), u)| \le \frac{p_2(t)}{1 - \lambda(t)} \psi(z(t)) \ \text{a.e. in } J. \quad (2.4.10)
$$

This and (g2) imply that

$$
|(\varphi \circ u)'(t)| \le |(\varphi \circ u)'(t) - g(t, u(t), u)| + |g(t, u(t), u)|
$$

$$
\le \frac{p_2(t)\psi(z(t))}{1 - \lambda(t)} + p_1(t)\psi(|\varphi(u(t))|) \quad \text{for a.e.} \ \ t \in J.
$$

Since $|L_2(u(t))| \le \frac{b_0 + d}{1 - c}$ by (2.4.8), it follows from (2.4.6) that

$$
|u(t)| \le |L_2 u(t)| + |B_0(u(t_0), u)| \le \frac{b_0 + d}{1 - c} + b_0 \le \varphi^{-1}(w_0), \ t \in J_0. \quad (2.4.11)
$$

Denoting $v(t) = |\varphi(u(t))|$, $t \in J$, then $v(t_0) = |\varphi(u(t_0))| \le w_0$ by the choice of w_0 in Lemma 2.4.2. The above inequalities imply that

$$
v'(t) \le \frac{p_2(t)\psi(z(t))}{1 - \lambda(t)} + p_1(t)\psi(v(t)) \quad \text{for a.e.} \ \ t \in J, \quad v(t_0) \le w_0. \quad (2.4.12)
$$

Denoting $y = \max\{v, z\}$, it follows from (2.4.7) and (2.4.12) by the monotonicity of ψ that

$$y'(t) \leq \left(\frac{p_2(t)}{1 - \lambda(t)} + p_1(t)\right) \psi(y(t)) \text{ for a.e. } t \in J, \quad y(t_0) = w_0.$$

In view of this, (2.4.7) and Lemma B.7.1 we see that $y(t) \leq z(t)$ on J. Thus $v(t) \leq z(t)$, so that

$$|\varphi(u(t))| \leq z(t), \quad t \in J.$$

Applying this result, (2.4.5), (2.4.10), and the hypothesis (f1) we get

$$|N_1 u(t)| = |f(t, u(t), u, L_1 u(t))| \leq p_2(t)\psi(|\varphi(u(t))|) + \lambda(t)\,|L_1 u(t)|$$

$$\leq p_2(t)\psi(z(t)) + \frac{\lambda(t)p_2(t)\psi(z(t))}{1 - \lambda(t)} = \frac{p_2(t)\psi(z(t))}{1 - \lambda(t)}$$

for a.e. $t \in J$. It follows from (2.4.11) and (B01) that for each $t \in J_0$,

$$|N_2 u(t)| = |B_1(t, u(t), u)| \leq c|u(t)| + d \leq c\frac{b_0 + d}{1 - c} + cb_0 + d \leq \frac{b_0 + d}{1 - c}.$$

The above inequalities and (2.4.8) imply that $h_- \leq Nu \leq h_-$. $\qquad\square$

Lemma 2.4.4. *If L, N, and V are defined by (2.4.5) and (2.4.8), then the following properties hold.*
a) If u, $v \in V$, $u \leq v$ and $Lu \leq Lv$, then $Nu \leq Nv$.
b) Monotone sequences of $N[V]$ converge in X with respect to the norm defined by (2.4.4).

Proof. a) Let u, $v \in V$ satisfy $u \leq v$ and $Lu \leq Lv$. It follows from (2.4.6) and from the hypotheses (f0) and (B1) that

$$N_1 u(t) = f(t, u(t), u, L_1 u(t)) \leq f(t, v(t), v, L_1 v(t)) = N_1 v(t)$$

for a.e. $t \in J$, and

$$N_2 u(t) = B_1(t, u(t), u) \leq B_1(t, v(t), v) = N_2 v(t), \quad t \in J_0.$$

These inequalities imply that $Nu \leq Nv$.

b) Assume that $(Nu_n)_{n=0}^{\infty}$ is a monotone sequence in $N[V]$. In view of Lemma 2.4.3 we have $-h \leq Nu_n \leq h$ for each $n \in \mathbb{N}$, so that the sequence $(N_1 u_n)$ is monotone and

$$|N_1 u_n(t)| \leq \frac{p_2(t)\psi(z(t))}{1 - \lambda(t)} \quad \text{for a.e. } t \in J.$$

By the monotone convergence theorem there is a function $h_1 \in L^1(J)$ such that

$$\int_J |N_1 u_n(t) - h_1(t)| dt \to 0 \quad \text{as} \quad n \to \infty.$$

Since $|N_2 u_n(t)| = |B_1(t, u_n(t), u)| \leq \frac{b_0 + d}{1 - c}$, and since $(N_2 u_n)$ is monotone, and also equicontinuous sequence in $C(J_0)$ by (B1), there exists $h_2 \in C(J_0)$ such that

$$\max_{J_0} |N_2 u_n(t) - h_2(t)| \to 0 \quad \text{as} \quad n \to \infty.$$

These two limes relations and (2.4.4) imply that

$$\|N u_n - (h_1, h_2)\| \to 0 \quad \text{as} \quad n \to \infty.$$

Since $(h_1, h_2) \in X = L^1(J) \times C(J_0)$, this proves the conclusion b).   □

2.4.2 The main existence and comparison results. As a consequence of Theorem 1.1.2 and the above results we now prove our main existence and comparison result.

Theorem 2.4.1. *Assume that the hypotheses (φ), (f0), (f1), (g0), (g1), (g2), (B0), (B1), and (B01) are satisfied. Then problem (2.4.1) has extremal solutions u_* and u^* in the sense that if $u \in Y$ is any solution of (2.4.1), then $u_*(t) \leq u(t) \leq u^*(t)$ for all $t \in J$. Moreover, u_* and u^* are increasing with respect to f and B_1.*

Proof. Let the operators L and N, the subset V of Y and the elements $h_\pm$ of X be defined by (2.4.5) and (2.4.8), respectively. It follows from Lemma 2.4.3, Proposition 2.3.1, and Lemma 2.4.4 that the following properties hold.

(I) If $u, v \in V$, $u \leq v$ and $Lu \leq Lv$, then $h_- \leq Nu \leq Nv \leq h_+$.
(II) Equation $Lu = h$ has for each $h \in [h_-, h_+]$ extremal solutions in V, and they are increasing with respect to h.
(III) Monotone sequences of $N[V]$ converge in X.

Thus equation $Lu = Nu$ has by Theorem 1.1.2 extremal solutions u_* and u^* in the set V, and they are increasing with respect to N. This result, Lemma 2.4.1, Lemma 2.4.2, and the definition (2.4.5) of N imply that u_* and u^* are extremal solutions of the BVP (2.4.1), and they are increasing with respect to f and B_1.   □

Remarks 2.4.1. In view of Proposition 1.1.2 the extremal solutions of equation $Lu = Nu$ can be obtained sometimes by the following method of successive approximations. Denote by u_0 the greatest solution of $Lu = h_+$, i.e., the greatest solution of problem (2.4.2) with $(h_1, h_0) = h_+$ defined by (2.4.8), and define a sequence $(u_n)_{n=0}^{\infty}$ recursively by choosing u_{n+1}, $n \in \mathbb{N}$ as the greatest solution of $Lu = Nu_n$, i.e., the greatest solution of problem

$$
\begin{cases}
\frac{d}{dt}\varphi(u(t)) = g(t, u(t), u) \\
+f(t, u_n(t), u_n, \frac{d}{dt}\varphi(u_n(t)) - g(t, u_n(t), u_n)) \quad \text{a.e. in } J, \quad (2.4.13) \\
u(t) = B_0(u(t_0), u) + B_1(t, u_n(t), u_n), \quad t \in J_0.
\end{cases}
$$

The hypotheses of Theorem 2.4.1 ensure that the properties (I) and (II) hold, whence the sequence (u_n) is decreasing. If $u_{n+1} = u_n$ for some $n \in \mathbb{N}$, it follows from Proposition 1.1.2 that u_n is the greatest solution of equation $Lu = Nu$, and hence the greatest solution of problem (2.4.1). The next possible candidate for the greatest solution is $\lim_{n \to \infty} u_n$. This is the case, e.g., if, in addition to the hypotheses of Theorem 2.4.1, we assume right continuity of f in its last three variables, g in its last two variables, B_1 in its last two variables, and the right continuity of B_0.

If u_0 is the least solution of (2.4.2) with $(h_1, h_0) = h_-$, and u_{n+1}, $n \in \mathbb{N}$, is the least solution of (2.4.13), then the above remarks hold for the least solution of (2.4.1) when right continuity is replaced by left continuity.

Example 2.4.1. Choose $J = [0, 1]$, $J_0 = [-1, 0]$ and $\mathcal{F} = C[-1, 1]$, and consider the problem

$$
u'(t) = H(u(t - 1) - 2t) + \frac{[u(t - 1) - t]}{1 + |[u(t - 1) - t]|}
$$
$$
+ \frac{[u'(t) - H(u(t - 1) - 2t)]}{1 + |[u'(t) - H(u(t - 1) - 2t)]|} \quad \text{a.e. in } J, \quad (2.4.14)
$$
$$
u(t) = \frac{2[u(0)]}{1 + |[u(0)]|} - \frac{[\int_{-1}^{1} u(s)ds]}{1 + |[\int_{-1}^{1} u(s)ds]|}t, \quad t \in J_0,
$$

where H is the Heaviside function and $[x]$ denotes the greatest integer less than or equal to x. Problem (2.4.14) is of the form (2.4.1) with

$$
\begin{cases}
g(t, x, u) = H(u(t - 1) - 2t), \quad t \in J, \\
f(t, x, u, y) = \frac{[u(t-1)-t]}{1+|[u(t-1)-t]|} + \frac{[y]}{1+|[y]|}, \quad t \in J, u \in \mathcal{F}, x \in \mathbb{R}, \\
B_0(x, u) = \frac{2[x]}{1+|[x]|}, \quad x \in \mathbb{R}, u \in \mathcal{F}, \quad (2.4.15) \\
B_1(t, x, u) = -\frac{[\int_{-1}^{1} u(s)ds]}{1+|[\int_{-1}^{1} u(s)ds]|}t, \quad t \in J_0, x \in \mathbb{R}, u \in \mathcal{F}.
\end{cases}
$$

It is easy to see that the hypotheses of Theorem 2.4.1 hold. Thus problem (2.4.14) has extremal solutions. Because H and $x \mapsto [x]$ are right-continuous, it follows from Remarks 2.4.1 that the greatest solution u^* of (2.4.14) is obtained as a limit of successive approximations.

By choosing $h_0(t) \equiv 3$ and $h_1(t) \equiv 2$ in (2.4.2), and calculating the successive approximations u_n by (2.4.13), we see that $u_4 = u_3$, whence $u^* = u_3$. The choices $h_0(t) \equiv -3$ and $h_1(t) \equiv -2$ in (2.4.2) imply that $u_5 = u_4$, whence $u_* = u_4$. Denoting by χ_W the characteristic function of $W \subset \mathbb{R}$, we get the following representations for u^* and u_*.

$$\begin{cases} u^*(t) = (1 - \frac{3t}{4})\chi_{[-1,0]}(t) + (1 + 2t)\chi_{[0,\frac{3}{7}]}(t) \\ +(\frac{10}{7} + t)\chi_{[\frac{3}{7},\frac{7}{11}]}(t) + \frac{159}{77}\chi_{[\frac{7}{11},1]}(t), \\ u_*(t) = (-\frac{4}{3} + \frac{4}{5}t)\chi_{[-1,0]}(t) - (\frac{4}{5} + \frac{17}{12}t)\chi_{[0,1]}(t). \end{cases}$$

Noticing that the possible values of $u(0)$ are $1, 0, -1$, and $-\frac{4}{3}$, similar methods can be used to obtain the following solutions of (2.4.14):

$$\begin{cases} u_0(t) = (1 - \frac{2t}{3})\chi_{[-1,0]}(t) + (1 + \frac{3t}{2})\chi_{[0,\frac{2}{5}]}(t) \\ +(\frac{7}{5} + \frac{t}{2})\chi_{[\frac{2}{5},\frac{5}{8}]}(t) + \frac{137}{80}\chi_{[\frac{5}{8},1]}(t), \\ u_1(t) = \frac{1}{2}t\chi_{[-1,0]}(t) - t\chi_{[0,1]}(t), \\ u_2(t) = \frac{1}{2}t\chi_{[-1,0]}(t) - \frac{7}{6}t\chi_{[0,1]}(t), \\ u_3(t) = (-1 + \frac{4}{5}t)\chi_{[-1,0]}(t) - (1 + \frac{17}{12}t)\chi_{[0,1]}(t). \end{cases}$$

Moreover, denoting

$$\begin{cases} A_1 = \{(t,x) \mid t \in [0,1], \, u_0(t) \leq x \leq u^*(t)\}, \\ A_2 = \{(t,x) \mid t \in [0,1], \, u_2(t) \leq x \leq u_1(t)\}, \end{cases}$$

it is easy to show that the points A_1 and A_2 are bifurcation points for solutions of (2.4.14). Thus between u_0 and u^*, and between u_2 and u_1 there is a continuum of chaotically behaving solutions of problem (2.4.14).

The hypotheses of Theorem 2.4.1 can be relaxed as follows.

Proposition 2.4.1. *The result of Theorem 2.4.1 holds if (f0) and (B1) are replaced by the following assumptions.*

(f2) *The function $t \mapsto f(t, u(t), u, v(t))$ is measurable for all $u \in \mathcal{F}$ and $v \in L^1(J)$, and there is a function $\alpha \in L^\infty_+(J)$ such that the function $f(t, x, u, y) + \alpha(t)y$ is increasing in x, u and y for a.e. $t \in J$.*

(B2) *$B_1(t, x, u) + \beta x$ is increasing in x and u for some $\beta \geq 0$.*

Proof. It is easy to see that problems (2.4.1) and

$$
\begin{cases}
\frac{d}{dt}\varphi(u(t)) = g(t, u(t), u) + \hat{f}(t, u(t), u, \frac{d}{dt}\varphi(u(t)) - g(t, u(t), u)) \\
\text{a.e. in } J, \quad u(t) = \hat{B}_0(u(t_0), u) + \hat{B}_1(t, u(t), u), \quad t \in J_0,
\end{cases}
$$

(2.4.16)

where $\hat{f}: J \times \mathbb{R} \times \mathcal{F} \times \mathbb{R} \to \mathbb{R}$, $\hat{B}_0: \mathbb{R} \times \mathcal{F} \to \mathbb{R}$ and $\hat{B}_1: J_0 \times \mathbb{R} \times \mathcal{C} \to \mathbb{R}$, are defined by

$$
\begin{cases}
\hat{f}(t, x, u, y) = \frac{f(t,x,u,y)+\alpha(t)y}{1+\alpha(t)}, \quad t \in J, \ x, \ y \in \mathbb{R}, \ u \in \mathcal{F}, \\
\hat{B}_0(x, u) = \frac{B_0(x,u)}{1+\beta}, \quad x \in \mathbb{R}, \ u \in \mathcal{F}, \\
\hat{B}_1(t, x, u) = \frac{B_1(t,x,u)+\beta x}{1+\beta}, \quad t \in J, \ x \in \mathbb{R}, \ u \in \mathcal{F},
\end{cases}
$$

(2.4.17)

have the same solutions. Moreover, the functions $\hat{f}$, $\hat{B}_0$, and $\hat{B}_1$ satisfy the hypotheses (f0), (f1), (B0), (B1), and (B01) with λ replaced by $\frac{\lambda+\alpha}{1+\alpha}$ and c by $\frac{c+\beta}{1+\beta}$. Thus problem (2.4.16), with $\hat{f}$ and $\hat{B}_0$ and $\hat{B}_1$ defined by (2.4.17), has by Theorem 2.4.1 extremal solutions u_* and u^*, and they are increasing with respect to $\hat{f}$ and $\hat{B}_1$. In view of (2.4.17), u_* and u^* are then extremal solutions of (2.4.1), and they are increasing with respect to f and B_1. $\qquad\square$

2.4.3 Special cases. Assume that $q: \mathbb{R} \to (0, \infty)$ has the property

(q0) q and $\frac{1}{q}$ belong to $L_{loc}^\infty(\mathbb{R})$, and $\int_0^{\pm\infty} \frac{dz}{q(z)} = \pm\infty$.

Applying Lemma 2.1.3 it can be shown that problem

$$
\begin{cases}
\frac{u'(t)}{q(u(t))} = g(t, u(t), u) + f(t, u, u(t), \frac{u'(t)}{q(u(t))} - g(t, u(t), u)) \quad \text{a.e. in } J, \\
u(t) = B_0(u(t_0), u) + B_1(t, u(t), u), \quad t \in J_0,
\end{cases}
$$

(2.4.18)

has same solutions as problem (2.4.1), where the function $\varphi: \mathbb{R} \to \mathbb{R}$ is defined by $\varphi(x) = \int_0^x \frac{dz}{q(z)}$, $x \in \mathbb{R}$. Moreover, this function φ has property $(\varphi 0)$. Thus the results of Theorem 2.4.1 and Proposition 2.4.1 are valid for problem (2.4.18) if, instead of $(\varphi 0)$ we assume that (q0) holds.

The function $\varphi: \mathbb{R} \to \mathbb{R}$, defined by $\varphi(x) = |x|^{p-2}x$, $x \in \mathbb{R}$, $p > 1$, has property $(\varphi 0)$. Thus the results of Theorem 2.4.1 and Proposition 2.4.1 hold for problem

$$
\begin{cases}
\frac{d}{dt}(|u(t)|^{p-2}u(t)) = g(t, u(t), u) + \\
f(t, u(t), u, \frac{d}{dt}(|u(t)|^{p-2}u(t)) - g(t, u(t), u)) \quad \text{a.e. in } J, \\
u(t) = B_0(u(t_0), u) + B_1(t, u(t), u), \quad t \in J_0,
\end{cases}
$$

(2.4.19)

if $(\varphi 0)$ is replaced by the assumption: $p > 1$.

When the functional dependence is omitted in problem (2.4.1) we obtain as a consequence of Proposition 2.4.1 the following result.

Proposition 2.4.2. *Assume that* $\varphi\colon \mathbb{R} \to \mathbb{R}$, $g\colon J\times\mathbb{R} \to \mathbb{R}$, $f\colon J\times\mathbb{R}\times\mathbb{R} \to \mathbb{R}$ *and* $B\colon \mathbb{R} \times \mathbb{R} \to \mathbb{R}$ *satisfy the following hypotheses:*

(φ0) φ *is an increasing homeomorphism.*

(fga) f *and* g *are Carathéodory functions, and there is* $\alpha \in L_+^\infty(J)$ *such that* $f(t,x,y) + \alpha(t)y$ *is increasing in* x *and* y *for a.e.* $t \in J$.

(fgb) *for a.e.* $t \in J$ *and all* x, $y \in \mathbb{R}$, $|f(t,x,y)| \le p_2(t)\psi(|\varphi(x)|) + \lambda(t)|y|$ *and* $|g(t,x)| \le p_1(t)\psi(|\varphi(x)|)$, *where* p_1, p_2, $\frac{p_2}{1-\lambda} \in L_+^1(J)$, *the function* $\psi\colon \mathbb{R}_+ \to (0,\infty)$ *is increasing and* $\int_0^\infty \frac{dx}{\psi(x)} = \infty$.

(Ba) $x \mapsto B(x,y) + \beta x$ *is increasing in* x *and* y *for some* $\beta \ge 0$.

(Bb) $|B(x,y)| \le c|x| + d$ *for all* x, $y \in \mathbb{R}$, *where* $c \in [0,1)$ *and* $d \in \mathbb{R}_+$.

Then the implicit initial value problem

$$\begin{cases} \frac{d}{dt}\varphi(u(t)) = g(t,u(t)) + f(t,u(t), \frac{d}{dt}\varphi(u(t)) - g(t,u(t))) \\ a.e. \ in \ \ J = [t_0,t_1], \quad u(t_0) = B(u(t_0),u(t_1)), \end{cases} \qquad (2.4.20)$$

has extremal solutions, and they are increasing with respect to f *and* B.

When the function f is deleted from problems (2.4.1), (2.4.18), (2.4.19), and (2.4.20) we get results for initial and boundary value problems of explicit differential equations.

The functional dependence can have many forms, some of which are presented in the following example.

Example 2.4.2. The function $q\colon \mathbb{R} \to (0,\infty)$, defined by

$$q(z) = \sum_{m=1}^\infty \sum_{k=1}^\infty \frac{(2 + [k^{\frac{1}{m}}z] - k^{\frac{1}{m}}z)}{(km)^2}\left(2 + \sin\left(\frac{1}{1 + [k^{\frac{1}{m}}z] - k^{\frac{1}{m}}z}\right)\right), \ z \in \mathbb{R},$$

where $[x]$ denotes the greatest integer $\le x$, has property (q0) (cf. Ex. 2.1.1).

Choose $J = [0,1]$, $r = 1$ and $\mathcal{F} = C([-1,1])$, and let $g\colon J \times \mathbb{R} \times \mathcal{F} \to \mathbb{R}$ and $B_0\colon \mathbb{R} \times \mathcal{F} \to \mathbb{R}$ be defined by (2.3.15) and (2.3.16). g has properties (g0), (g1), and (g2), and B_0 is bounded and has property (B0).

The function $f\colon J \times \mathbb{R} \times \mathcal{F} \times \mathbb{R} \to \mathbb{R}$, defined by

$$f(t,x,u,y) = \sum_{n=1}^\infty \frac{\arctan([n(u(1-t) + x + y - t)])}{n^2}, \quad t \in J, \ x, y \in \mathbb{R},$$

has properties (f0) and (f1), and the function $B_1\colon J_0 \times \mathbb{R} \times \mathcal{F} \to \mathbb{R}$, defined by

$$B_1(t,x,u) = \sum_{m=-\infty}^\infty \sum_{n=1}^\infty \frac{H(\int_{-1}^1 u(s)ds - \frac{m}{n})}{2^{|m|+n}} + \sin(t+x),$$

where H is the Heaviside function, has properties (B1) and (B01). Thus problem (2.4.18) has for these functions q, g, f, B_0, and B_1 extremal solutions.

Remarks 2.4.2. We have assumed that (g2) and (f1) hold with the same ψ. If ψ is replaced by $\hat{\psi}$ in (f1), property $\int_0^\infty \frac{dx}{\max\{\psi(x),\hat{\psi}(x)\}} = \infty$ is required. This and all the other properties assumed for ψ:s in (g2) and (f1) hold when ψ:s are any of the functions:

$$\psi_0(x) = ax + b, \quad x \geq 0, \ a \geq 0, \ b > 0,$$

and

$$\psi_n(x) = (x + 1) \ln(x + e) \cdots \ln_n(x + exp_n(1)), \quad x \geq 0, \ n = 1, 2, \ldots,$$

where $\ln_n$ and $\exp_n$ denote the n-fold iterated logarithm and exponential function, respectively.

By Remarks 2.1.1 we can replace $\psi(|\varphi(x)|)$ by $\psi(|x|)$ in (g2), (f1) and (fgb) if φ is Lipschitz continuous.

Problem

$$\begin{cases} F(t, u(t), u, \frac{d}{dt}\varphi(u(t))) - g(t, u(t), u)) = 0 \ \text{ a.e. in } \ J = [t_0, t_1], \\ B(t, u(t), u) = 0, \quad t \in J_0 = [t_0 - r, t_0], \end{cases} \quad (2.4.21)$$

where $g \colon J \times \mathbb{R} \times \mathcal{F} \to \mathbb{R}$, $F \colon J \times \mathbb{R} \times \mathcal{F} \times \mathbb{R} \to \mathbb{R}$, $B \colon J_0 \times \mathbb{R} \times \mathcal{F} \to \mathbb{R}$ and $\varphi \colon \mathbb{R} \to \mathbb{R}$, has the same solutions as the BVP (2.4.1), if the functions $B_0 \colon \mathbb{R} \times \mathcal{F} \to \mathbb{R}$, $f \colon J \times \mathbb{R} \times \mathcal{F} \times \mathbb{R} \to \mathbb{R}$ and $B_1 \colon J_0 \times \mathbb{R} \times \mathcal{F} \to \mathbb{R}$ are defined by $B_0(x, u) \equiv 0$,

$$\begin{cases} f(t, x, u, y) = y - \mu(t, x, u, y)F(t, x, u, y), \quad t \in J, \ x, y \in \mathbb{R}, \ u \in \mathcal{F}, \\ B_1(t, x, u) = x - \nu(t, x, u)B(t, x, u), \quad t \in J_0, \ x \in \mathbb{R}, \ u \in \mathcal{F}, \end{cases}$$
$$(2.4.22)$$

for any functions $\mu \colon J \times \mathbb{R} \times \mathcal{F} \times \mathbb{R} \to (0, \infty)$ and $\nu \colon J_0 \times \mathbb{R} \times \mathcal{F} \to (0, \infty)$. Hence, if μ and ν can be chosen so that the hypotheses of Theorem 2.4.1 hold when f and B_1 are defined by (2.4.22), then problem (2.4.21) has extremal solutions.

2.5 Notes and comments

Existence and comparison results derived for discontinuous first order explicit initial value problems in section 2.1, and for corresponding boundary

value problems in section 2.2, are taken from [4]. As for recent results in
the special case when φ is the identity function, see, e.g., [25], [26], [40],
[42], [89],[125], [128], [130], [131], [138], [139], [141], [142], [143], [162], [194],
[203], [206]. Finite systems are considered in [125], [127], [162], and in [141],
where also infinite systems are studied. As for first order explicit differential equations in ordered Banach spaces see, e.g., [141] and the references
therein.

Extremality results presented for explicit functional problems in section
2.3, and for implicit functional problems in section 2.4, are adapted from
[84]. First order implicit differential equations are considered, for instance,
in [13], [16], [17], [32], [42], [76], [80], [132], [133], [137], [148], [149], [150],
[153], [157], [159], [183], [195], [196], [199], [200], [201], [202], [204], [210],
and [220].

Finite implicit systems are studied, e.g., in [114]. As for implicit differential equations in an ordered Banach space X, see, e.g., [73], [78], and
[87]. In [135] equations are functional, and only the existence of weak limits
of bounded and monotone sequences of X is assumed. Implicit impulsive
differential equations in ordered Banach spaces are considered in [82] and
[86].

Chapter 3

Uniqueness, comparison, and well-posedness results for quasilinear differential equations

The purpose of this chapter is to present comparison, uniqueness, and well-posedness results for initial and boundary value problems of first and second order quasilinear differential equations. We begin section 3.1 with a comparison principle, which is then applied to prove comparison and uniqueness results for a boundary value problem of the differential equation

$$\frac{d}{dt}\varphi(t, u(t)) = g(t, u(t)).$$

In section 3.2 we apply an analogous procedure to provide comparison and uniqueness results for initial value problems of the above-mentioned differential equation and the implicit differential equation

$$\frac{d}{dt}\varphi(t, u(t)) = g(t, u(t)) + f(t, u(t), \frac{d}{dt}\varphi(t, u(t)) - g(t, u(t))).$$

The assumptions imposed on the functions φ, g, and f allow them to be discontinuous in all their variables. The so-obtained results, combined with existence results derived in chapter 2, yield also existence and uniqueness results for the above problems with $\varphi(t, u(t))$ replaced by $\varphi(u(t))$.

Section 3.3 is devoted to well-posedness study of initial value problems of the above differential equations when φ does not depend on its first variable. In section 3.4 we derive maximum principles for a differential operator which consists of a generalized phi-Laplacian operator and lower order terms. No sup-measurability hypotheses are required. These maximum principles are then applied to prove uniqueness and comparison results for boundary value problems of the differential equation

$$-\frac{d}{dt}\varphi(t, u'(t)) = g(t, u(t), u'(t)).$$

3.1　First order boundary value problems

In this section we generalize some known comparison and uniqueness results derived for first order boundary value problems of the differential equation $u'(t) = g(t, u(t))$ (cf., e.g., [138]), replacing the term $u'(t)$ by $\frac{d}{dt}\varphi(t, u(t))$, and allowing both g and φ to be discontinuous. After treating some special cases, including periodic boundary value problems, we combine results of this section and chapter 2 to obtain existence and uniqueness results.

3.1.1　Comparison and uniqueness results. Given a closed interval $J = [t_0, t_1]$, functions $\varphi\colon J \times \mathbb{R} \to \mathbb{R}$ and $g\colon J \times \mathbb{R} \to \mathbb{R}$, and positive constants a and b, we derive first a comparison principle for the operators A and B, defined by

$$\begin{cases} Au(t) := \frac{d}{dt}\varphi(t, u(t)) - g(t, u(t)), & t \in J, \\ Bu := au(t_0) - bu(t_1), \\ u \in Y := \{u \in C(J) \mid \varphi(\cdot, u(\cdot)) \in AC(J)\}. \end{cases} \qquad (3.1.1)$$

Assuming that φ and g satisfy the following hypotheses for some $k, K > 0$:

(φ1)　$\varphi(t, \cdot)$ is strictly increasing for all t in a dense subset of J, and if $x < y$, then $k(\varphi(t_0, y) - \varphi(t_0, x)) \leq y - x \leq K(\varphi(t_1, y) - \varphi(t_1, x))$,

($g\varphi$1)　there exists a $p \in L^1(J)$ with $\int_{t_0}^{t_1} p(t)dt > \ln(\frac{bK}{ak})$ such that the function $x \mapsto g(t, x) + p(t)\,\varphi(t, x)$ is decreasing in x for a.e. $t \in J$,

we shall prove our first comparison result.

Lemma 3.1.1. *Assume that the hypotheses (φ1) and ($g\varphi$1) hold, and let the operators A and B be defined by (3.1.1). If the functions $v, w \in Y$ satisfy inequalities*

$$Av(t) \leq Aw(t) \quad \text{a.e. in } J, \quad \text{and} \quad Bv \leq Bw, \qquad (3.1.2)$$

then $v(t) \leq w(t)$ on J.

Proof. If the claim is wrong, then either

(i)　$w(t) < v(t)$ on J,

or there exist a_1, a_2 in J, $a_1 < a_2$, such that one of the following cases hold.

(ii)　$v(a_1) = w(a_1)$ and $w(t) < v(t)$ on (a_1, a_2).
(iii)　$v(a_2) = w(a_2)$ and $w(t) < v(t)$ on (a_1, a_2).

Suppose that (i) holds, and define

$$x(t) := \varphi(t, v(t)) - \varphi(t, w(t)), \quad t \in J. \tag{3.1.3}$$

Applying this definition, the first of inequalities (3.1.2), and the hypothesis $(g\varphi 1)$ we obtain

$$x'(t) = \frac{d}{dt}\varphi(t, v(t)) - \frac{d}{dt}\varphi(t, w(t)) \leq g(t, v(t)) - g(t, w(t))$$
$$\leq -p(t)(\varphi(t, v(t)) - \varphi(t, w(t))) = -p(t)x(t) \text{ for a.e. } t \in J.$$

This implies that

$$x(t_1) e^{\int_{t_0}^{t_1} p(t)dt} \leq x(t_0). \tag{3.1.4}$$

It follows from the inequality $Bv \leq Bw$ and $(\varphi 1)$ that

$$akx(t_0) \leq a(v(t_0) - w(t_0)) \leq b(v(t_1) - w(t_1)) \leq bKx(t_1). \tag{3.1.5}$$

The inequalities (3.1.4) and (3.1.5) imply that

$$x(t_1)(ak \, e^{\int_{t_0}^{t_1} p(t)dt} - bK) \leq 0.$$

Since $ake^{\int_{t_0}^{t_1} p(t)dt} > bK$ by $(g\varphi 1)$, then $x(t_1) \leq 0$, which contradicts the fact that $x(t_1) > 0$ by (3.1.3), (i) and $(\varphi 1)$.

If (ii) holds, we get by similar reasoning as above,

$$x'(t) \leq -p(t)x(t) \text{ for a.e. } t \in (a_1, a_2), \quad x(a_1) = 0,$$

so that

$$x(t) \leq x(a_1) e^{-\int_{t_0}^{t} p(s)ds} = 0, \quad t \in (a_1, a_2).$$

But this is impossible, because $x(t) > 0$ for some $t \in (a_1, a_2)$ by (ii), (3.1.3) and $(\varphi 1)$.

Finally, if (iii) holds, then $w(t) < v(t)$ for all $t \in [t_0, a_2)$, for otherwise there is a $t_2 \in [t_0, a_2)$ such that $v(t_2) = w(t_2)$ and $w(t) < v(t)$ on (t_2, a_2). But then (ii) holds with a_1 replaced by t_2, which is impossible. In particular, $w(t_0) < v(t_0)$. This and the inequalities (3.1.5) imply that $w(t_1) < v(t_1)$. Since $w(a_2) = v(a_2)$ and $v - w$ is continuous, there is a $t_3 \in [a_2, t_1)$ such that $v(t_3) = w(t_3)$ and $w(t) < v(t)$ on (t_3, t_1). But then (ii) holds with a_1, a_2 replaced by t_3, t_1, which is impossible. Thus (iii) cannot hold either. This concludes the proof.   □

Next we apply the result of Lemma 3.1.1 to prove uniqueness and comparison results for the boundary value problem

$$\frac{d}{dt}\varphi(t, u(t)) = g(t, u(t)) \text{ a.e. in } J, \quad au(t_0) - bu(t_1) = c. \tag{3.1.6}$$

Definition 3.1.1. A function $u \in Y$ is said to be a *lower solution* of the BVP (3.1.6) if

$$\frac{d}{dt}\varphi(t, u(t)) \leq g(t, u(t)) \text{ for a.e. } t \in J, \text{ and } au(t_0) - bu(t_1) \leq c, \quad (3.1.7)$$

and an *upper solution* if the reversed inequalities are satisfied. If equalities hold in (3.1.7), we say that u is a *solution* of (3.1.6).

As an immediate consequence of Lemma 3.1.1 we obtain the following comparison and uniqueness result.

Theorem 3.1.1. *Assume that the hypotheses $(\varphi 1)$ and $(g\varphi 1)$ hold. If $v \in Y$ is a lower solution of (3.1.6), and if $w \in Y$ is an upper solution of (3.1.6), then $v(t) \leq w(t)$ on J. In particular, (3.1.6) can have at most one solution.*

Proof. If v and w are lower and upper solutions of (3.1.6), then the inequalities (3.1.2) hold when the operators A and B are defined by (3.1.1). Thus $v(t) \leq w(t)$ on J by Lemma 3.1.1. The last conclusion follows from the first one.   ☐

Obvious modifications to the proof of Lemma 3.1.1 yield our next result.

Proposition 3.1.1. *The results of Lemma 3.1.1 and Theorem 3.1.1 hold when the functions φ and g have the following properties for some k, $K > 0$:*

$(\varphi 11)$ $\varphi(t, \cdot)$ *is increasing for all t in a dense subset of J, and if $x < y$, then $k(\varphi(t_0, y) - \varphi(t_0, x)) \leq y - x \leq K(\varphi(t_1, y) - \varphi(t_1, x))$.*

$(g\varphi 11)$ *There is a $p \in L^1(J)$ with $\int_{t_0}^{t_1} p(t)dt > \ln(\frac{bK}{ak})$ such that the function $x \mapsto g(t, x) + p(t)\,\varphi(t, x)$ is strictly decreasing in x for a.e. $t \in J$.*

3.1.2 Special cases. In this section we present some special cases of the results derived in subsection 3.1.1. In the case when $a = b = 1$, $(g\varphi 1)$ is reduced to the form

$(g\varphi 2)$ *There is a $p \in L^1(J)$ with $\int_{t_0}^{t_1} p(t)dt > \ln(\frac{K}{k})$ such that the function $x \mapsto g(t, x) + p(t)\,\varphi(t, x)$ is decreasing in x for a.e. $t \in J$.*

Since the periodic boundary value problem (PBVP)

$$\frac{d}{dt}\varphi(t, u(t)) = g(t, u(t)) \text{ a.e. in } J, \quad u(t_0) = u(t_1), \quad\quad (3.1.8)$$

is a special case of the BVP (3.1.6) when $a = b = 1$ and $c = 0$, we obtain the following consequence of Theorem 3.1.1.

Corollary 3.1.1. *Assume that the hypotheses (φ1) and ($g\varphi$2) hold. Then the comparison and uniqueness results of Theorem 3.1.1 are valid for (3.1.8).*

Example 3.1.1. Denote by $[x]$ the greatest integer $\leq x$, and by χ_U the characteristic function of a nonempty subset U of $(0,1)$. The PBVP

$$\begin{cases} \frac{d}{dt}(u(t) + [\chi_U(t)u(t)]) = -u(t) - [\chi_U(t)u(t)] \text{ a.e. in } J = [0,1], \\ u(0) = u(1), \end{cases} \quad (3.1.9)$$

is a special case of problem (3.1.8) with

$$\varphi(t,x) = x + [\chi_U(t)x], \ g(t,x) = -\varphi(t,x), \quad t \in J, \ x \in \mathbb{R}.$$

It is easy to see that these functions φ and g satisfy the hypotheses (φ1) and ($g\varphi$2) when $k = K = 1$ and $p(t) \equiv 1$. Thus Corollary 3.1.1 implies that the PBVP (3.1.9) can have only one solution. Obviously, $u(t) \equiv 0$ is the solution of (3.1.9). The functions φ and g are discontinuous in all their variables, and even nonmeasurable in t if U is a nonmeasurable subset of J.

In the special case when the BVP (3.1.6) can be rewritten as

$$\frac{d}{dt}\varphi(u(t)) = g(t,u(t)) \text{ a.e. in } J, \quad au(t_0) - bu(t_1) = c, \quad (3.1.10)$$

the hypothesis (φ1) is reduced to the form

(φ12) $k(\varphi(y) - \varphi(z)) \leq y - z \leq K(\varphi(y) - \varphi(z))$ for all $y, z \in \mathbb{R}$, $z < y$.

Thus we get the following result.

Proposition 3.1.2. *The results of Theorem 3.1.1 hold for the BVP (3.1.10) if the function $\varphi \colon \mathbb{R} \to \mathbb{R}$ has property (φ12) and if $g \colon J \times \mathbb{R} \to \mathbb{R}$ satisfies the following hypothesis for some $k, K > 0$.*

($g\varphi$12) *There is a $p \in L^1(J)$ with $\int_{t_0}^{t_1} p(t)dt > \ln(\frac{bK}{ak})$ such that the function $x \mapsto g(t,x) + p(t)\varphi(x)$ is decreasing for a.e. $t \in J$.*

The hypothesis (φ12) implies that φ is an increasing homeomorphism. Thus the results of Theorem 2.2.1 and Proposition 3.1.2 imply the following existence and uniqueness result.

Proposition 3.1.3. *The BVP (3.1.10) has a unique solution if the hypotheses of Proposition 3.1.2 and the following hypotheses are valid.*

(g0) *For each $x \in \mathbb{R}$ the function $g(\cdot, x)$ is measurable, and*
$$\limsup_{y \to x-} g(t,y) \leq g(t,x) \leq \liminf_{y \to x+} g(t,y) \quad \text{for a.e. } t \in J.$$

(A) *(3.1.10) has a lower solution $\underline{u}$ and an upper solution $\overline{u}$ such that $\underline{u} \leq \overline{u}$, and g is L^1-bounded in $\Omega = \{(t,x) \in J \times \mathbb{R} \mid \underline{u}(t) \leq x \leq \overline{u}(t)\}$.*

Consider next the BVP

$$u'(t) = q(u(t))g(t, u(t)) \quad \text{a.e. in } J, \quad au(t_0) - bu(t_1) = c, \qquad (3.1.11)$$

where

(q2) $q \colon \mathbb{R} \to (0, \infty)$ and $\frac{1}{q}$ are measurable and essentially bounded.

It follows from Lemma 2.1.3 that (3.1.11) can be reduced to the BVP (3.1.10), where $\varphi \colon \mathbb{R} \to \mathbb{R}$ is defined by

$$\varphi(x) = \int_0^x \frac{dz}{q(z)}, \qquad x \in \mathbb{R}. \qquad (3.1.12)$$

As a direct consequence of (q2) we see that the function φ, defined by (3.1.12), has the following property.

(φ13) $\|\frac{1}{q}\|_\infty^{-1}(\varphi(y) - \varphi(z)) \le y - z \le \|q\|_\infty(\varphi(y) - \varphi(z))$ whenever $z < y$.

This property and Proposition 3.1.2 yield the following results.

Proposition 3.1.4. *Assume that the function $\varphi \colon \mathbb{R} \to \mathbb{R}$ is defined by (3.1.12), where $q \colon \mathbb{R} \to (0, \infty)$ has property (q2), and that $g \colon J \times \mathbb{R} \to \mathbb{R}$ satisfies the following hypothesis.*

($g\varphi$13) *There is a $p \in L^1(J)$ with $\int_{t_0}^{t_1} p(t)dt > \ln(\frac{b}{a}\|\frac{1}{q}\|_\infty\|q\|_\infty)$ such that the function $x \mapsto g(t, x) + p(t)\,\varphi(x)$ is decreasing for a.e. $t \in J$.*

If $v \in Y$ is a lower solution and $w \in Y$ an upper solution of (3.1.11), then $v(t) \le w(t)$ on J. In particular, (3.1.11) can have only one solution.

When $a = b = 1$ and $c = 0$ we get the following result for the PBVP

$$u'(t) = q(u(t))g(t, u(t)) \quad \text{a.e. in } J, \quad u(t_0) = u(t_1). \qquad (3.1.13)$$

Corollary 3.1.2. *Assume that a function $\varphi \colon \mathbb{R} \to \mathbb{R}$ is defined by (3.1.12), where $q \colon \mathbb{R} \to (0, \infty)$ has property (q2), and that $g \colon J \times \mathbb{R} \to \mathbb{R}$ has the following property.*

($g\varphi$14) *There is a $p \in L^1(J)$ for which $\int_{t_0}^{t_1} p(t)dt > \ln(\|\frac{1}{q}\|_\infty\|q\|_\infty)$ such that the function $x \mapsto g(t, x) + p(t)\,\varphi(x)$ is decreasing for a.e. $t \in J$.*

If $v \in Y$ is a lower solution and $w \in Y$ an upper solution of (3.1.13), then $v(t) \le w(t)$ on J. In particular, (3.1.13) can have only one solution.

Combining the results of Propositions 2.2.4 and 3.1.4 we get the following existence and uniqueness result.

Proposition 3.1.5. *The BVP (3.1.11) has a unique solution if the hypotheses of Proposition 3.1.4 and the following hypotheses are valid.*

(g0) *For each $x \in \mathbb{R}$ the function $g(\cdot, x)$ is measurable, and*
$$\limsup_{y \to x-} g(t, y) \le g(t, x) \le \liminf_{y \to x+} g(t, y) \quad \text{for a.e. } t \in J.$$

(g2) $|g(t, x)| \le M(t) + p_1(t)|x|$ *for a.e. $t \in J$ and for all $x \in \mathbb{R}$, where $M, p_1 \in L^1_+(J)$, and $\int_{t_0}^{t_1} p_1(s)\, ds < \frac{1}{\|q\|_\infty} \ln \frac{a}{b}$.*

3.2 First order initial value problems

In this section we provide comparison and uniqueness results for the initial value problem of the first order implicit differential equation

$$\frac{d}{dt}\varphi(t, u(t)) = g(t, u(t)) + f(t, u(t), \frac{d}{dt}\varphi(t, u(t)) - g(t, u(t))). \qquad (3.2.1)$$

No continuity or measurability conditions are imposed on the functions φ, g, and f.

3.2.1 Comparison and uniqueness results for implicit IVPs. Given a real interval $J = [t_0, t_1]$ and functions $\varphi \colon J \times \mathbb{R} \to \mathbb{R}$, $g \colon J \times \mathbb{R} \to \mathbb{R}$ and $f \colon J \times \mathbb{R} \times \mathbb{R} \to \mathbb{R}$, we derive first a comparison principle for the differential operator D, defined by

$$\begin{cases} Du(t) := \frac{d}{dt}\varphi(t, u(t)) - g(t, u(t)) - f(t, u(t), \frac{d}{dt}\varphi(t, u(t)) - g(t, u(t))), \\ u \in Y := \{u \in C(J) \mid \varphi(\cdot, u(\cdot)) \in AC(J)\}, \quad t \in J. \end{cases}$$
$$(3.2.2)$$

We assume that φ, g, and f satisfy the following hypotheses.

($\varphi 2$) $\varphi(t, \cdot)$ is strictly increasing for a.e. $t \in J$.

(fg2) For a.e. $t \in J$ and for all $x, y \in \mathbb{R}$, $x < y$, and $u, v \in \mathbb{R}$, $v < u$,

$$f(t, y, u) - f(t, x, v) \le p(t)\phi(\varphi(t, y) - \varphi(t, x)) + \lambda(t)(u - v), \qquad (3.2.3)$$

and

$$g(t, y) - g(t, x) \le \bar{p}(t)\phi(\varphi(t, y) - \varphi(t, x)), \qquad (3.2.4)$$

where $\bar{p}, \frac{p}{1-\lambda} \in L^1_+(J)$, $\phi \colon \mathbb{R}_+ \to \mathbb{R}_+$ is increasing and $\int_{0+}^{1} \frac{dx}{\phi(x)} = \infty$.

Our main comparison result reads as follows.

Lemma 3.2.1. *Assume that the hypotheses ($\varphi 2$) and (fg2) hold, and let the operator D be defined by (3.2.2). If $v, w \in Y$ satisfy inequalities*

$$Dv(t) \leq Dw(t) \quad a.e. \ in \ J, \quad v(t_0) \leq w(t_0), \tag{3.2.5}$$

then $v(t) \leq w(t)$ on J.

Proof. If the claim is wrong, there exist $a \in [t_0, t_1)$ and $b \in (a, t_1]$ such that

$$v(a) = w(a) \quad \text{and} \quad w(t) < v(t) \quad \text{for each} \ t \in (a, b]. \tag{3.2.6}$$

Denote

$$Au(t) = \frac{d}{dt}\varphi(t, u(t)) - g(t, u(t)), \quad u \in Y, \ t \in J, \tag{3.2.7}$$

and

$$x(t) = \begin{cases} 0, & t_0 \leq t \leq a, \\ \varphi(t, v(t)) - \varphi(t, w(t)), & a \leq t \leq b, \\ x(b), & b \leq t \leq t_1. \end{cases} \tag{3.2.8}$$

Since $Dv(t) \leq Dw(t)$ a.e. in J, then by applying (3.2.2), (3.2.3), (3.2.6), and (3.2.7) we see that for a.e. $t \in [a, b]$ for which $Av(t) - Aw(t) > 0$,

$$
\begin{aligned}
Av(t) - Aw(t) &\leq f(t, v(t), Av(t)) - f(t, w(t), Aw(t)) \\
&\leq p(t)\phi(\varphi(t, v(t)) - \varphi(t, w(t))) + \lambda(t)(Av(t) - Aw(t)).
\end{aligned}
$$

Hence,

$$Av(t) - Aw(t) \leq \frac{p(t)}{1 - \lambda(t)}\phi(x(t)) \tag{3.2.9}$$

for a.e. $t \in [a, b]$ satisfying $Av(t) - Aw(t) > 0$. Since the right-hand side of (3.2.9) is nonnegative for a.e. $t \in J$, it holds also when $Av(t) - Aw(t) \leq 0$, so that (3.2.9) holds a.e. in $[a, b]$. It then follows from (3.2.4), (3.2.7), (3.2.8), and (3.2.9) that

$$
\begin{aligned}
x'(t) &\leq g(t, v(t)) - g(t, w(t)) + \frac{p(t)}{1 - \lambda(t)}\phi(x(t)) \\
&\leq \bar{p}(t)\phi(x(t)) + \frac{p(t)}{1 - \lambda(t)}\phi(x(t)),
\end{aligned}
$$

or equivalently,

$$x'(t) \leq \left(\bar{p}(t) + \frac{p(t)}{1 - \lambda(t)}\right)\phi(x(t)) \tag{3.2.10}$$

for a.e. $t \in [a, b]$. Since $x'(t) = 0$ when $t \in J \backslash [a, b]$, and since the right-hand side of (3.2.10) is nonnegative for a.e. $t \in J$, it holds a.e. in J. Because $x(t_0) = 0$, then $x(t) = 0$ on J by Lemma B.6.1, which contradicts the fact that $x(t) > 0$ a.e. in $(a, b]$ by (3.2.6) and ($\varphi 2$). This concludes the proof. $\square$

Next we shall apply the result of Lemma 3.2.1 to the IVP

$$
\begin{cases}
\frac{d}{dt}\varphi(t, u(t)) = g(t, u(t)) + f(t, u(t), \frac{d}{dt}\varphi(t, u(t)) - g(t, u(t))) \\
\qquad\qquad \text{for a.e. } t \in J, \\
u(t_0) = x_0.
\end{cases}
\tag{3.2.11}
$$

A function $u \in Y$ is said to be a *lower solution* of the IVP (3.2.11) if

$$
\begin{cases}
\frac{d}{dt}\varphi(t, u(t)) \leq g(t, u(t)) + f(t, u(t), \frac{d}{dt}\varphi(t, u(t)) - g(t, u(t))) \\
\qquad\qquad \text{for a.e. } t \in J, \\
u(t_0) \leq x_0,
\end{cases}
$$

and an *upper solution* if the reversed inequalities are satisfied. If equalities hold, we say that u is a *solution* of (3.2.11).

As an immediate consequence of Lemma 3.2.1 we obtain the following comparison and uniqueness result.

Theorem 3.2.1. *Assume that the hypotheses ($\varphi 2$) and (fg2) are valid, and let $x_0 \in \mathbb{R}$ be given. If $v \in Y$ is a lower solution of (3.2.11) and $w \in Y$ is an upper solution of (3.2.11), then $v(t) \leq w(t)$ on J. In particular, (3.2.11) can have at most one solution.*

Proof. If v and w are lower and upper solutions of (3.2.11), then the inequalities (3.2.3) hold when the operator D is defined by (3.2.2). Thus $v(t) \leq w(t)$ on J by Lemma 3.2.1. The last conclusion follows from the first one. $\square$

If $g(t, x)$ and $f(t, x, u)$ are decreasing in x and u, the hypothesis (fg2) is valid when $\bar{p}$, p, and λ are zero functions. Thus we obtain the following result.

Corollary 3.2.1. *The results of Lemma 3.2.1 and Theorem 3.2.1 hold if for a.e. $t \in J$, $\varphi(t, \cdot)$ is strictly increasing, $g(t, \cdot)$ is decreasing, and both $f(t, \cdot, x)$ and $f(t, x, \cdot)$ are decreasing for all $x \in \mathbb{R}$.*

In the next result we modify somewhat the hypotheses ($\varphi 2$) and (fg2).

Proposition 3.2.1. *The results of Lemma 3.2.1 and Theorem 3.2.1 hold if ($\varphi2$) and (fg2) are replaced by the following hypotheses.*

($\varphi3$) *For each bounded subinterval I_1 of $\mathbb{R}$ there exists such a $K > 0$ that $y - z \le K(\varphi(t, y) - \varphi(t, z))$ for a.e. $t \in J$ and for all $y, z \in I_1$, $z < y$.*

(fg3) *For a.e. $t \in J$ and for all $y, z \in \mathbb{R}$, $z < y$, and $u, v \in \mathbb{R}$, $v < u$,*

$$f(t, y, u) - f(t, z, v) \le p(t)\phi(y - z) + \lambda(t)(u - v), \quad and \qquad (3.2.12)$$

$$g(t, y) - g(t, z) \le \bar{p}(t)\phi(y - z), \qquad (3.2.13)$$

where $\bar{p}, \frac{p}{1-\lambda} \in L^1_+(J)$, $\phi: \mathbb{R}_+ \to \mathbb{R}_+$ is an increasing function and $\int_{0+}^1 \frac{dx}{\phi(x)} = \infty$.

Proof. Obviously ($\varphi3$) implies that ($\varphi2$) holds. Moreover, it follows from ($\varphi3$), (3.2.12), and (3.2.13) that (fg2) is valid when x, y are restricted to a bounded interval I_1 and ϕ is replaced by $x \mapsto \phi(Kx)$. The conclusions follow when I_1 is chosen in the proof of Lemma 3.2.1 so that it contains the ranges of v and w.   □

Remark 3.2.1. The hypotheses given for ϕ in (fg2) and (fg3) hold, for instance, if ϕ is the continuous extension on $\mathbb{R}_+$ of the function

$$\phi(x) = x \ln \frac{1}{x} \cdots \ln_n \frac{1}{x}, \quad 0 < x \le \frac{1}{exp_n(1)},$$

where $\ln_n$ and $\exp_n$ denote the n-fold iterated logarithm and exponential functions, respectively. In the case when $\phi(x) = x$, $x \in \mathbb{R}_+$, (3.2.13) is reduced to an L^1-Lipschitz condition.

A special feature of the above hypotheses is that they allow the functions φ, g, and f to be discontinuous, and even nonmeasurable. However, the hypotheses (fg2) and (fg3) don't allow upward jumps to functions $x \mapsto g(t, x)$ and $x \mapsto f(t, x, y)$. The next theorem shows that such jumps are possible when $\varphi(\cdot, x)$ is decreasing and the functions g and f are nonnegative-valued.

Theorem 3.2.2. *The results of Theorem 3.2.1 hold if $\varphi: J \times \mathbb{R} \to \mathbb{R}$ has properties ($\varphi2$) and*

($\varphi4$) *$\varphi(\cdot, z)$ is decreasing for each $z \in \mathbb{R}$,*

and if $g: J \times \mathbb{R} \to \mathbb{R}$ and $f: J \times \mathbb{R} \times \mathbb{R} \to \mathbb{R}$ have the following properties.

(fg4) *f and g are nonnegative-valued, and for each $(s, z) \in [t_0, t_1) \times \mathbb{R}$ there exist positive numbers δ and ϵ, functions $\bar{p}, p, \lambda \in L^1_+(J)$ with*

$\frac{p}{1-\lambda} \in L^1_+(J)$, and an increasing function $\phi \colon \mathbb{R}_+ \to \mathbb{R}_+$ for which $\int_{0+}^1 \frac{dx}{\phi(x)} = \infty$ such that (3.2.3) and (3.2.4) hold for a.e. $t \in [s, s+\delta)$ and when $x, y \in \mathbb{R} \cap [z, z+\epsilon)$ and $u, v \in \mathbb{R}$, $x < y$ and $u < v$.

Proof. Let $v \in Y$ be a lower solution of (3.2.11) and $w \in Y$ an upper solution of (3.2.11). If $v(t) \le w(t)$ on J does not hold, we may choose a subinterval $[a, b]$ of J such that (3.2.6) holds. Because w is an upper solution of (3.2.11), and since f and g are nonnegative-valued by (fg4), it follows from ($\varphi 2$) and ($\varphi 4$) that w is increasing on $[a, b]$. Choose $(s, z) = (a, w(a))$, and let $\delta > 0$ and $\epsilon > 0$ be as in (fg4). Since $v, w \in Y$, we can choose b in (3.2.6) so that $z \le w(t) < v(t) < z + \epsilon$ for each $t \in (a, b]$. Thus we can use the same reasoning as in the proof of Lemma 3.2.1 to get a contradiction, which implies the assertions. $\qquad\square$

3.2.2 Comparison and uniqueness results for explicit IVPs. When f is a zero-function, the operator D, defined by (3.2.2) is reduced to A, given by

$$Au(t) = \frac{d}{dt}\varphi(t, u(t)) - g(t, u(t)), \quad u \in Y, \ t \in J, \qquad (3.2.7)$$

and the IVP (3.2.11) can be rewritten as

$$\frac{d}{dt}\varphi(t, u(t)) = g(t, u(t)) \text{ a.e. in } J, \quad u(t_0) = x_0. \qquad (3.2.14)$$

In this case we get the following result.

Proposition 3.2.2. *Assume that* $\varphi \colon J \times \mathbb{R} \to \mathbb{R}$ *satisfies* ($\varphi 2$), *and that* $g \colon J \times \mathbb{R} \to \mathbb{R}$ *has the following property.*

(g3) $g(t, y) - g(t, x) \le l(t, \varphi(t, y) - \varphi(t, x))$ *for a.e.* $t \in J$ *and for all* $x, y \in \mathbb{R}$, $x < y$, *where* $l \colon J \times \mathbb{R}_+ \to \mathbb{R}_+$, *and zero-function is the only lower solution of the IVP*

$$x'(t) = l(t, x(t)) \text{ a.e. in } J, \quad x(t_0) = 0. \qquad (3.2.15)$$

a) *If* $v, w \in Y$, $Av \le Aw$ *and* $v(t_0) \le w(t_0)$, *then* $v(t) \le w(t)$ *on* J.
b) *If* v *and* w *are lower and upper solutions of (3.2.14) in* Y, *then* $v \le w$.
c) *The IVP (3.2.14) can have only one solution.*

Proof. a) Assume that $v, w \in Y$, $Av \le Aw$ and $v(t_0) \le w(t_0)$, and let $x \colon J \to \mathbb{R}_+$ be defined by (3.2.8). If the claim $v(t) \le w(t)$ on J is wrong, there exist $a \in [t_0, t_1)$ and $b \in (a, t_1]$ such that (3.2.6) holds. Since $Av(t) \leq Aw(t)$ a.e. in J, it then follows from (3.2.6), (3.2.7), (3.2.8), and (g3) that

$$x'(t) \leq l(t, x(t)) \qquad (3.2.16)$$

for a.e. $t \in [a, b]$. Since $x'(t) = 0$ when $t \in J \setminus [a, b]$, and since the right-hand side of (3.2.16) is nonnegative for a.e. $t \in J$, it holds a.e. in J. Because $x(t_0) = 0$, then $x(t) \equiv 0$ in J by (g3), which contradicts the fact that $x(t) > 0$ a.e. in $(a, b]$ by (3.2.6) and (φ2). This concludes the proof of a). The assertions b) and c) follow from a) by a reasoning used in the proof of Theorem 3.2.1.   □

The result of Theorem 3.2.2 has the following analog for the IVP (3.2.14).

Proposition 3.2.3. *The existence and comparison results of Theorem 3.2.2 hold for problem (3.2.14) if $\varphi \colon J \times \mathbb{R} \to \mathbb{R}$ has properties (φ2) and (φ4), and if $g \colon J \times \mathbb{R} \to \mathbb{R}$ satisfies the following hypothesis.*

(g4) *g is nonnegative-valued, and for each $(s, z) \in [t_0, t_1) \times \mathbb{R}$ there exist positive numbers δ and ϵ and a function $l \colon [s, s + \delta) \times \mathbb{R}_+ \to \mathbb{R}_+$ such that (g2) holds when J is replaced by $J \cap [s, s + \delta)$ and $\mathbb{R}$ by $\mathbb{R} \cap [z, z + \epsilon)$.*

When φ is increasing and g is decreasing in their second arguments, and one of these monotonicities is strict, we obtain the following result.

Proposition 3.2.4. *The results of Proposition 3.2.3 hold if there is such a dense subset D of J that one of the following hypotheses is satisfied.*

(1) *$\varphi(t, \cdot)$ is strictly increasing for each $t \in D$, and $g(t, \cdot)$ is decreasing for a.e. $t \in J$.*

(2) *$\varphi(t, \cdot)$ is increasing for each $t \in D$, and $g(t, \cdot)$ is strictly decreasing for a.e. $t \in J$.*

Proof. Let $v, w \in Y$ be lower and upper solutions of (3.2.14). Assume on the contrary the existence of $a \in [t_0, t_1)$ and $b \in (a, t_1]$ such that (3.2.6) holds. Denoting $x(t) = \varphi(t, v(t)) - \varphi(t, w(t))$, $a \leq t \leq b$, we obtain

$$x'(t) = \frac{d}{dt}\varphi(t, v(t)) - \frac{d}{dt}\varphi(t, w(t)) \leq g(t, v(t)) - g(t, w(t))$$

for a.e. $t \in [a, b]$. Then $x(a) = 0$, and the above inequality and the hypotheses given for g imply that $x'(t) \leq 0$ a.e. in $[a, b]$, i.e., $x(t) \leq 0$ on $(a, b]$ in case (1), and $x'(t) < 0$ a.e. in $[a, b]$, i.e., $x(t) < 0$ on $(a, b]$ in case (2). But the hypotheses given for φ in (1) and (2) imply that $x(t) > 0$ in a dense subset of $(a, b]$ in case (1) and $x(t) \geq 0$ in a dense subset of $(a, b]$ in case (2). These contradictions show that the asserted results hold.   □

Example 3.2.1. Denote by $[x]$ the greatest integer $\leq x$, and by χ_U the characteristic function of a nonempty subset U of $(0,1)$. The IVP

$$\begin{cases} \frac{d}{dt}(u(t) + [\chi_U(t)u(t)]) = u(t) - [\chi_U(t)u(t)] - t + 1 \text{ a.e. in } [0,1], \\ u(0) = 0, \end{cases}$$

$$(3.2.17)$$

is a special case of problem (3.2.14) with

$$\varphi(t,x) = x + [\chi_U(t)x], \ g(t,x) = x - [\chi_U(t)x] - t + 1, \quad t \in [0,1], \ x \in \mathbb{R}.$$

It is easy to see that these functions φ and g satisfy the hypotheses $(\varphi 2)$ and (g3), whence Proposition 3.2.2 implies that the IVP (3.2.17) can have only one solution. Obviously, $u(t) = t$ is the solution of (3.2.17). The functions φ and g are discontinuous in all their variables, and even nonmeasurable in t if U is nonmeasurable.

Example 3.2.2. The IVP $u'(t) = g(t,u(t))$ a.e. in J, $x(0) = 0$, has many solutions on each interval $J = [0,T]$, $T > 0$, if g is one of the following functions.

$$g(t,x) = \begin{cases} tx, & x \geq 0, \\ -t, & x < 0, \end{cases} \quad g(t,x) = \begin{cases} tx^{\frac{1}{2}}, & x \geq 0, \\ 0, & u < 0. \end{cases}$$

The first of these functions satisfies (g4), except that g is nonnegative-valued. In the second case (g2) does not hold.

3.2.3 Special cases. When $\varphi(t,x) = \mu(t)|x|^{c-2}x$, where $\mu\colon J \to \mathbb{R}$, the hypotheses $(\varphi 2)$-$(\varphi 4)$ are consequences of the respective properties $(c\mu 2)$-$(c\mu 4)$, where

$(c\mu 2)$ $c > 1$ and $\mu(t) > 0$ for a.e. $t \in J$;

$(c\mu 3)$ $c \in (1,2]$ and there is such a $K > 0$ that $K\mu(t) \geq 1$ a.e. in J;

$(c\mu 4)$ μ is decreasing.

The function $\varphi(t,x) = \frac{\mu(t)x}{\sqrt{1+x^2}}$ has property $(\varphi 2)$ if $\mu\colon J \to \mathbb{R}$ is a.e. positive-valued, whence the comparison and uniqueness results of Lemma 3.2.1 and Theorem 3.2.1 hold when $(\varphi 2)$ is replaced by the hypothesis: $\mu(t) > 0$ for a.e. $t \in J$.

In the special case when the IVP (3.2.11) can be rewritten as

$$\begin{cases} \frac{d}{dt}\varphi(u(t)) = g(t,u(t)) + f(t,u(t), \frac{d}{dt}\varphi(u(t)) - g(t,u(t))), \\ u(t_0) = x_0, \end{cases}$$

$$(3.2.18)$$

the hypotheses $(\varphi 2)$ and $(\varphi 4)$ hold if φ is strictly increasing. Thus we obtain the following result.

Proposition 3.2.5. *The comparison and uniqueness results of Theorems 3.2.1 and 3.2.2 hold for the IVP (3.2.18) if (φ2) and (φ4) are replaced by the assumption that $\varphi\colon \mathbb{R} \to \mathbb{R}$ is strictly increasing.*

The results of Lemma 2.1.2 and Lemma 2.1.3 ensure that the results of Proposition 3.2.5 and Theorem 3.2.3 hold for the IVP

$$\begin{cases} \frac{u'(t)}{q(u(t))} = g(t, u(t)) + f(t, u(t), \frac{u'(t)}{q(u(t))} - g(t, u(t))) & \text{a.e. in } J, \\ u(t_0) = x_0, \end{cases} \quad (3.2.19)$$

if (φ0) is replaced by the following assumption.

(q0) $q\colon \mathbb{R} \to (0, \infty)$ and $\frac{1}{q}$ belong to $L^\infty_{loc}(\mathbb{R})$, and $\int_0^{\pm\infty} \frac{dz}{q(z)} = \pm\infty$.

Example 3.2.2. The points

$$c(n_0, \ldots, n_m) = 1 - 2^{-m-1} - \sum_{k=0}^{m} 2^{-k-m-2} \prod_{j=0}^{k} 2^{-n_j} - 2^{-2m-2} \prod_{j=0}^{m} 2^{-n_j},$$

where $m, n_0, \ldots, n_m \in \mathbb{N}$, form a well-ordered set C of rational numbers with $\min C = 0$ and $\sup C = 1$. Define

$$h(z) = \frac{c(n_0, \ldots, n_m + 1 - z)}{c(n_0, \ldots, n_m + 1) - c(n_0, \ldots, n_m)}, \quad \text{where}$$

$$c(n_0, \ldots, n_m) \le z < c(n_0, \ldots, n_m + 1), \quad m, n_0, \ldots, n_m \in \mathbb{N},$$

and

$$g(t, x) = h(t + x - [t + x]), \quad t \in J = [0, 1], \ x \in \mathbb{R}.$$

It is easy to see that g is bounded and has properties (g0) and (g4). It then follows from Theorem 2.1.4 and Proposition 3.2.3 that the IVP

$$\frac{d}{dt}(|u(t)|^{c-2}u(t)) = h(t + u(t) - [t + u(t)]) \text{ a.e. in } J, \quad u(t_0) = x_0, \quad (3.2.20)$$

has for all choices of $c > 1$ and $x_0 \in \mathbb{R}$ a unique solution.

Consider finally the following initial value problem

$$F(t, u(t), \frac{d}{dt}\varphi(t, u(t)) - g(t, u(t))) = 0, \text{ a.e. in } J, \quad u(t_0) = x_0, \quad (3.2.21)$$

where $\varphi\colon J \times \mathbb{R} \to \mathbb{R}$, $g\colon J \times \mathbb{R} \to \mathbb{R}$ and $F\colon J \times \mathbb{R} \times \mathbb{R} \to \mathbb{R}$. A function $u \in Y$ is called a *lower solution* of the IVP (3.2.21) if

$$F(t, u(t), \frac{d}{dt}\varphi(t, u(t)) - g(t, u(t))) \le 0 \text{ for a.e. } t \in J, \text{ and } u(t_0) \le x_0,$$

and an *upper solution* if the reversed inequalities are satisfied. If equalities hold, we say that u is a *solution* of (3.2.21).

As a consequence of Theorems 3.2.1 and 3.2.2 and Proposition 3.2.1 we obtain the following result.

Proposition 3.2.6. *Let* $\varphi\colon J \times \mathbb{R} \to \mathbb{R}$, $g\colon J \times \mathbb{R} \to \mathbb{R}$, $F\colon J \times \mathbb{R} \times \mathbb{R} \to \mathbb{R}$ *and* $\mu\colon J \times \mathbb{R} \times \mathbb{R} \to (0, \infty)$ *be given, and let* $f\colon J \times \mathbb{R} \times \mathbb{R} \to \mathbb{R}$ *be defined by*

$$f(t, x, y) = y - \mu(t, x, y)F(t, x, y), \quad t \in J, \ x \in \mathbb{R}, \ y \in \mathbb{R}. \qquad (3.2.22)$$

Assume that φ, f, *and* g *have properties* $(\varphi 2)$ *and* $(fg2)$, *or* $(\varphi 3)$ *and* $(fg3)$, *or* $(\varphi 4)$ *and* $(fg4)$, *and let* $x_0 \in \mathbb{R}$ *be given. If* $v \in Y$ *is a lower solution of* (3.2.21) *and* $w \in Y$ *is an upper solution of* (3.2.21), *then* $v(t) \leq w(t)$ *on* J. *In particular,* (3.2.21) *can have at most one solution.*

Proof. It is easy to see that the IVPs (3.2.21) and (3.2.11) with f defined by (3.2.22) have the same lower solutions, solutions and upper solutions, respectively, whence the assertions follow from Theorems 3.2.1 and 3.2.2 and Proposition 3.2.1. $\square$

3.3 Well-posedness results

In this section we derive well-posedness results for explicit and implicit initial value problems.

3.3.1 Hypotheses and main results. Consider the following initial value problem

$$\begin{cases} \frac{d}{dt}\varphi(u(t)) = g(t, u(t)) + f(t, u(t), \frac{d}{dt}\varphi(u(t)) - g(t, u(t))) \\ \text{a.e. in } J = [t_0, t_1], \quad u(t_0) = C(u(t_0), x_0). \end{cases} \qquad (3.3.1)$$

First we shall prove that (3.3.1) has at most one solution if the functions $\varphi\colon \mathbb{R} \to \mathbb{R}$, $g\colon J \times \mathbb{R}$, $f\colon J \times \mathbb{R} \times \mathbb{R} \to \mathbb{R}$ and $C\colon \mathbb{R} \times \mathbb{R} \to \mathbb{R}$ have the following properties.

(φa) φ is an increasing homeomorphism, and φ^{-1} is Lipschitz-continuous.
(fga) For a.e. $t \in J$ and for all $x, y, u, v \in \mathbb{R}$,

$$|f(t, x, u) - f(t, y, v)| \leq \bar{p}(t)\phi(|x - y|) + \lambda(t)\,|u - v| \qquad (3.3.2)$$

and

$$|g(t, x) - g(t, y)| \leq p(t)\phi(|x - y|), \qquad (3.3.3)$$

where p, $\bar{p}$, $\frac{\bar{p}}{1-\lambda} \in L^1_+(J)$, the function $\phi\colon \mathbb{R}_+ \to \mathbb{R}_+$ is increasing and $\int_{0+}^1 \frac{dx}{\phi(x)} = \infty$.

(C) $|C(x, x_0) - C(y, y_0)| \leq c|x - y| + k|x_0 - y_0|$ for all $x, y, x_0, y_0 \in \mathbb{R}$, where $c \in [0, 1)$ and $k \geq 0$.

Moreover, by making the further hypotheses:

(gb) $g(\cdot, x)$ is measurable for all $x \in \mathbb{R}$ and $g(\cdot, 0) \in L^1(J)$,

(fb) $f(\cdot, x, y)$ is measurable for all $x, y \in \mathbb{R}$, $f(\cdot, 0, 0) \in L^1(J)$ and $f(t, \cdot, y)$ is increasing for a.e. $t \in J$ and for all $y \in \mathbb{R}$,

we prove by the above results and an existence result derived in section 2.4 that (3.3.1) has for each $x_0 \in \mathbb{R}$ a unique solution $u = u(\cdot, x_0)$ which is increasing with respect to f and C, and depends continuously on x_0. As a special case we demonstrate how to apply the results derived for (3.3.1) to the IVP

$$\begin{cases} \frac{u'(t)}{q(u(t))} = g(t, u(t) + f(t, u(t), \frac{u'(t)}{q(u(t))} - g(t, u(t)))) \text{ a.e. in } J, \\ u(t_0) = C(u(t_0), x_0), \end{cases}$$

where $q : \mathbb{R} \to (0, \infty)$ has the property

(qa) q is measurable and essentially bounded, and $\frac{1}{q}$ is locally essentially bounded.

Finally we shall prove a result concerning continuity of the solution of the IVP

$$\frac{d}{dt} \varphi(u(t)) = g(t, u(t)) + h(t) \quad \text{a.e. in} \quad J, \quad u(t_0) = x_0$$

with respect to x_0 and h in the case when g is a Carathéodory function.

3.3.2 A uniqueness result. Our main uniqueness result is based on the following Lemma.

Lemma 3.3.1. *Let the hypotheses (φa) and (fga) hold, and let $K > 0$ be a Lipschitz-constant of φ^{-1}. If $u, v \in Y$ satisfy the differential equation of the IVP (3.3.1), then the function*

$$x(t) = K |\varphi(u(t)) - \varphi(v(t))|, \quad t \in J, \qquad (3.3.4)$$

satisfies the differential inequality

$$x'(t) \leq \hat{p}(t) \phi(x(t)) \qquad a.e. \ in \ J, \qquad (3.3.5)$$

where

$$\hat{p}(t) = K \left(\frac{\overline{p}(t)}{1 - \lambda(t)} + p(t) \right), \quad t \in J.$$

Proof. For each $u \in Y = \{u \in C(J) \mid \varphi \circ u \in AC(J)\}$ we denote

$$Au(t) = (\varphi \circ u)'(t) - g(t, u(t)), \quad t \in J. \tag{3.3.6}$$

If u and v are solutions of (3.3.1), we get by (3.3.2) and (3.3.6)

$$|Au(t) - Av(t)| = |f(t, u(t), Au(t)) - f(t, v(t), Av(t))|$$
$$\leq \overline{p}(t)\phi(|u(t) - v(t)|) + \lambda(t)|Au(t) - Av(t)|,$$

or equivalently,

$$|Au(t) - Av(t)| \leq \frac{\overline{p}(t)}{1 - \lambda(t)}\phi(|u(t) - v(t)|) \tag{3.3.7}$$

for a.e. $t \in J$. In view of the definition of A we have

$$(\varphi \circ u)'(t) - (\varphi \circ v)'(t) = g(t, u(t)) - g(t, v(t)) + Au(t) - Av(t) \tag{3.3.8}$$

for a.e. $t \in J$. Hence, it follows from (3.3.3) (3.3.4), (3.3.7), and (3.3.8) that for a.e. $t \in J$,

$$x'(t) \leq K|(\varphi \circ u)'(t) - (\varphi \circ v)'(t)|$$
$$\leq K|g(t, u(t)) - g(t, v(t))| + K|Au(t) - Av(t)|$$
$$\leq K\left(p(t) + \frac{\overline{p}(t)}{1 - \lambda(t)}\right)\phi(|u(t) - v(t)|) = \hat{p}(t)\phi(|u(t) - v(t)|).$$

Since ϕ is increasing and since K is a Lipschitz-constant of φ^{-1}, we have

$$\phi(|u(t) - v(t)|) \leq \phi(K|\varphi(u(t)) - \varphi(v(t))|) = \phi(x(t)), \quad t \in J.$$

The above inequalities imply that (3.3.5) holds.   □

Lemma 3.3.1 implies the following uniqueness result.

Proposition 3.3.1. *Assume that the hypotheses (φa), (fga), and (C) are valid. Then the IVP (3.3.1) can have only one solution.*

Proof. Let $x_0 \in \mathbb{R}$ be given, and let u, $v \in Y$ be solutions of the IVP (3.3.1). Applying the hypothesis (C) we get

$$|u(t_0) - v(t_0)| = |C(u(t_0), x_0) - C(v(t_0), x_0))| \leq c|u(t_0) - v(t_0)|.$$

Since $c \in [0, 1)$, this implies that $u(t_0) - v(t_0) = 0$. Because φ is a homeomorphism, then $x(t_0) = K|\varphi(u(t_0)) - \varphi(v(t_0))| = 0$. In view of this result and (3.3.4) we see that x, defined by (3.3.4), has properties

$$x'(t) \leq \hat{p}(t)\phi(x(t)) \text{ a.e. in } J, \quad x(t_0) = 0. \tag{3.3.9}$$

Thus the hypotheses given for ϕ imply by Lemma B.6.1 that $x(t) \equiv 0$. Since φ is strictly increasing and $K > 0$, it follows from (3.3.4) that $u = v$.   □

3.3.3 A well-posedness result. As a consequence of Proposition 2.4.2 we have the following result.

Lemma 3.3.2. *Assume that* $\varphi \colon \mathbb{R} \to \mathbb{R}$, $g \colon J \times \mathbb{R} \to \mathbb{R}$, $f \colon J \times \mathbb{R} \times \mathbb{R} \to \mathbb{R}$ *and* $B \colon \mathbb{R} \to \mathbb{R}$ *satisfy the following hypotheses:*

- $(\varphi 0)$ φ *is an increasing homeomorphism.*
- (fgb) *f and g are Carathéodory functions, and there is a function $\alpha \in L^1_+(J)$ such that $f(t,x,y) + \alpha(t)y$ is increasing in x and y for a.e. $t \in J$.*
- (fgc) *for a.e. $t \in J$ and all x, $y \in \mathbb{R}$, $|g(t,x)| \le p_1(t)\psi(|\varphi(x)|)$ and $|f(t,x,y)| \le p_2(t)\psi(|\varphi(x)|) + \lambda(t)|y|$, where p_1, p_2, $\frac{p_2}{1-\lambda} \in L^1_+(J)$, $\psi \colon \mathbb{R}_+ \to (0,\infty)$ is increasing and $\int_0^\infty \frac{dx}{\psi(x)} = \infty$.*
- (Ba) *$x \mapsto B(x) + \beta x$ is increasing for some $\beta \ge 0$.*
- (Bb) *$|B(x)| \le c|x| + d$ for all $x \in \mathbb{R}$, where $c \in [0,1)$ and $d \in \mathbb{R}_+$.*

Then the implicit boundary value problem

$$\begin{cases} \frac{d}{dt}\varphi(u(t)) = g(t,u(t)) + f(t,u(t), \frac{d}{dt}\varphi(u(t)) - g(t,u(t))) \\ a.e. \ in \ J, \quad u(t_0) = B(u(t_0)), \end{cases} \tag{3.3.10}$$

has extremal solutions, and they are increasing with respect to f and B.

Now we are ready to prove our main result of this section.

Theorem 3.3.1. *Assume that the hypotheses (φa), (fga), (fb), (gb), and (C) are valid. Then the IVP (3.3.1) has a unique solution, it is increasing with respect to f and C, and it depends continuously on x_0.*

Proof. It follows from (fga) that the functions $g(t, \cdot)$ and $f(t, \cdot, \cdot)$ are continuous for a.e. $t \in J$. These properties and the hypotheses (gb) and (fb) imply that g and f are Carathéodory functions. The hypotheses (fga) and (fb) imply also that $f(t,x,y) + \lambda(t)y$ is increasing in x and in y for a.e. $t \in J$. Thus the hypothesis (fgb) of Lemma 3.3.2 holds when $\alpha = \lambda$.

To prove that (fgc) holds, let x, $y \in \mathbb{R}$ be given. In view of (fga) we have

$$|f(t,x,y) - f(t,0,0)| \le |f(t,x,y) - f(t,0,y)| + \lambda(t)\,|y| \tag{3.3.11}$$

for a.e. $t \in J$. Choose next $n \in \mathbb{N}$ such that $|x| \le n < |x| + 1$. Denoting $x_i = \frac{i}{n}x$, $i = 0, \ldots, n$, we have $|x_i - x_{i-1}| \le 1$ for each $i = 1, \ldots, n$, whence (fga) implies that

$$\begin{aligned} |f(t,x,y) - f(t,0,y)| &\le \sum_{i=1}^{n} |f(t,x_i,y) - f(t,x_{i-1},y)| \\ &\le \sum_{i=1}^{n} \overline{p}(t)\phi(1) \le \overline{p}(t)\phi(1)(|x|+1) \end{aligned} \tag{3.3.12}$$

for a.e. $t \in J$ and for all x, $u \in \mathbb{R}$, and

$$
\begin{aligned}
|g(t, x) - g(t, 0)| &\leq \sum_{i=1}^{n} |g(t, x_i) - g(t, x_{i-1})| \\
&\leq \sum_{i=1}^{n} p(t)\phi(1) \leq p(t)\phi(1)(|x| + 1)
\end{aligned} \tag{3.3.13}
$$

for a.e. $t \in J$ and for all $x \in \mathbb{R}$. Noticing that $|x| \leq K|\varphi(x) - \varphi(0)|$ for all $x \in \mathbb{R}$, it follows from (3.3.11), (3.3.12), and (3.3.13) that (fgc) is valid when

$$
\begin{cases}
p_1(t) = |g(t, 0)| + p(t)\phi(1)(|\varphi(0)| + 1), \ t \in J, \\
p_2(t) = |f(t, 0, 0)| + \overline{p}(t)\phi(1)(|\varphi(0)| + 1), \ t \in J, \\
\psi(z) = z + 1, \quad z \geq 0.
\end{cases}
$$

Let $x_0 \in \mathbb{R}$ be given. Denoting

$$
B(x) = C(x, x_0), \quad x \in \mathbb{R},
$$

it follows from (C) that the hypotheses (Ba) and (Bb) are valid when

$$
\beta = c \quad \text{and} \quad d = |C(0, x_0)|.
$$

These properties imply by Lemma 3.3.2 that the IVP (3.3.1) has a solution $u = u(\cdot, x_0)$. The uniqueness follows from Proposition 3.3.1, and monotonicity assertions from Lemma 3.3.2.

To prove continuous dependence of $u(\cdot, x_0)$ on x_0, denote $u = u(\cdot, x_0)$ and $v = u(\cdot, y_0)$. Since $u(t_0) = C(u(t_0), x_0)$ and $v(t_0) = C(v(t_0), y_0)$, we get by (C),

$$
|u(t_0) - v(t_0)| \leq c|u(t_0) - v(t_0)| + k|x_0 - y_0|.
$$

Thus

$$
|u(t_0) - v(t_0)| \leq \frac{k}{1 - c}|x_0 - y_0|. \tag{3.3.14}
$$

On the other hand, it follows from (φa) and (3.3.4) that

$$
|u(t) - v(t)| \leq K|\varphi(u(t)) - \varphi(v(t))| = x(t), \quad t \in J. \tag{3.3.15}
$$

According to Lemma 3.3.2 we have

$$
x'(t) \leq \hat{p}(t)\phi(x(t)) \quad \text{a.e. in } J. \tag{3.3.5}
$$

If $x(t_0) > 0$ we then have

$$\int_{x(t_0)}^{x(t)} \frac{dx}{\phi(x)} \le \int_{t_0}^t \hat{p}(s)ds \le \int_{t_0}^{t_1} \hat{p}(s)ds, \quad t \in J, \qquad (3.3.16)$$

Since $\int_{0+}^1 \frac{dx}{\phi(x)} = \infty$, it follows from (3.3.16) that $x(t) \to 0$ uniformly on J if $x(t_0) \to 0+$. This result and (3.3.4) imply, because φ^{-1} is continuous, that $v(t) \to u(t)$ uniformly on J if $v(t_0) \to u(t_0)$, or by (3.3.14), if $y_0 \to x_0$. This proves the continuous dependence of $u(\cdot, x_0)$ on x_0. □

3.3.4 Special cases. Consider first the IVP

$$\begin{cases} \frac{u'(t)}{q(u(t))} = g(t, u(t) + f(t, u(t), \frac{u'(t)}{q(u(t))} - g(t, u(t)))) \quad \text{a.e. in } J, \\ u(t_0) = C(u(t_0), x_0), \end{cases} \qquad (3.3.17)$$

where the function $q \colon \mathbb{R} \to (0, \infty)$ has the following property.

(qa) q is measurable and essentially bounded and $\frac{1}{q}$ is locally essentially bounded.

The equation

$$\varphi(x) := \int_0^x \frac{ds}{q(s)}, \qquad t \in \mathbb{R}, \qquad (3.3.18)$$

defines an increasing homeomorphism from $\mathbb{R}$ onto $\mathbb{R}$, and φ^{-1} is Lipschitz-continuous with $\|q\|_\infty$ as a Lipschitz-constant. Thus the function φ, defined by (3.3.18), satisfies (φa). As in the proof of Lemma 2.1.3 it can be shown that u is a solution of (3.3.17) if and only if u is a solution of (3.3.1) in Y with φ defined by (3.3.18). Thus we obtain the following result.

Proposition 3.3.2. *The results of Theorem 3.3.1 are valid for the IVP (3.3.17) when (φa) is replaced by (qa).*

When the last variable of f is deleted we get results for the explicit IVP

$$u'(t) = q(u(t))(g(t, u(t)) + f(t, u(t))) \quad \text{a.e. in } J, \quad u(t_0) = C(u(t_0), x_0).$$

Next we shall prove a result concerning continuity of the solution of the IVP

$$\frac{d}{dt}\varphi(u(t)) = g(t, u(t)) + h(t) \quad \text{a.e. in } J, \quad u(t_0) = x_0, \qquad (3.3.19)$$

with respect to x_0 and h.

Proposition 3.3.3. *Let $\varphi\colon \mathbb{R} \to \mathbb{R}$ be an increasing homeomorphism, and let $g\colon J \times \mathbb{R} \to \mathbb{R}$ be a Carathéodory function which satisfies the following hypotheses.*

(g1) *for a.e. $t \in J$ and all $x,\, y \in \mathbb{R}$, $|g(t,x)| \le p_1(t)\psi(|\varphi(x)|)$ and where $p_1 \in L^1_+(J)$, $\psi\colon \mathbb{R}_+ \to (0,\infty)$ is increasing and $\int_0^\infty \frac{dx}{\psi(x)} = \infty$.*

(g3) *$g(t,y) - g(t,x) \le l(t,\varphi(t,y) - \varphi(t,x))$ for a.e. $t \in J$ and for all $x,\, y \in \mathbb{R}$, $x < y$, where $l\colon J \times \mathbb{R}_+ \to \mathbb{R}_+$, and zero-function is the only lower solution of the IVP*

$$x'(t) = l(t,x(t)) \quad a.e. \ in \ \ J, \quad x(t_0) = 0.$$

If $(x_n)_{n=1}^\infty$ is a sequence of real numbers which converges to x_0, and (h_n) is a sequence which converges to h in $L^1(J)$, then the IVP

$$\frac{d}{dt}\varphi(u(t)) = g(t,u(t)) + h_n(t) \quad a.e. \ in \ \ J, \quad u(t_0) = x_n, \qquad (3.3.20)$$

has for each $n = 1,2,\dots$ a unique solution u_n, and the sequence $(u_n)_{n=1}^\infty$ converges uniformly on J to the solution of the IVP (3.3.19).

Proof. It follows from Propositions 2.1.1 and 3.2.2 that the IVPs (3.3.20) have unique solutions. Thus the IVP

$$v'(t) = g(t,\varphi^{-1}(v(t))) + h_n(t) \quad \text{a.e. in} \ \ J, \quad v(t_0) = \varphi(x_n), \qquad (3.3.21)$$

has for each $n = 1,2,\dots$ a unique solution v_n by Lemma 2.1.1. Denote $w_0 = \sup\{|\varphi(x_n)| \mid n = 1,2,\dots\}$, by h_0 an L^1-bound of the functions h_n, $n = 1,2,\dots$, and by w the solution of the IVP

$$w'(t) = (p_1(t) + h_0(t))(\psi(w(t)) + 1) \quad \text{a.e. in} \ \ J, \qquad w(t_0) = w_0.$$

As in the proof of Proposition 2.1.2 it can be shown that

$$v_n(t) \in [-w(t_1), w(t_1)] \ \text{ for all } n = 1,2,\dots \text{ and } t \in J, \qquad (3.3.22)$$

and that for each $n = 1,2,\dots$,

$$|v_n(b) - v_n(a)| \le w(b) - w(a) \ \text{ if } t_0 \le a \le b \le t_1.$$

Thus the sequence $(v_n)_{n=1}^\infty$ and each its subsequence is uniformly bounded and equicontinuous. It then follows from Arzela-Ascoli Theorem that if

$(v_{n_k})_{k=1}^{\infty}$ is a subsequence of $(v_n)_{n=1}^{\infty}$, it has a subsequence $(v_{n_{k_j}})_{j=1}^{\infty}$ which converges uniformly on J to a function v which has property

$$|v(b) - v(a)| \leq |w(b) - w(a)|, \quad a, b \in J.$$

In particular, $v \in AC(J)$. It follows from (3.3.22) that

$$v(t) \in [-w(t_1), w(t_1)] \quad \text{for all} \ \ t \in J.$$

Each $v_{n_{k_j}}$ satisfies the integral equation

$$v_{n_{k_j}}(t) = \varphi(x_{n_{k_j}}) + \int_{t_0}^{t} (g(s, \varphi^{-1}(v_{n_{k_j}}(s))) + h_{n_{k_j}}(s)) \, ds, \quad t \in J. \quad (3.3.23)$$

Noticing that the function $g(t, \cdot)$ is continuous for a.e. $t \in J$, that the sequence $(\varphi^{-1} \circ v_{n_{k_j}})$ converges uniformly $\varphi^{-1} \circ v$, that $x_{n_{k_j}} \to x_0$ and $v_{n_{k_j}}(t) \to v(t)$, and that both φ and φ^{-1} are continuous, it follows from (3.3.23) when $j \to \infty$, applying also the dominated convergence theorem, that

$$v(t) = \varphi(x_0) + \int_{t_0}^{t} (g(s, \varphi^{-1}(v(s))) + h(s)) \, ds, \quad t \in J.$$

This implies that v is a solution of the IVP (3.3.21).

The above proof shows that each subsequence of $(v_n)_{n=1}^{\infty}$ has a subsequence which converges uniformly to a solution v of (3.3.21). Because v is uniquely determined, then the whole sequence $(v_n)_{n=1}^{\infty}$ converges uniformly to v. Thus the results of Lemma 2.1.1 and the hypotheses given for φ imply that the function $u_n = \varphi^{-1} \circ v_n$ is for each $n = 1, 2, \ldots$, a unique solution of the IVP (3.3.20), and that the sequence $(u_n)_{n=1}^{\infty}$ converges uniformly to the solution of the IVP (3.3.19). This concludes the proof.   □

Consider finally the initial value problem

$$\begin{cases} F(t, u(t), \frac{d}{dt}\varphi(u(t)) - g(t, u(t))) = 0 \ \ \text{a.e. in} \ \ J, \\ B(u(t_0), x_0) = 0, \end{cases} \quad (3.3.24)$$

where $\varphi \colon \mathbb{R} \to \mathbb{R}$, $g \colon J \times \mathbb{R} \to \mathbb{R}$, $F \colon J \times \mathbb{R} \times \mathbb{R} \to \mathbb{R}$ and $B \colon \mathbb{R} \times \mathbb{R} \to \mathbb{R}$.

As a consequence of Theorems 3.3.1 and Proposition 3.3.1 we obtain the following result.

Proposition 3.3.4. *Given functions* $g\colon J \times \mathbb{R} \to \mathbb{R}$, $F\colon J \times \mathbb{R} \times \mathbb{R} \to \mathbb{R}$, $\varphi\colon J \times \mathbb{R} \to \mathbb{R}$ *and* $B\colon \mathbb{R} \times \mathbb{R} \to \mathbb{R}$, *assume there exist* $\eta\colon J \times \mathbb{R} \times \mathbb{R} \to (0, \infty)$ *and* $\gamma\colon \mathbb{R} \times \mathbb{R} \to (0, \infty)$ *such that the functions* φ *and* g, *and the functions* $f\colon J \times \mathbb{R} \times \mathbb{R} \to \mathbb{R}$ *and* $C\colon \mathbb{R} \times \mathbb{R} \to \mathbb{R}$, *defined by*

$$\begin{cases} f(t,x,y) = y - \eta(t,x,y)F(t,x,y), & t \in J,\ x,\ y \in \mathbb{R}, \\ C(x,y) = y - \gamma(x,y)B(x,y), & x,\ y \in \mathbb{R}, \end{cases} \tag{3.3.25}$$

have properties (φa), (fga), *and* C. *Then the IVP (3.3.24) can have only one solution. The solution exists and depends continuously on* x_0 *if also the hypotheses* (fb) *and* (gb) *hold.*

Proof. The IVPs (3.3.1) and (3.3.24) with f and C defined by (3.3.25) have same solutions. Thus the assertions follow from Theorem 3.3.1 and Proposition 3.3.1. □

3.4 Second order problems

In this section we first derive maximum principles for a generalized phi-Laplacian operator combined with a part containing lower order terms. No continuity or measurability hypotheses are required. These maximum principles are then applied to prove uniqueness and comparison results for second order quasilinear differential equations with separated, periodic, Neumann, or Dirichlet boundary conditions. Special cases including phi-Laplacian, p-Laplacian, and Sturm-Liouville problems are also considered. Examples and counterexamples are given to illustrate the obtained results.

3.4.1 Maximum principles. Given intervals I_0 and $J = [t_0, t_1]$, and functions $\varphi\colon J \times I_0 \to \mathbb{R}$ and $g\colon J \times \mathbb{R} \times \mathbb{R} \to \mathbb{R}$, we derive maximum principles for the differential operator L_1, defined by

$$\begin{cases} L_1 u(t) := -\frac{d}{dt}\varphi(t, u'(t)) - g(t, u(t), u'(t)), & t \in J, \\ u \in Y := \{u \in C^1(J) \mid u'[J] \subseteq I_0 \text{ and } \varphi(\cdot, u'(\cdot)) \in AC(J)\}, \end{cases} \tag{3.4.1}$$

where $C^1(J)$ is the space of continuously differentiable functions $u\colon J \to \mathbb{R}$. In the proof of our main maximum principle we assume that the functions φ and g satisfy the following hypotheses.

- (φa) To each choice of s_1, $s_2 \in I_0$, $s_0 < s_1$, there corresponds such an $M > 0$ that $\varphi(t,y) - \varphi(t,z) \geq M(y - z)$ whenever $t \in J$ and $s_0 \leq z < y \leq s_1$.
- (ga) $g(t,x,z) \leq g(t,y,z)$ for a.e. $t \in J$ and for all $x,\ y,\ z \in \mathbb{R}$, $x \geq y$.
- (gb) $|g(t,x,y) - g(t,x,z)| \leq p(t)\phi(|y - z|)$ for a.e. $t \in J$ and for all $x, y,\ z \in \mathbb{R}$, where $p \in L^1_+(J)$, $\phi\colon \mathbb{R}_+ \to \mathbb{R}_+$ is increasing and $\int_{0+}^{1} \frac{dz}{\phi(z)} = \infty$.

Lemma 3.4.1. *Assume that* $g: J \times \mathbb{R} \times \mathbb{R} \to \mathbb{R}$ *and* $\varphi: J \times I_0 \to \mathbb{R}$ *satisfy the hypotheses* (φa), (ga), *and* (gb). *Let* u, $w \in Y$ *satisfy* $L_1 u(t) \leq L_1 w(t)$ *for a.e.* $t \in J$, *where the operator* $L_1: Y \to \mathbb{R}$ *is defined by* (3.4.1). *If* $u - w$ *attains a positive maximum* c *in* (t_0, t_1), *then* $u(t) - w(t) \equiv c$.

Proof. Assume that $u - w$ attains a positive maximum c at $t_2 \in (t_0, t_1)$. Let $p \in L^1_+(J)$ and $\phi: \mathbb{R}_+ \to \mathbb{R}_+$ be as in (gb). The proof is divided into two steps.

(i) Let t_3 be the greatest number on $(t_2, t_1]$ such that $u(t) \geq w(t)$ for each $t \in [t_2, t_3]$. To prove that $w'(t) \leq u'(t)$ for each $t \in [t_2, t_3]$, assume on the contrary: there is a subinterval $[a, b]$ of $[t_2, t_3]$ such that

$$0 < w'(t) - u'(t), \ t \in (a, b], \quad w'(a) - u'(a) = 0.$$

By (φa) there exists a $K > 0$ such that

$$w'(t) - u'(t) \leq K(\varphi(t, w'(t)) - \varphi(t, u'(t))) \tag{3.4.2}$$

for all $t \in [a, b]$. Denote $x(t) = K(\varphi(t, w'(t)) - \varphi(t, u'(t)))$, $t \in J$. Since $u, w \in Y$, then $x \in AC(J)$. Using the assumption $L_1 u(t) \leq L_1 w(t)$ a.e. in J, we then have by (3.4.1), (3.4.2), (ga), and (gb),

$$
\begin{aligned}
x'(t) &= K \frac{d}{dt} \varphi(t, w'(t)) - K \frac{d}{dt} \varphi(t, u'(t)) \\
&\leq K(g(t, u(t), u'(t)) - g(t, w(t), w'(t))) \\
&\leq K|g(t, w(t), u'(t)) - g(t, w(t), w'(t))| \leq K p(t) \phi(|u'(t) - w'(t)|) \\
&= K p(t) \phi(w'(t) - u'(t)) \leq K p(t) \phi(K(\varphi(t, w'(t)) - \varphi(t, u'(t)))) \\
&= K p(t) \phi(x(t))
\end{aligned}
$$

for a.e. $t \in (a, b]$. Thus we have

$$x'(t) \leq K p(t) \phi(x(t)) \text{ a.e. in } (a, b], \quad x(a) = 0.$$

Applying this inequality and Lemma C.3.2 we obtain

$$\int_{0+}^{x(b)} \frac{dx}{\phi(x)} = \lim_{s \to a+} \int_s^b \frac{x'(t) dt}{\phi(x(t))} \leq \int_a^b K p(t) dt < \infty.$$

This contradicts the hypotheses given for ϕ in (gb). Consequently, $w'(t) \leq u'(t)$ on $[t_2, t_3]$, whence

$$u(t) - w(t) = u(t_2) - w(t_2) + \int_{t_2}^t (u'(s) - w'(s)) ds \geq u(t_2) - w(t_2), \ t \in [t_2, t_3].$$

Because t_2 was the maximum point of $u(t) - w(t)$, then $u(t) - w(t) \equiv c$ on $[t_2, t_3]$. This and the choice of t_3 imply that $t_3 = t_1$. Thus $u(t) - w(t) \equiv c$ on $[t_2, t_1]$.

(ii) Choose next t_4 to be the least number on $[t_0, t_2)$ such that $u(t) \geq w(t)$ for each $t \in [t_4, t_2]$. To prove that $u'(t) \leq w'(t)$ for each $t \in [t_4, t_2]$, assume on the contrary: there is a subinterval $[a, b]$ of $[t_4, t_2]$ such that

$$0 < u'(t) - w'(t), \ t \in [a, b), \quad u'(b) = w'(b).$$

Choose $K > 0$ so that

$$u'(t) - w'(t) \leq K(\varphi(t, u'(t)) - \varphi(t, w'(t))) \qquad (3.4.3)$$

holds for all $t \in [a, b]$. Denoting $x(t) = K(\varphi(u'(t)) - \varphi(w'(t)))$, $t \in J$, and noticing that $L_1 u(t) \leq L_1 w(t)$ a.e. in J we obtain, by applying (3.4.1), (3.4.3), (ga), and (gb),

$$
\begin{aligned}
-x'(t) &= K\frac{d}{dt}\varphi(t, w'(t)) - K\frac{d}{dt}\varphi(t, u'(t)) \\
&\leq K(g(t, u(t), u'(t)) - g(t, w(t), w'(t))) \\
&\leq K|g(t, w(t), u'(t)) - g(t, w(t), w'(t))| \\
&\leq Kp(t)\phi(|u'(t) - w'(t)|) = Kp(t)\phi(u'(t) - w'(t)) \\
&\leq Kp(t)\phi(K(\varphi(t, u'(t)) - \varphi(t, w'(t)))) = Kp(t)\phi(x(t))
\end{aligned}
$$

for a.e. $t \in [a, b)$. Because $x(b) = 0$, we obtain

$$-x'(t) \leq Kp(t)\phi(x(t)) \text{ a.e. in } [a, b), \quad x(b) = 0,$$

which implies a contradiction

$$\infty = \int_{0+}^{x(a)} \frac{dx}{\phi(x)} = \int_{b-}^{a} \frac{x'(t)dt}{\phi(x(t))} = \int_{a}^{b-} \frac{-x'(t)dt}{\phi(x(t))} \leq \int_{a}^{b} Kp(t)dt < \infty.$$

Thus $u'(t) \leq w'(t)$ on $[t_4, t_2]$, whence

$$u(t_2) - w(t_2) = u(t) - w(t) + \int_{t}^{t_2} (u'(s) - w'(s))ds \leq u(t) - w(t), \ t \in [t_4, t_2].$$

Because t_2 was the maximum point of $u(t) - w(t)$, then $u(t) - w(t) \equiv c$ in $[t_4, t_2]$. This and the choice of t_4 imply that $t_4 = t_0$. Thus $u(t) - w(t) \equiv c$ on $[t_0, t_2]$.

The results of (i) and (ii) imply that $u(t) - w(t) \equiv c$. $\square$

Given a_j, $b_j \in \mathbb{R}_+$, $j = 0, 1$ and $u \in C^1(J)$, denote

$$L_2 u = a_0 u(t_0) - b_0 u'(t_0), \quad L_3 u = a_1 u(t_1) + b_1 u'(t_1). \qquad (3.4.4)$$

As an application of Lemma 3.4.1 we obtain the following result.

Lemma 3.4.2. *Let the hypotheses of Lemma 3.4.1 hold, and assume that functions $u, w \in Y$ satisfy inequalities*

$$L_1 u(t) \leq L_1 w(t) \quad a.e. \ in \ J, \quad L_2 u \leq L_2 w, \quad L_3 u \leq L_3 w, \qquad (3.4.5)$$

where L_1, L_2, and L_3 are defined by (3.4.1) and (3.4.4) with $a_j, b_j \in \mathbb{R}_+$ and $a_j + b_j > 0$, $j = 0, 1$. If $c = \max\{u(t) - w(t) \mid t \in J\}$ is positive, then $u(t) - w(t) \equiv c$ and $a_0 = a_1 = 0$.

Proof. Let $u(t) - w(t)$ attain its positive maximum c at a point $t_2 \in J$. Assume first that $t_0 < t_2 < t_1$. From Lemma 3.4.1 we get $u(t) - w(t) \equiv c$, so that

$$\begin{cases} u(t_0) = w(t_0) + c, & \text{and } u'(t_0) = w'(t_0), \\ u(t_1) = w(t_1) + c, & \text{and } u'(t_1) = w'(t_1). \end{cases}$$

Thus $L_2 u = L_2 w + a_0 c$ and $L_3 u = L_3 w + a_1 c$, which imply by (3.4.5) that $a_j c \leq 0$, $j = 0, 1$, i.e. $a_0 = a_1 = 0$.

Assume next that the positive maximum c of $u(t) - w(t)$ is attained at t_0. Then $u'(t_0) \leq w'(t_0)$, so that

$$L_2 u = a_0 u(t_0) - b_0 u'(t_0) \geq a_0(w(t_0) + c) - b_0 w'(t_0) = L_2 w + a_0 c.$$

In view of this result and (3.4.5) we have $a_0 = 0$ and $b_0 u'(t_0) = b_0 w'(t_0)$. Because $a_0 + b_0 > 0$, then $b_0 \neq 0$, whence $u'(t_0) = w'(t_0)$. Thus we can choose $t_2 = t_0$ in part (i) of the proof of Lemma 3.4.1, which yields $u(t) - w(t) \equiv c$ on J, and hence $L_3 u = L_3 w + a_1 c$. This and (3.4.5) imply that $a_1 = 0$.

In the case when $u(t) - w(t)$ is assumed to obtain its positive maximum at t_1 we have $u'(t_1) \geq w'(t_1)$. This and (3.4.5) imply that $a_1 = 0$ and $w'(t_1) = u'(t_1)$. Thus the choice $t_2 = t_1$ in part (ii) of the proof of Lemma 3.4.1 yields $u(t) - w(t) \equiv c$. Hence $L_2 u = L_2 w + a_0 c$, so that $a_0 = 0$ by (3.4.5). $\qquad\qquad \square$

The proof of Lemma 3.4.2 contains also the proof of the following result.

Lemma 3.4.3. *Let the hypotheses of Lemma 3.4.1 hold, and let $u, w \in Y$ satisfy inequalities*

$$L_1 u(t) \leq L_1 w(t) \quad a.e. \ in \ J, \quad u'(t_0) \geq w'(t_0), \ u'(t_1) \leq w'(t_1), \qquad (3.4.6)$$

where L_1 is defined by (3.4.1). If $c = \max\{u(t) - w(t) \mid t \in J\}$ is positive, then $u(t) - w(t) \equiv c$.

3.4.2 Comparison and uniqueness results. The maximum principles derived in subsection 3.4.1 will now be applied to prove comparison and uniqueness results for second order boundary value problems. The following consequence of Lemma 3.4.2 is used in the proofs.

Proposition 3.4.1. *Assume that $g: J \times \mathbb{R} \times \mathbb{R} \to \mathbb{R}$ and $\varphi: J \times I_0 \to \mathbb{R}$ satisfy the hypotheses (φa), (ga), and (gb), and that functions u, $w \in Y$ satisfy inequalities (3.4.5). Then $u(t) \le w(t)$ for each $t \in J$ in the following cases.*

a) a_j, $b_j \in \mathbb{R}_+$, $j = 0, 1$ and $a_0 a_1 + a_0 b_1 + a_1 b_0 > 0$.

b) a_j, $b_j \in \mathbb{R}_+$, $j = 0, 1$, $b_0 b_1 > 0$, and there is a $\hat{J}$ of J with positive measure such that $g(t, x, z) < g(t, y, z)$ for all $t \in \hat{J}$ and x, y, $z \in \mathbb{R}$, $x > y$.

c) a_j, $b_j \in \mathbb{R}_+$, $j = 0, 1$, $b_0 b_1 > 0$, and there is such a linear functional $Q: C(J) \to \mathbb{R}$ that $Qv > 0$ whenever $v(t) \equiv c > 0$, and that $Qu \le Qw$.

Proof. Assume on the contrary that $c = \max\{u(t) - w(t) \mid t \in J\}$ is positive. In all the cases a)–c) the hypotheses of Lemma 3.4.2 are satisfied, whence $u(t) - w(t) \equiv c > 0$ and $a_0 = a_1 = 0$. But then $a_0 a_1 + a_0 b_1 + a_1 b_0 = 0$ which contradicts the hypotheses of a). Because $u(t) = w(t) + c$, $t \in J$, it follows from (3.4.1) that

$$L_1 u(t) - L_1 w(t) = g(t, w(t), w'(t)) - g(t, w(t) + c, w'(t)) \quad \text{a.e. in } J.$$

Since $c > 0$, then $L_1 u(t) > L_1 w(t)$ a.e. in $\hat{J}$ by the hypotheses given in b), which contradicts (3.4.5). If Q is as in c), then $Q(u - w) = Qc > 0$, contradicting $Qu \le Qw$. Thus the given hypotheses do not allow that $u(t) = w(t) + c$, $c > 0$ for all $t \in J$, which concludes the proof.   □

Remarks 3.4.1. The only continuity assumption for φ and g is imposed on $g(t, x, \cdot)$.

Linear functionals $Q: C(J) \to \mathbb{R}$, defined by

$$Qv = \int_J v(s)\, ds \quad \text{and} \quad Qv = v(\bar{t}) \quad (\bar{t} \in J \text{ is fixed})$$

satisfy $Qv > 0$ if $v(t) \equiv c > 0$, assumed in Proposition 3.4.1.c.

The results of Proposition 3.4.1 will now be applied to the quasilinear differential equation

$$-\frac{d}{dt}\varphi(t, u'(t)) = g(t, u(t), u'(t)) \quad \text{a.e. in } J, \tag{3.4.7}$$

associated with separated boundary conditions

$$a_0 u(t_0) - b_0 u'(t_0) = c_0, \quad a_1 u(t_1) + b_1 u'(t_1) = c_1. \tag{3.4.8}$$

We say that a function $u \in Y$ is a *lower solution* of (3.4.7),(3.4.8) if

$$\begin{cases} -\frac{d}{dt}\varphi(t, u'(t)) \le g(t, u(t), u'(t)) & \text{a.e. in } J, \\ a_0 u(t_0) - b_0 u'(t_0) \le c_0, \quad a_1 u(t_1) + b_1 u'(t_1) \le c_1, \end{cases} \qquad (3.4.9)$$

and an *upper solution* if the reversed inequalities hold. If equalities hold in (3.4.9), we say that u is a *solution* of (3.4.7),(3.4.8). As a consequence of Proposition 3.4.1.a we get the following comparison and uniqueness results.

Theorem 3.4.1. *Let the functions $g: J \times \mathbb{R} \times \mathbb{R} \to \mathbb{R}$ and $\varphi: J \times I_0 \to \mathbb{R}$ have properties (φa), (ga), and (gb), and let the constants $c_j \in \mathbb{R}$, a_j, $b_j \ge 0$, $j = 0, 1$, satisfy $a_0 a_1 + a_0 b_1 + a_1 b_0 > 0$. If $u \in Y$ is a lower solution and $w \in Y$ an upper solution of (3.4.7),(3.4.8), then $u(t) \le w(t)$ on J. In particular, the separated problem (3.4.7),(3.4.8) can have at most one solution.*

Proof. If u and w are lower and upper solutions of (3.4.7),(3.4.8), then the inequalities (3.4.5) hold for the operators L_1, L_2, and L_3, defined by (3.4.1) and (3.4.4). Thus $u(t) \le w(t)$ on J by Proposition 3.4.1.a. The last conclusion follows from the first one. $\qquad \square$

The special case where $a_0 = a_1 = 0$ and $b_0 b_1 > 0$ in (3.4.8) is not covered by Theorem 3.4.1. In this case the boundary conditions (3.4.8) are reduced to the Neumann conditions

$$u'(t_0) = c_0, \quad u'(t_1) = c_1. \qquad (3.4.10)$$

The hypothesis (ga) will now be replaced by the following stronger property.

(ga') There exists a subset $\hat{J}$ of J which is of positive measure such that $g(t, x, z) < g(t, y, z)$ for all $t \in \hat{J}$ and x, y, $z \in \mathbb{R}$, $x > y$.

The proof of the following result is the same as that of Theorem 3.4.1 when we apply part b) of Proposition 3.4.1, instead of part a).

Theorem 3.4.2. *Assume that $g: J \times \mathbb{R} \times \mathbb{R} \to \mathbb{R}$ and $\varphi: J \times I_0 \to \mathbb{R}$ have properties (φa), (ga') and (gb). If $u \in Y$ is a lower solution and $w \in Y$ an upper solution of (3.4.7),(3.4.10), then $u(t) \le w(t)$ on J. In particular, the Neumann problem (3.4.7),(3.4.10) can have at most one solution.*

Consider next (3.4.7) equipped with periodic boundary conditions

$$u(t_0) = u(t_1), \quad u'(t_0) = u'(t_1). \qquad (3.4.11)$$

We say that a function $u \in Y$ is a *lower solution* of problem (3.4.7),(3.4.11) if

$$\begin{cases} -\frac{d}{dt}\varphi(t, u'(t)) \leq g(t, u(t), u'(t)) & \text{a.e. in } J, \\ u(t_0) = u(t_1), \quad u'(t_0) \geq u'(t_1), \end{cases}$$

an *upper solution* if the reversed inequalities hold, and a *solution* of (3.4.7), (3.4.11) if equalities hold.

The following comparison and uniqueness results are consequences of Lemmas 3.4.1 and 3.4.3.

Theorem 3.4.3. *Let* $g: J \times \mathbb{R} \times \mathbb{R} \to \mathbb{R}$ *and* $\varphi: J \times I_0 \to \mathbb{R}$ *have properties* (φa), (ga'), *and* (gb). *If* $u \in Y$ *is a lower solution and* $w \in Y$ *an upper solution of (3.4.7),(3.4.11), then* $u(t) \leq w(t)$ *for each* $t \in J$. *In particular, the periodic problem (3.4.7),(3.4.11) can have at most one solution.*

Proof. Assume on the contrary that $c = \max\{u(t) - w(t) \mid t \in J\}$ is positive. If $u(t_2) - w(t_2) = c$ for some $t_2 \in (t_0, t_1)$, then $u(t) - w(t) \equiv c$ by Lemma 3.4.1. Assume next that $u(t_0) - w(t_0) = c$. This and the definition of upper and lower solution of (3.4.7),(3.4.11) imply that $u(t_1) - w(t_1) = c$. Thus $u - w$ attains its positive maximum at t_0 and t_1, whence $u'(t_0) \leq w'(t_0)$ and $u'(t_1) \geq w'(t_1)$. These inequalities and the definition of upper and lower solutions of (3.4.7),(3.4.11) imply that

$$\begin{cases} u'(t_1) \leq u'(t_0) \leq w'(t_0) \leq w'(t_1), & \text{and} \\ u'(t_0) \geq u'(t_1) \geq w'(t_1) \geq w'(t_0). \end{cases}$$

But then $u(t) - w(t) \equiv c$ by Lemma 3.4.3.

The above proof shows that $u(t) - w(t) \equiv c > 0$, which leads to contradiction with (ga') (cf. the proof of Proposition 3.4.1.b). This concludes the proof. $\square$

The above results can be proved also by slightly different assumptions.

Corollary 3.4.1. *The results of Lemmas 3.4.1–3.4.3, Theorems 3.4.1–3.4.3, and Proposition 3.4.1 are valid also when* (φa) *and* (gb) *are replaced by the following hypotheses.*

(φ) $\varphi(t, \cdot)$ *is strictly increasing for each* $t \in J$;
(gφ) $|g(t, x, y) - g(t, x, z)| \leq p(t)\phi(|\varphi(t, y) - \varphi(t, z)|)$ *for a.e.* $t \in J$ *and for all* $x \in \mathbb{R}$, $y, z \in I_0$, *where* $p \in L^1_+(J)$, $\phi: \mathbb{R} \to \mathbb{R}$ *is increasing and satisfies* $\int_{0+}^1 \frac{dz}{\phi(z)} = \infty$.

Proof. In the case when (φ), (gφ), and (ga) hold we do not need inequalities (3.4.2) and (3.4.3), and the proof of Lemma 3.4.1 is almost the same as above with $K = 1$. $\square$

In the case of Dirichlet boundary conditions

$$u(t_0) = c_0, \qquad u(t_1) = c_1, \qquad\qquad (3.4.12)$$

instead of the two-sided Osgood condition $(g\varphi)$ we will require only the following one-sided condition

(gc) $g(t, x, z) - g(t, x, y) \le l(t, \varphi(t, y) - \varphi(t, z))$ for a.e. $t \in J$ and for all $x \in \mathbb{R}$, $y, z \in I_0$, $y > z$, $0 < \varphi(t, y) - \varphi(t, z) \le r$, where $r > 0$, $l: J \times [0, r] \to \mathbb{R}_+$, and $x(t) \equiv 0$ is the only function in $AC(J)$ which satisfies

$$x'(t) \le l(t, x(t)) \quad \text{a.e. in } J, \quad x(t_0) = 0.$$

Theorem 3.4.4. *Given functions $g: J \times \mathbb{R} \times \mathbb{R} \to \mathbb{R}$ and $\varphi: J \times I_0 \to \mathbb{R}$ having properties (φ), (ga), and (gc), assume that $u, w \in Y$ satisfy*

$$L_1 u(t) \le L_1 w(t) \quad \text{a.e. in } J, \ u(t_0) \le w(t_0), \ u(t_1) \le w(t_1). \qquad (3.4.13)$$

Then $u(t) \le w(t)$ for each $t \in J$. In particular, the Dirichlet problem (3.4.7),(3.4.12) can have at most one solution.

Proof. If the first claim is wrong, then $w - u$ has a minimum point t_2 in (t_0, t_1), and $w(t_2) - u(t_2) < 0$ and $w'(t_2) - u'(t_2) = 0$. Define $t_* = \min\{t \ge t_2 \mid w(t) - u(t) \ge 0\}$. It follows that $t_* \in (t_2, t_1]$, that $w(t_*) - u(t_*) = 0$, and that $w(t) - u(t) < 0$ for each $t \in [t_2, t_*)$. Hence there exists a $b \in (t_2, t_*)$ such that $w'(b) - u'(b) > 0$. Finally, define $a = \max\{t \in [t_2, b) \mid w'(t) - u'(t) \le 0\}$.

The above construction gives a subinterval $[a, b]$ of J such that

$$w'(a) - u'(a) = 0, \ w(t) - u(t) < 0 \ \text{ and } \ w'(t) - u'(t) > 0 \text{ for all } \ t \in (a, b].$$

Moreover, the end-point b of this interval can be chosen so that the function

$$x(t) = \begin{cases} 0, & t_0 \le t \le a, \\ \varphi(t, w'(t)) - \varphi(t, u'(t)), & a \le t \le b, \\ x(b), & b \le t \le t_1, \end{cases}$$

satisfies $0 \le x(t) \le r$ on J. Since $L_1 u(t) \le L_1 w(t)$ a.e. in J, we obtain by (3.4.1), (ga), and (gc),

$$\begin{aligned} x'(t) &= \frac{d}{dt}\varphi(t, w'(t)) - \frac{d}{dt}\varphi(t, u'(t)) \le g(t, u(t), u'(t)) - g(t, w(t), w'(t)) \\ &\le g(t, w(t), u'(t)) - g(t, w(t), w'(t)) \le l(t, \varphi(t, w'(t)) - \varphi(t, u'(t))) \\ &= l(t, x(t)) \end{aligned}$$

for a.e. $t \in [a, b]$. Thus we have proved that

$$x'(t) \leq l(t, x(t)) \quad \text{a.e. in } [a, b].$$

This inequality holds also when $t \in J \setminus [a, b]$. Because $x(t_0) = 0$, it then follows from the hypothesis (gc) that $x(t) \equiv 0$ on J, which contradicts the fact that $x(t) = \varphi(t, w'(t)) - \varphi(t, u'(t)) > 0$ on $(a, t_1]$. This concludes the proof of the first assertion, and hence also the second one. □

Remarks 3.4.2. The following one-sided Osgood condition

(gc') $g(t, x, z) - g(t, x, y) \leq p(t)\phi(\varphi(t, y) - \varphi(t, z))$ for a.e. $t \in J$ and for all $x \in \mathbb{R}$, $y, z \in I_0$, $y > z$, $0 < \varphi(t, y) - \varphi(t, z) \leq r$, where $r > 0$, $p \in L^1_+(J)$, $\phi: (0, r] \to (0, \infty)$ is increasing and satisfies $\int_{0+}^r \frac{dz}{\phi(z)} = \infty$

is a special case of condition (gc), as shown in [144]. The considerations of [144] imply also that (gc) can be replaced by the following condition.

(gd) $g(t, x, z) - g(t, x, y) \leq \frac{p(t)}{P(t)}(\varphi(t, y) - \varphi(t, z))$ for a.e. $t \in J$ and for all $x \in \mathbb{R}$, $y, z \in I_0$, $y > z$, $0 < \varphi(t, y) - \varphi(t, z) \leq r$, where $r > 0$, $p \in L^1(J, \mathbb{R}_+)$ with $P(t) = \int_{t_0}^t p(s)\, ds > 0$ for $t \in (t_0, t_1]$, and $\sup\{[g(t, x, y) - g(t, x, z)]_+ \mid 0 \leq y - z \leq r,\ x \in \mathbb{R}\} = o(p(t))$ as $t \to t_0+$, where $[a]_+ = \max\{a, 0\}$.

For each $n = 1, 2, \ldots$ the continuous extension of the function

$$\phi_n(z) = z \ln \frac{1}{z} \cdots \ln_n \frac{1}{z}, \quad 0 < z \leq \frac{1}{\exp_n 1},$$

to $\mathbb{R}_+$ satisfies the hypotheses given for ϕ in (gb), (gc'), and (gφ). This is true also when $\phi(z) = z$, $z \in \mathbb{R}_+$.

By assuming mixed monotonicity of g in its last two arguments we can relax the strict monotonicity of $\varphi(t, \cdot)$, assumed in (φ).

Theorem 3.4.5. *The results of Theorem 3.4.4 are valid if $\varphi: J \times \mathbb{R} \to I_0$ and $g: J \times \mathbb{R} \times \mathbb{R} \to \mathbb{R}$ satisfy the following hypotheses.*

(φ') $\varphi(t, y) \leq \varphi(t, z)$ whenever $t \in J$, $y, z \in I_0$ and $y \leq z$.

(g) $g(t, x, z) < g(t, y, s)$ for a.e. $t \in J$ and for all $x, y, z \in \mathbb{R}$, $x > y$ and $z < s$;

Proof. Let $u, w \in Y$ satisfy (3.4.13), and assume on the contrary that $u(t) > w(t)$ for some $t \in J$. As in the proof of Theorem 3.4.4 we can choose a subinterval $[a, b]$ of J such that

$$w'(a) - u'(a) = 0, \quad w(t) - u(t) < 0 \quad \text{and} \quad w'(t) - u'(t) > 0 \quad \text{for all} \quad t \in (a, b],$$

and that

$$x(t) = \varphi(t, w'(t)) - \varphi(t, u'(t)) \leq r, \quad a \leq t \leq b.$$

It then follows from (3.4.13), and (g) that

$$x'(t) = \frac{d}{dt}\varphi(t, w'(t)) - \frac{d}{dt}\varphi(t, u'(t)) \leq g(t, u(t), u'(t)) - g(t, w(t), w'(t)) < 0$$

for a.e. $t \in [a, b]$. Because $x(a) = 0$, then $x(t) < 0$ on $(a, b]$, which contradicts the fact that $x(t) = \varphi(t, w'(t)) - \varphi(t, u'(t)) \geq 0$ on $(a, b]$ by (φ'). This concludes the proof. $\square$

When g is constant with respect to its last argument, the comparison results of Theorems 3.4.4 and 3.4.5 can be restated in the following form.

Proposition 3.4.2. *Given* $\varphi \colon J \times \mathbb{R} \to \mathbb{R}$ *and* $g \colon J \times \mathbb{R} \to \mathbb{R}$, *assume that* $\varphi(t, \cdot)$ *is increasing and* $g(t, \cdot)$ *is decreasing, one of them being strict for all* $t \in J$. *If* $u, w \in Y$ *satisfy*

$$\begin{cases} -\frac{d}{dt}\varphi(t, u(t)) - g(t, u(t)) \leq -\frac{d}{dt}\varphi(t, w(t)) - g(t, w(t)) & \text{a.e. in } J, \\ u(t_0) \leq w(t_0), \ u(t_1) \leq w(t_1), \end{cases}$$

then $u(t) \leq w(t)$ *on* J.

Example 3.4.1. Denote by $[x]$ the greatest integer $\leq x$, and by χ_U the characteristic function of a subset U of $J = [0, 1]$. The periodic boundary value problem

$$\begin{cases} \frac{d}{dt}(u'(t) + [\chi_U(t)u'(t)]) = [1 + 2t](u'(t) + [\chi_U(t)u'(t)]) - u(t) + 1 \\ \text{a.e. in } J, \quad u(0) = u(1), \quad u'(0) = u'(1), \end{cases}$$

$$(3.4.14)$$

is a special case of problem (3.4.7),(3.4.11) with

$$\varphi(t, x) = x + [\chi_U(t)x], \ g(t, x, z) = [1 + 2t](z + [\chi_U(t)z]) - x + 1,$$

$t \in J$, $x, y \in \mathbb{R}$. It is easy to see that these functions φ and g satisfy the hypotheses of Theorem 3.4.3. Thus problem (3.4.14) can have only one solution. Obviously, $u(t) \equiv 1$ is the solution of (3.4.14). Notice that the functions φ and g are discontinuous, and even nonmeasurable in t if U is a nonmeasurable subset of J.

Example 3.4.2. The BVP

$$\begin{cases} \frac{d}{dt}(u'(t) + [\chi_U(t)u'(t)]) = u'(t) + [\chi_U(t)u'(t)] - [\chi_V(t)u(t)] - \frac{1}{2} \\ \text{a.e. in } \ J = [0,1], \quad u(0) = 0, \ u(1) = \frac{1}{2}, \end{cases}$$

$$(3.4.15)$$

where U, $V \subseteq J$, is a special case of problem (3.4.7),(3.4.12) with

$$\varphi(t,x) = x + [\chi_U(t)x], \ g(t,x,z) = z + [\chi_U(t)z] - [\chi_V(t)x] - \frac{1}{2},$$

$t \in J$, x, $y \in \mathbb{R}$. It is easy to see that the hypotheses of Theorem 3.4.4 hold. Thus $u(t) = \frac{1}{2}$, $t \in J$, is the only solution of (3.4.15). Also in this example the functions φ and g are discontinuous in all their variables and may be nonmeasurable in t.

3.4.3 Special cases. In the *phi-Laplacian* case where the differential equation (3.4.7) is reduced to

$$-\frac{d}{dt}\varphi(u'(t)) = g(t, u(t), u'(t)),$$

$$(3.4.16)$$

the hypotheses (φ) and (φa) can be rewritten as follows.

$(\varphi 0)$ $\varphi \colon I_0 \to \mathbb{R}$ is strictly increasing.

$(\varphi 1)$ If s_1, $s_2 \in I_0$ and $s_1 < s_2$, there exists a positive constant M such that $\varphi(y) - \varphi(z) \geq M(y - z)$ whenever $s_1 \leq z < y \leq s_2$.

Next we are looking for uniqueness and comparison results for the differential equation

$$-\frac{d}{dt}\varphi(u'(t)) = q(u'(t))g(t, u(t), u'(t)),$$

$$(3.4.17)$$

associated with one of the boundary conditions (3.4.8), (3.4.10), (3.4.11), and (3.4.12). By assuming that $q \colon \mathbb{R} \to (0, \infty)$ has the following property:

(q) q, $\frac{1}{q} \in L_{loc}^\infty(\mathbb{R})$ and $q \circ \varphi^{-1} \colon I_0 \to \mathbb{R}$ is measurable,

we prove the following result.

Proposition 3.4.3. *The comparison and uniqueness results of Theorems 3.4.1–3.4.4 hold for the corresponding boundary value problems of the differential equation (3.4.17) if property (q) is added to the hypotheses.*

Proof. Define $\overline{\varphi} \colon I_0 \to \mathbb{R}$ for a fixed $y_0 \in I_0$ by

$$\overline{\varphi}(x) = \int_{\varphi(y_0)}^{\varphi(x)} \frac{dz}{q(\varphi^{-1}(z))}, \quad x \in I_0.$$

If $u \in Y$, then $u'[J] \subseteq I_0$ and $\varphi \circ u' \in AC(J)$. The hypothesis (q) ensures that $\frac{1}{q \circ \varphi^{-1}}$ is measurable and locally essentially bounded. Thus an
application of Lemma C.3.2 yields

$$\overline{\varphi}(u'(t)) = \overline{\varphi}(u'(t_0)) + \int_{\varphi(u'(t_0))}^{\varphi(u'(t))} \frac{dz}{q(\varphi^{-1}(z))}$$

$$= \overline{\varphi}(u'(t_0)) + \int_{t_0}^{t} \frac{\frac{d}{ds}\varphi(u'(s))ds}{q(u'(s))}$$

for each $t \in J$. This implies that $\overline{\varphi} \circ u' \in AC(J)$, and that

$$\frac{d}{dt}\overline{\varphi}(u'(t)) = \frac{d}{dt}\int_{t_0}^{t} \frac{\frac{d}{ds}\varphi(u'(s))ds}{q(u'(s))} = \frac{\frac{d}{dt}\varphi(u'(t))}{q(u'(t))} \quad \text{a.e. in } J.$$

Hence, if $u \in Y$ is a lower solution, an upper solution or solution of (3.4.17)
with boundary conditions (3.4.8), (3.4.10), (3.4.11), or (3.4.12), then u is
a lower solution, an upper solution, or solution of corresponding problems
where the differential equation (3.4.17) is replaced by

$$-\frac{d}{dt}(\overline{\varphi}(u'(t)) = g(t, u(t), u'(t)) \quad \text{a.e. in } J. \tag{3.4.18}$$

Obviously, $\overline{\varphi}$ satisfies the hypothesis $(\varphi 1)$ so that also $(\varphi 1)$ holds. Thus the
results of Theorems 3.4.1–3.4.4 hold for equation (3.4.18) with correspond-
ing boundary conditions, which implies the assertions. $\qquad\square$

Remark 3.4.3. The conditions imposed in the hypothesis (q) on the function q allow the right-hand side of (3.4.17) to have discontinuous dependence
also on u'.

When $\varphi(t, x) = \mu(t)|x|^{p-2}x$, the differential equation (3.4.7) is reduced
to equation containing the *p-Laplacian* operator, i.e.,

$$-\frac{d}{dt}(\mu(t)|u'(t)|^{p-2}u'(t)) = g(t, u(t), u'(t)) \quad \text{a.e. in } J. \tag{3.4.19}$$

If $\mu: J \to [a, b]$, $0 < a < b < \infty$, then (φ) holds for all $p > 1$, and (φa) holds
when $1 < p \leq 2$. Thus we get the following corollary.

Corollary 3.4.2. *If $\mu: J \to [a, b]$, $0 < a < b < \infty$, then the comparison
and uniqueness results of Theorems 3.4.1–3.4.4 hold when (3.4.7) is replaced
by (3.4.19), property (φ) by $p > 1$, and property (φa) by $1 < p \leq 2$.*

If $\varphi(t, x) = \frac{\mu(t)x}{\sqrt{1+x^2}}$, we have the following result.

Corollary 3.4.3. *If* $\mu\colon J \to [a,b]$, $0 < a < b < \infty$, *then the comparison and uniqueness results of Theorems 3.4.1–3.4.4 hold when* $\varphi(t,x) = \frac{\mu(t)x}{\sqrt{1+x^2}}$.

In the case when $\varphi(t,x) = \mu(t)x$ we get the following consequence of Corollary 3.4.2.

Corollary 3.4.4. *The comparison and uniqueness results of Theorem 3.4.1 hold for the separated Sturm-Liouville problem*

$$\begin{cases} -\frac{d}{dt}(\mu(t)u'(t)) = g(t, u(t), u'(t)) & a.e. \ in \ J, \\ a_0 u(t_0) - b_0 u'(t_0) = c_0, \quad a_1 u(t_1) + b_1 u'(t_1) = c_1, \end{cases} \tag{3.4.20}$$

if $\mu \in C(J, (0, \infty))$, *and if* $g\colon J \times \mathbb{R} \times \mathbb{R} \to \mathbb{R}$ *has properties (ga) and (gb).*

3.4.4 Remarks, examples, and counterexamples. In Lemma 3.4.1 it suffices to assume that p in (gb) and (gφ) is locally Lebesgue integrable in (t_0, t_1). For instance, if a solution u of

$$-u''(t) = \frac{3}{t}u'(t), \quad t \in (0,1),$$

has a positive maximum c, then $u(t) \equiv c$. In particular, p does not need to be bounded, as assumed, e.g., in [197]. On the other hand, the local integrability of p is needed, since the Dirichlet problem

$$-u''(t) = p(t)u'(t), \quad \text{a.e. in } J = [-1,1], \ u(-1) = u(1) = 0,$$

where $p(t) = \begin{cases} -\frac{3}{t}, \ t \neq 0, \\ 0, \ t = 0, \end{cases}$, has solutions $u(t) \equiv 0$ and $u(t) = 1 - t^4$.

Assumption $\int_{0+}^{1} \frac{dx}{\phi(x)} = \infty$ in (gb) and (gc') is also needed because the BVP

$$-u''(t) = 3(2u'(t))^{\frac{2}{3}}, \ t \in J = [-1,1], \quad u(-1) = u(1) = 0 \tag{3.4.21}$$

has solutions $u(t) \equiv 0$ and $u(t) = 1 - t^4$. However, if we are interested in such solutions of (3.4.21) whose derivatives are nonzero, we can rewrite (3.4.21) in the following forms:

$$\begin{cases} -\frac{d}{dt}u'(t)^{\frac{1}{3}} = 2^{\frac{2}{3}}, \ t \in J = [-1,1], \quad u(-1) = u(1) = 0, \\ -\frac{d}{dt}(|u'(t)|^{\frac{4}{3}-2}u'(t)) = 2^{\frac{2}{3}}, \ t \in J = [-1,1], \quad u(-1) = u(1) = 0. \end{cases}$$

The first one is in a phi-Laplacian form with $\varphi(x) = x^{\frac{1}{3}}$, whereas the second one is in a p-Laplacian form with $p = \frac{4}{3}$. In the former case the hypotheses of Proposition 3.4.3 hold, and in the latter case the hypotheses of Corollary 3.4.2 hold, whence these problems can have only one solution, which is $u(t) = 1 - t^4$.

The periodic boundary value problem

$$\begin{cases} \frac{d}{dt}(u'(t) + [tu'(t)]) = [1 + 2t](u'(t) + [tu'(t)]) - [\sin(t)u(t)] \\ \text{a.e. in } J = [0, 1], \quad u(0) = u(1), \quad u'(0) = u'(1) \end{cases}$$

has a continuum of solutions of the form $x(t) \equiv c$, $c \in [0, \frac{1}{\sin(1)}]$. In this example hypotheses of Theorem 3.4.3 are not valid. On the other hand, the hypotheses of Theorem 3.4.4 hold for the BVP

$$\begin{cases} \frac{d}{dt}(u'(t) + [tu'(t)]) = [1 + 2t](u'(t) + [tu'(t)]) - [\sin(t)u(t)] \\ \text{a.e. in } J = [0, 1], \quad u(0) = c_0, \quad u(1) = c_1, \end{cases}$$

whence this problem can have only one solution. In fact $u(t) \equiv c$ is the solution when $c_0 = c_1 = c \in [0, \frac{1}{\sin(1)}]$.

The Dirichlet problem

$$-u''(t) = \mu u(t), \ t \in J = [0, \frac{\pi}{\sqrt{\mu}}], \quad u(0) = u(\frac{\pi}{\sqrt{\mu}}) = 0$$

has for each $\mu > 0$ solutions $u(t) \equiv 0$ and $u(t) = \sin(\sqrt{\mu}t)$. Thus the results of Proposition 3.4.1 and Theorems 3.4.1 and 3.4.4 do not hold in general if condition (ga) is not satisfied.

The Dirichlet problem

$$-u''(t) = \mu u(t), \ t \in J = [0, \frac{\pi}{\sqrt{\mu}}], \quad u(0) = u(\frac{\pi}{\sqrt{\mu}}) = 0,$$

has for each $\mu > 0$ solutions $u(t) \equiv 0$ and $u(t) = \sin(\sqrt{\mu}t)$. Thus the results of Proposition 3.4.1 and Theorem 3.4.1 do not hold in general if (ga) is not satisfied.

If g, $h \in L^1(J)$, and if $g(t) = 0$ a.e. in J, then the Neumann problem

$$-u''(t) - g(t)u(t) = h(t) \ \text{a.e. in } J, \quad u'(t_0) = 1, \ u'(t_1) = -1,$$

has solutions if and only if $\int_{t_0}^{t_1} h(t)dt = 0$. In this case the solutions are of the form $u(t) = -\int_{t_0}^{t}(\int_{t_0}^{s} h(\tau)d\tau)ds + t + C$. This example shows that the result of Theorem 3.4.2 does not hold in general if (ga') is replaced by (ga), and that the results of Proposition 3.4.1.b and c do not hold in general if we only assume that a_j, $b_j \in \mathbb{R}_+$, $j = 0, 1$, and $b_0 b_1 > 0$.

3.5 Notes and comments

The uniqueness and well-posedness results presented in sections 3.1, 3.2, and 3.3 for explicit and implicit first order initial and boundary value problems are adopted from [136]. As for uniqueness and well-posedness results for explicit initial and boundary value problems in the case when $\varphi(t, x) \equiv x$, see, e.g., [141], [144], [162], [167], [197], [211]. Uniqueness results in this special case for first order implicit problems are derived, e.g., in [145], [158], [207], [210].

Maximum and comparison principles, and the uniqueness results derived in section 3.4 for second order generalized phi-Laplacian boundary value problems are taken from [146]. As for related results in special cases, see, e.g., [39], [71], [74], [129], [141], [212].

Chapter 4

Second order functional differential equations

In this chapter we introduce existence and comparison results for explicit and implicit boundary value problems of second order differential equations. Both differential equations and boundary conditions may involve discontinuities and functional terms.

In section 4.1 we study explicit functional Sturm-Liouville differential equations with separated boundary conditions, by proving existence and comparison results for their extremal solutions. Applying also an uniqueness criterion proved in section 3.4 we obtain existence and uniqueness results.

The results of section 4.1 and those derived for the equation $Lu = Nu$ in section 1.1 are applied in section 4.2 to prove existence and comparison theorems for extremal solutions of implicit Sturm-Liouville problems. Special cases where the existence of extremal solutions can be proved by the method of successive approximations are considered in section 4.3.

In section 4.4 we study the existence of extremal solutions of explicit functional phi-Laplacian differential equations with functional initial conditions. We shall first convert the problem under consideration to a pair of first order problems. Applying results of chapter 2 to the so-obtained system we then derive extremality results for the original problem between assumed upper and lower solutions. With the help of growth conditions we finally prove the existence of extremal solutions and study their dependence on the data.

In section 4.5 we provide extremality results for implicit functional phi-Laplacian differential equations equipped with implicit functional initial conditions. We reduce them first to operator equations of the form $Lu = Nu$ in suitable ordered function spaces. The results of sections 4.4 and 1.1 are then applied to prove existence and comparison results for extremal solutions of given implicit problems. Special cases where extremal solutions are obtained by the method of successive approximations are also presented.

Theoretical and concrete examples are given to illustrate the results.

4.1 Explicit Sturm-Liouville boundary value problems

In this section we prove existence results for extremal solutions of the BVP

$$\begin{cases} -\frac{d}{dt}(\mu(t)u'(t)) = \lambda g(t, u, u(t), u'(t)) & \text{a.e. in } J = [t_0, t_1], \\ a_0 u(t_0) - b_0 u'(t_0) = c_0, \quad a_1 u(t_1) + b_1 u'(t_1) = c_1, \end{cases} \qquad (4.1.1)$$

in the set

$$Y = \{u \in C^1(J) \mid \mu\, u' \in AC(J)\},$$

and study their dependence on data. In this section we assume that

$$\lambda,\, a_j,\, b_j \in \mathbb{R}_+,\, a_0 a_1 + a_0 b_1 + a_1 b_0 > 0,\, c_0,\, c_1 \in \mathbb{R}, \text{ and } \mu \in C(J,(0,\infty)),$$

that $C(J)$ is ordered pointwise, and the function $g\colon J \times C(J) \times \mathbb{R} \times \mathbb{R} \to \mathbb{R}$ satisfies the following hypotheses.

(g0) $(t, x, y) \mapsto g(t, v, x, y)$ is a Carathéodory function for each $v \in C(J)$.

(g1) $|g(t, v, x, y)| \le p_1(t) \max\{|x|, |y|\} + m(t)$ for all $x,\, y \in \mathbb{R}$ and for a.e. $t \in J$, where $p_1,\, m \in L^1_+(J)$.

(g2) $g(t, v, x, y)$ is increasing in v and decreasing in x for a.e. $t \in J$ and for all $y \in \mathbb{R}$.

(g3) $|g(t, v, x, y) - g(t, v, x, z)| \le p(t)\phi(|y - z|)$ for a.e. $t \in J$ and for all $x, y, z \in \mathbb{R}$ and $v \in C(J)$, where $p \in L^1_+(J)$, $\phi\colon \mathbb{R}_+ \to \mathbb{R}_+$ is increasing and $\int_{0+}^1 \frac{dx}{\phi(x)} = \infty$.

Notice that g can be discontinuous in its first and second arguments.

4.1.1 Auxiliary results. For the sake of completeness we shall prove

Lemma 4.1.1. *If $q \in L^1(J)$, then the BVP*

$$\begin{cases} -\frac{d}{dt}(\mu(t)u'(t)) = q(t) & \text{a.e. in } J, \\ a_0 u(t_0) - b_0 u'(t_0) = c_0, \quad a_1 u(t_1) + b_1 u'(t_1) = c_1, \end{cases} \qquad (4.1.2)$$

has a unique solution u in Y, which is increasing with respect to q, c_0, and c_1, and can be represented as

$$u(t) = \frac{c_0 y_1(t) + c_1 y_0(t)}{D} + \int_{t_0}^{t_1} k(t, s) q(s)\, ds, \quad t \in J, \qquad (4.1.3)$$

where

$$\begin{cases} y_0(t) = \int_{t_0}^t \frac{a_0}{\mu(s)} ds + \frac{b_0}{\mu(t_0)}, \quad D = \int_{t_0}^{t_1} \frac{a_0 a_1}{\mu(s)} ds + \frac{a_0 b_1}{\mu(t_1)} + \frac{a_1 b_0}{\mu(t_0)}, \\ y_1(t) = \int_t^{t_1} \frac{a_1}{\mu(s)} ds + \frac{b_1}{\mu(t_1)}, \quad k(t, s) = \begin{cases} \frac{y_1(t) y_0(s)}{D}, & t_0 \le s \le t, \\ \frac{y_0(t) y_1(s)}{D}, & t \le s \le t_1. \end{cases} \end{cases} \qquad (4.1.4)$$

Proof. In view of (4.1.4) we can rewrite (4.1.3) as

$$u(t) = \frac{y_1(t)}{D}\left(c_0 + \int_{t_0}^t y_0(s)q(s)ds\right) + \frac{y_0(t)}{D}\left(c_1 + \int_t^{t_1} y_1(s)q(s)ds\right).$$

This implies by differentiation that for a.e. $t \in J$,

$$\mu(t)u'(t) = \frac{-a_1}{D}\left(c_0 + \int_{t_0}^t y_0(s)q(s)ds\right) + \frac{a_0}{D}\left(c_1 + \int_t^{t_1} y_1(s)q(s)ds\right). \quad (4.1.5)$$

Since the right-hand side of (4.1.5) is absolutely continuous and μ is continuous, then (4.1.5) holds for all $t \in J$. In particular, $u \in C^1(J)$ and $\mu u' \in AC(J)$, whence $u \in Y$. It follows from (4.1.5) that

$$\frac{d}{dt}(\mu(t)u'(t)) = -\frac{a_0 y_1(t) + a_1 y_0(t)}{D}q(t) = -q(t) \quad \text{a.e. in } J.$$

The above representations of u and $\mu u'$ imply also that u satisfies the boundary conditions of (4.1.2). Thus u is a solution of the BVP (4.1.2). Uniqueness follows from Corollary 3.4.4, and (4.1.3) implies that u is increasing with respect to q, c_0, and c_1. □

Consider next the BVP

$$\begin{cases} -\frac{d}{dt}(\mu(t)u'(t)) = \lambda g(t, v, u(t), u'(t)) & \text{for a.e. } t \in J, \\ a_0 u(t_0) - b_0 u'(t_0) = c_0, \quad a_1 u(t_1) + b_1 u'(t_1) = c_1, \end{cases} \quad (4.1.6)$$

where $\lambda \geq 0$, and $v \in C(J)$ are fixed. As an application of Lemma 4.1.1 we get the following result.

Lemma 4.1.2. *If the hypotheses (g0) and (g1) are satisfied, then $u \in Y$ is a solution of the BVP (4.1.6) if and only if*

$$u(t) = \frac{c_0 y_1(t) + c_1 y_0(t)}{D} + \lambda \int_{t_0}^{t_1} k(t, s)g(s, v, u(s), u'(s))ds, \quad t \in J. \quad (4.1.7)$$

Proof. The given hypotheses ensure that if $u \in Y$, the function $q: J \to \mathbb{R}$, defined by $q(t) = \lambda g(t, v, u(t), u'(t))$, $t \in J$, is Lebesgue integrable. Thus the assertion follows from Lemma 4.1.1. □

Denote

$$z_0(t) = \max\{y_0(t), \frac{a_0}{\mu(t)}\}, \quad z_1(t) = \max\{y_1(t), \frac{a_1}{\mu(t)}\}, \quad (4.1.8)$$

and define an operator $A: C(J) \to C(J)$ by

$$Au(t) = \int_{t_0}^{t_1} \ell(t, s)p_1(s)u(s)ds, \quad \ell(t, s) = \begin{cases} \frac{z_1(t)y_0(s)}{D}, & t_0 \leq s \leq t, \\ \frac{z_0(t)y_1(s)}{D}, & t \leq s \leq t_1. \end{cases} \quad (4.1.9)$$

Lemma 4.1.3. *If the hypothesis (g1) holds, and if $\lambda \in [0, \lambda_1)$, where λ_1 is the least positive eigenvalue of A, then the integral equation*

$$b(t) = \frac{|c_0|z_1(t) + |c_1|z_0(t)}{D} + \lambda \int_{t_0}^{t_1} \ell(t, s)(p_1(s)b(s) + m(s))\, ds \quad (4.1.10)$$

has a unique solution $b \in C(J)$. Moreover, if $w \in C(J)$ and

$$w(t) \le \frac{|c_0|z_1(t) + |c_1|z_0(t)}{D} + \lambda \int_{t_0}^{t_1} \ell(t, s)(p_1(s)w(s) + m(s))\, ds, \quad t \in J,$$

then $w \le b$.

Proof. The integral equation (4.1.10) can be rewritten as

$$b = v_1 + \lambda A b,$$

where $v_1(t) = \frac{|c_0|z_1(t) + |c_1|z_0(t)}{D} + \lambda \int_{t_0}^{t_1} \ell(t, s)m(s)\, ds$, $t \in J$. Since the spectral radius $\rho(A)$ of A equals to $\frac{1}{\lambda_1}$, then $0 \le \lambda\, r(A) < 1$, so that the function $b = \sum_{n=0}^{\infty} (\lambda A)^n v_1$ is the unique solution of (4.1.10) in $C(J)$. The operator A is linear, continuous, and increasing, whence the inequality

$$w \le v_1 + \lambda A w$$

implies by the Abstract Gronwall Lemma (cf. B.8) that $w \le b$. This proves the last conclusion. $\square$

4.1.2 Existence and comparison results for (4.1.6). We shall first prove an existence result for the BVP (4.1.6) by using Schauder's fixed point theorem and Lemma 4.1.2.

Proposition 4.1.1. *Let the hypotheses (g0) and (g1) hold, assume that $\lambda \in [0, \lambda_1)$, where λ_1 is the least positive eigenvalue of the operator A, defined by (4.1.9), and let b be the solution of (4.1.10). Then for each $v \in [-b, b]$ the BVP (4.1.6) has a solution in the set $Y \cap B$, where*

$$B = \{u \in C^1(J) \mid \max\{|u(t)|, |u'(t)|\} \le b(t),\ t \in J\}. \quad (4.1.11)$$

Proof. Let $v \in [-b, b]$ be given. The hypotheses (g0) and (g1) imply that the function $q_u \colon J \to \mathbb{R}$, defined by

$$q_u(t) = \lambda g(t, v, u(t), u'(t)), \quad t \in J, \quad (4.1.12)$$

is Lebesgue integrable for each $u \in B$. Thus we can define a mapping F on B by

$$Fu(t) = \frac{c_0 y_1(t) + c_1 y_0(t)}{D} + \int_{t_0}^{t_1} k(t,s) g_u(s) ds, \quad t \in J. \qquad (4.1.13)$$

It follows from the proof of Lemma 4.1.1 that

$$(Fu)'(t) = \frac{-a_1}{\mu(t)D}\left(c_0 + \int_{t_0}^{t} y_0(s) q_u(s) ds\right) + \frac{a_0}{\mu(t)D}\left(c_1 + \int_{t}^{t_1} y_1(s) q_u(s) ds\right)$$
$$(4.1.14)$$

for all $t \in J$, that $Fu \in Y$, and that

$$\frac{d}{dt}(\mu(t)(Fu)'(t)) = -q_u(t) \quad \text{a.e. in } J. \qquad (4.1.15)$$

To show that $F[B] \subseteq B$, let $u \in B$ be given. It follows from (4.1.2), (4.1.8)–(4.1.14), and (g1), and from the choice of v that

$$|Fu(t)| \leq b(t), \qquad |(Fu)'(t)| \leq b(t), \quad t \in J. \qquad (4.1.16)$$

In view of (4.1.11) and (4.1.16) we then have $Fu \in B$.

Define a norm in $C^1(J)$ by

$$\|u\| = \max\{|u(t)|, |u'(t)| \mid t \in J\}. \qquad (4.1.17)$$

Obviously, B is closed and convex subset of $C^1(J)$. To prove that F is continuous in B, assume that $u_n, u \in B$, $n \in \mathbb{N}$. Denoting $k_0 = \max\{\ell(t,s) \mid t, s \in J\}$, it follows from (4.1.13) and (4.1.14) that for each $t \in J$,

$$\begin{cases} |(Fu_n)'(t) - (Fu)'(t)| \leq k_0 \int_{t_0}^{t_1} |q_{u_n}(s) - q_u(s)| ds, \\ |Fu_n(t) - Fu(t)| \leq k_0 \int_{t_0}^{t_1} |q_{u_n}(s) - q_u(s)| ds. \end{cases}$$

Noticing that $(x,y) \mapsto g(t,v,x,y)$ is by (g0) continuous for a.e. $t \in J$, the above inequalities and the dominated convergence theorem imply that $\|Fu_n - Fu\| \to 0$ as $\|u_n - u\| \to 0$. Thus F is continuous in B.

It follows from (4.1.13) and (4.1.16) that the set $\{Fu \mid u \in B\}$ is uniformly bounded and equicontinuous. In view of (4.1.15) we obtain

$$(Fu)'(t) = -\frac{1}{\mu(t)} \int_{t_0}^{t} q_u(s) ds + \frac{\mu(t_0)(Fu)'(t_0)}{\mu(t)}, \quad t \in J.$$

This implies by (4.1.12) and (g1) that also the functions $(Fu)'$, $u \in B$, form an uniformly bounded and equicontinuous set. These results ensure by Arzela-Ascoli theorem that given any sequence (u_n) of B there is a subsequence of (Fu_n) which converges with respect to the norm given by (4.1.17).

The above proof implies that F is a compact mapping in a closed and convex subset B of $C^1(J)$, whence F has by Schauder's theorem a fixed point u in B. This result, (4.1.12), and (4.1.13) ensure that u is a solution of the integral equation (4.1.7). It then follows from Lemma 4.1.2 that u is a solution of (4.1.6) in $Y \cap B$. $\qquad\qquad\qquad\qquad\qquad\qquad\qquad\square$

Remark 4.1.1. The BVP

$$-u''(t) = \lambda u(t), \ t \in J = [0, \pi], \quad u(0) = 0, \ u(\pi) = 1, \qquad (4.1.18)$$

does not have any solution when $\lambda = 1$. Thus the result of Proposition 4.1.1 is not valid in general if $\lambda \notin [0, \lambda_1)$.

As a consequence of Corollary 3.4.4 we get the following comparison result for lower and upper solutions of the BVP (4.1.6).

Proposition 4.1.2. *Assume that $g: J \times C(J) \times \mathbb{R} \times \mathbb{R} \to \mathbb{R}$ has properties (g2) and (g3), and that $v \in C(J)$. If u is a lower solution and w an upper solution of (4.1.6), then $u(t) \leq w(t)$ for each $t \in J$. In particular, for each $v \in C(J)$ there exists at most one solution of (4.1.6).*

Proof. The hypotheses (g2) and (g3) ensure that the hypotheses given in Corollary 3.4.4 hold for the function $(t, x, y) \to \lambda g(t, v, x, y)$, which implies the assertions. $\qquad\qquad\qquad\qquad\qquad\qquad\qquad\qquad\qquad\square$

4.1.3 Extremality results for the BVP (4.1.1). Now we are ready to prove the main existence theorem of this section. In the proof we use the results of subsection 4.1.2 and Corollary 1.1.1.

Theorem 4.1.1. *Assume that g has properties (g0)–(g3), and that $\lambda \in [0, \lambda_1)$, where λ_1 is the least positive eigenvalue of the operator A, defined by (4.1.9). Then the BVP (4.1.1) has extremal solutions u_* and u^* in Y, i.e., if $u \in Y$ is a solution of (4.1.1), then $u_*(t) \leq u(t) \leq u^*(t)$ on J.*

Proof. Let $b \in C(J)$ be the solution of (4.1.10), and let B be defined by (4.1.11). In view of Propositions 4.1.1 and 4.1.2 the relation

$$\begin{cases} Gv := \text{ the solution of the BVP (4.1.6),} \\ v \in [-b, b] = \{v \in C(J) \mid -b(t) \leq v(t) \leq b(t), \ t \in J\}, \end{cases} \qquad (4.1.19)$$

defines a mapping G from $[-b, b]$ to B. Since $B \subset [-b, b]$, then G maps $[-b, b]$ into itself. Let $v, \hat{v} \in [-b, b]$ be given, and let $u, \hat{u}$ be the corresponding solutions of (4.1.6). If $\hat{v} \leq v$, it follows from (g2) that $\hat{u}$ is a lower solution of the BVP (4.1.6). Since u, as the solution, is an upper solution of (4.1.6), then $\hat{u} \leq u$ by Proposition 4.1.2, which means that $G\hat{v} \leq Gv$, i.e., G is increasing. Lemma 4.1.2 implies that $u = Gv$, as the solution of (4.1.6), satisfies the integral equation

$$Gv(t) = \frac{c_0 y_1(t) + c_1 y_0(t)}{D} + \lambda \int_{t_0}^{t_1} k(t, s) g(s, Gv(s), v(s), (Gv)'(s)) \, ds.$$

Since $FGv = Gv \in B$ for each $v \in [-b, b]$, it follows from the proof of Proposition 4.1.1 that $G[-b, b]$ is an equicontinuous subset of $[-b, b]$. This implies that each monotone sequence of $G[-b, b]$ converges uniformly on J to a function $u \in [-b, b]$.

The above proof implies that the hypotheses of Corollary 1.1.1 hold when $[a, b] = J$ and $[\underline{x}, \overline{x}] = [-b, b]$. Thus the operator G has the least fixed point u_* and the greatest fixed point u^*. The definition (4.1.19) of G and Lemma 4.1.2 imply that u_* and u^* are the least and the greatest solutions of the BVP (4.1.1) in $[-b, b]$.

Assume now that $u \in Y$ is a solution of the BVP (4.1.1). The hypotheses (g0) and (g1) ensure that $g(\cdot, u, u(\cdot), u'(\cdot)) \in L^1(J)$. This implies by Lemma 4.1.2 that u satisfies the integral equation

$$u(t) = \frac{c_0 y_1(t) + c_1 y_0(t)}{D} + \lambda \int_{t_0}^{t_1} k(t, s) g(s, u, u(s), u'(s)) \, ds. \qquad (4.1.20)$$

To show that u belongs to $[-b, b]$, denote $w(t) = \max\{|u(t)|, |u'(t)|\}$, $t \in J$. It follows from (4.1.20), (4.1.8), (4.1.9), and (g1) that for each $t \in J$,

$$|u(t)| \leq \frac{|c_0| z_1(t) + |c_1| z_0(t)}{D} + \lambda \int_{t_0}^{t_1} \ell(t, s)(p_1(s) w(s) + m(s)) \, ds. \qquad (4.1.21)$$

From (4.1.20) we obtain by differentiation that for each $t \in J$,

$$u'(t) = \frac{-a_1}{\mu(t) D} \left(c_0 + \int_{t_0}^{t} y_0(s) g(s, u, u(s), u'(s)) ds \right.$$
$$+ \frac{a_0}{\mu(t) D} \left(c_1 + \int_{t}^{t_1} y_1(s) g(s, u, u(s), u'(s)) ds. \right.$$

Applying this equation, (4.1.8), (4.1.9), and (g1) we get

$$|u'(t)| \leq \frac{|c_0|z_1(t) + |c_1|z_0(t)}{D} + \lambda \int_{t_0}^{t_1} \ell(t,s)(p_1(s)w(s) + m(s))\, ds \quad (4.1.22)$$

for each $t \in J$. In view of (4.1.21) and (4.1.22) we then have

$$w(t) \leq \frac{|c_0|z_1(t) + |c_1|z_0(t)}{D} + \lambda \int_{t_0}^{t_1} \ell(t,s)(p_1(s)w(s) + m(s))\, ds.$$

Thus $w \leq b$ by Lemma 4.1.3, or equivalently, $\max\{|u(t)|, |u'(t)|\} \leq b(t)$, $t \in J$, so that $u \in B$. In particular, $u \in [-b, b]$. Since u_* and u^* are the least and greatest solutions of the BVP (4.1.1) in $[-b, b]$, then $u_* \leq u \leq u^*$. This concludes the proof. $\qquad\square$

Remark 4.1.2. In the proof of Theorem 4.1.1 it is shown that under the given hypotheses each solution of the BVP (4.1.1) belongs to the set B, defined by (4.1.11).

Next we shall derive results for the dependence of extremal solutions of (4.1.1) on g, c_0, and c_1.

Proposition 4.1.3. *If the hypotheses of Theorem 4.1.1 are satisfied, then the extremal solutions of the BVP (4.1.1) are increasing with respect to g, c_0, and c_1.*

Proof. Assume that functions g, $\hat{g} \colon J \times C(J) \times \mathbb{R} \times \mathbb{R} \to \mathbb{R}$ have properties (g0)–(g3), that

$$g(t, v, x, y) \leq \hat{g}(t, v, x, y) \quad (4.1.23)$$

for a.e. $t \in J$ and for all $x, y \in \mathbb{R}$ and $v \in C(J)$, and that

$$c_j, \hat{c}_j \in \mathbb{R}, \quad c_j \leq \hat{c}_j, \quad j = 0, 1. \quad (4.1.24)$$

Let b be the solution of the integral equation

$$\begin{cases} b(t) = u_0(t) + \lambda \int_{t_0}^{t_1} \ell(t,s)(p_1(s)b(s) + m(s))\, ds, \\ u_0(t) = \frac{(|c_0| + |\hat{c}_0|)z_1(t) + (|c_1| + |\hat{c}_1|)z_0(t)}{D}. \end{cases}$$

Let Gv and $\hat{G}v$ denote for each $v \in [-b, b]$ the solutions of the BVPs (4.1.6) and

$$\begin{cases} -\frac{d}{dt}(\mu(t)u'(t)) = \lambda\, \hat{g}(t, v, u(t), u'(t)) \quad \text{a.e. in } J, \\ a_0 u(t_0) - b_0 u'(t_0) = \hat{c}_0, \quad a_1 u(t_1) + b_1 u'(t_1) = \hat{c}_1, \end{cases} \quad (4.1.25)$$

respectively. It follows from (4.1.23), (4.1.24), and (4.1.25) that $\hat{G}v$ is an upper solution of (4.1.6), which implies by Proposition 4.1.2 that $Gv \leq \hat{G}v$ for each $v \in [-b, b]$. Denoting $\underline{x} = -b$ and $\overline{x} = b$, we obtain operators $G, \hat{G} \colon [\underline{x}, \overline{x}] \to [\underline{x}, \overline{x}]$ which satisfy the hypotheses of Proposition 1.1.1. Let u_* be the least solution of (4.1.1), and $\hat{u}$ the least solution of the BVP

$$\begin{cases} \frac{d}{dt}(\mu(t)\hat{u}'(t)) = \lambda\,\hat{g}(t, \hat{u}, \hat{u}(t), \hat{u}'(t)) \quad \text{a.e. in } J, \\ a_0\hat{u}(t_0) - b_0\hat{u}'(t_0) = \hat{c}_0, \quad a_1\hat{u}(t_1) + b_1\hat{u}'(t_1) = \hat{c}_1. \end{cases}$$

The proof of Theorem 4.1.1 implies that $u_* = Gu_*$ and $\hat{u} = \hat{G}\hat{u}$, so that $G\hat{u} \leq \hat{G}\hat{u} = \hat{u}$. Because $u_* = \min\{u \in [\underline{x}, \overline{x}] \mid Gu \leq u\}$ by Proposition 1.1.1, it follows that $u_* \leq \hat{u}$. This proves the assertion for least solutions. The proof for greatest solutions is similar.   □

As a consequence of Theorem 4.1.1, Proposition 4.1.2, and Proposition 4.1.3 we obtain the following result.

Proposition 4.1.4. *Assume that* $q\colon J \times \mathbb{R} \times \mathbb{R} \to \mathbb{R}$ *has the following properties.*

(qa) $q(t, \cdot, z)$ *is continuous and decreasing for a.e.* $t \in J$ *and for all* $z \in \mathbb{R}$, *and there exists a* $p \in L^1_+(J)$ *and an increasing function* $\phi\colon \mathbb{R}_+ \to \mathbb{R}_+$ *with* $\int_{0+}^{1} \frac{dz}{\phi(z)} = \infty$ *such that* $|q(t, x, z) - q(t, x, y)| \leq p(t)\phi(|z - y|)$ *for a.e.* $t \in J$ *and all* $x, y, z \in \mathbb{R}$.

(qb) $|q(t, x, y)| \leq p_1(t)\max\{|x|, |y|\} + m(t)$ *for all* $x, y \in \mathbb{R}$ *and for a.e.* $t \in J$, *where* $p_1, m \in L^1_+(J)$, *and* $\|A^n\|^{\frac{1}{n}} < 1$ *for some* $n = 1, 2, \dots$, *with* $A\colon C(J) \to C(J)$ *given by (4.1.9).*

If $h\colon J \times \mathbb{R} \to \mathbb{R}$ *is* L^1*-bounded and sup-measurable, and if* $h(t, \cdot)$ *is increasing for a.e.* $t \in J$, *then the BVP*

$$\begin{cases} -\frac{d}{dt}(\mu(t)u'(t)) = q(t, u(t), u'(t)) + h(t, u(t)) \quad \text{for a.e. } t \in J, \\ a_0\,u(t_0) - b_0\,u'(t_0) = c_0, \quad a_1\,u(t_1) + b_1\,u'(t_1) = c_1, \end{cases} \tag{4.1.26}$$

has extremal solutions which are increasing with respect to q, h, c_0, *and* c_1. *Moreover, if* h *is constant with respect to its second argument, then the solution of the BVP (4.1.26) is uniquely determined.*

Proof. The function $g(t, u, x, y) = q(t, x, y) + h(t, u(t))$ satisfies the hypotheses of Theorem 4.1.1.   □

The following result is a consequence of Proposition 4.1.4.

Corollary 4.1.1. *Assume that $f: \mathbb{R} \to \mathbb{R}$ is of bounded variation, and that $f(x-) \leq f(x) \leq f(x+)$ for each $x \in \mathbb{R}$. Then for each $h_0 \in L^1(J)$ the BVP*

$$\begin{cases} -\frac{d}{dt}(\mu(t)u'(t)) = f(u(t)) + h_0(t) & \text{for a.e. } t \in J, \\ a_0 \, u(t_0) - b_0 \, u'(t_0) = c_0, \quad a_1 \, u(t_1) + b_1 \, u'(t_1) = c_1, \end{cases} \qquad (4.1.27)$$

has extremal solutions which are increasing with respect to h_0, c_0, and c_1.

Proof. The hypotheses given for f imply that it can be represented as a sum $f = f_1 + f_2$ where f_1 is increasing and f_2 is decreasing and continuous. Denoting

$$q(t, x, y) = f_2(x), \quad h(t, x) = h_0(t) + f_1(x), \quad t \in J, \ x, \ y \in \mathbb{R},$$

we get functions $q: J \times \mathbb{R} \times \mathbb{R} \to \mathbb{R}$ and $h: J \times \mathbb{R} \to \mathbb{R}$ which satisfy the hypotheses of Proposition 4.1.4 with $p_1(t) = p(t) \equiv 0$.   □

As a special case of Proposition 4.1.4 we obtain the following existence, uniqueness, and comparison results for a linear BVP.

Corollary 4.1.2. *Assume that $p_1, h \in L^1(J)$, and that $\|A^n\|^{\frac{1}{n}} < 1$ for some $n \geq 1$, where $A: C(J) \to C(J)$ is defined by (4.1.9). Then the BVP*

$$\begin{cases} -\frac{d}{dt}(\mu(t)u'(t)) = p_1(t)u(t) + h(t) & \text{for a.e. } t \in J, \\ a_0 \, u(t_0) - b_0 \, u'(t_0) = c_0, \quad a_1 \, u(t_1) + b_1 \, u'(t_1) = c_1, \end{cases} \qquad (4.1.28)$$

has a solution. If $p_1(t) \leq 0$ for a.e. $t \in J$, then (4.1.28) has a unique solution which is increasing with respect to h, c_0, and c_1.

Example 4.1.1. Define $q: J \times \mathbb{R} \times \mathbb{R} \to \mathbb{R}$ and $h: J \times \mathbb{R} \times \mathbb{R} \to \mathbb{R}$ by

$$q(t, x, y) = \frac{\cos(y)}{1 + e^x} \sum_{j=1}^{\infty} \sum_{k=1}^{\infty} \frac{2 + \sin((1 + [k^{\frac{1}{j}}t] - k^{\frac{1}{j}}t)^{-1})}{(kj)^2}, \quad t \in J, \ x, y \in \mathbb{R},$$

$$h(t, x) = \arctan\left(\sum_{j=1}^{\infty} \sum_{k=1}^{\infty} \frac{[2 + [k^{\frac{1}{j}}t] - k^{\frac{1}{j}}t] + [k^{\frac{1}{j}}x]}{(kj)^2}\right), \quad t \in J, \ x \in \mathbb{R},$$

where $[z]$ denotes the greatest integer $\leq z$. It is easy to see that the functions q and h satisfy the hypotheses of Proposition 4.1.4. Thus the BVP (4.1.26) has for each choices of λ, a_j, $b_j \in \mathbb{R}_+$, $j = 0, 1$, $a_0 a_1 + a_0 b_1 + a_1 b_0 > 0$, and c_0, $c_1 \in \mathbb{R}$ extremal solutions which are increasing with respect to q, h, c_0, and c_1.

4.2 Implicit Sturm-Liouville boundary value problems

In this section we consider the following implicit boundary value problem

$$\begin{cases} L_1 u(t) = f(t, u, L_1 u(t)) \quad \text{for a.e. } t \in J, \\ L_i u = C_i(L_2 u, L_3 u, u), \quad i = 2, 3, \end{cases} \tag{4.2.1}$$

where

$$\begin{cases} L_1 u(t) = -\frac{d}{dt}(\mu(t) u'(t)) - \lambda\, g(t, u, u(t), u'(t)), \quad t \in J, \\ L_2 u = a_0 u(t_0) - b_0 u'(t_0), \; L_3 u = a_1 u(t_1) + b_1 u'(t_1), \\ a_i, b_i \in \mathbb{R}_+, \; i = 0, 1, \; \text{and} \; a_0 a_1 + a_0 b_1 + a_1 b_0 > 0. \end{cases} \tag{4.2.2}$$

We derive existence and comparison results for extremal solutions of (4.2.1) by assuming that $\mu \in C(J, (0, \infty))$, that $g \colon J \times C(J) \times \mathbb{R} \times \mathbb{R} \to \mathbb{R}$ has properties (g0)–(g3), that $C(J)$ is ordered pointwise, and that the functions $f \colon J \times C(J) \times \mathbb{R} \to \mathbb{R}$ and $C_i \colon \mathbb{R} \times \mathbb{R} \times C(J) \to \mathbb{R}$, $i = 2, 3$, satisfy the following hypotheses.

(f0) $t \mapsto f(t, u, v(t))$ is measurable for all $u \in C(J)$ and $v \in L^1(J)$, and $f(t, u, y)$ is increasing in u and in y for a.e. $t \in J$.

(f1) $|f(t, u, y)| \le M(t) + \gamma(t)|y|$ for a.e. $t \in J$ and for all $u \in C(J)$ and $y \in \mathbb{R}$, where $M \in L^1_+(J)$, $\gamma \colon J \to [0, 1)$ and $\frac{1}{1-\gamma} \in L^\infty(J)$.

(C0) The functions C_2 and C_3 are increasing in all their variables.

(C1) $|C_i(x_2, x_3, u)| \le c_i |x_i| + d_i$ for all $x_2, x_3 \in \mathbb{R}$ and $u \in C(J)$, where $d_i \ge 0$, and $c_i \in [0, 1)$, $i = 2, 3$.

Throughout this section we also assume that $\lambda \in [0, \lambda_1)$, where λ_1 is the least positive eigenvalue of the operator $A \colon C(J) \to C(J)$, defined by (4.1.9).

4.2.1 Preliminaries. To reduce problem (4.2.1) to an operator equation of the form $Lu = Nu$ we denote

$$X = L^1(J) \times \mathbb{R} \times \mathbb{R} \quad \text{and} \quad Y = \{u \in C^1(J) \mid \mu\, u' \in AC(J)\}. \tag{4.2.3}$$

Define a partial ordering and a norm on X by

$$(h_0, x_0, y_0) \le (h_1, x_1, y_1) \quad \text{iff} \quad h_0 \le h_1, \; x_0 \le x_1 \text{ and } y_0 \le y_1, \tag{4.2.4}$$

and

$$\|(h, x, y)\| = \|h\|_1 + |x| + |y|. \tag{4.2.5}$$

Lemma 4.2.1. *Let L_i, $i = 1, 2, 3$, be given by (4.2.2). Denoting*

$$N_1 u = f(\cdot, u, L_1 u(\cdot)), \quad N_i u = C_i(L_2 u, L_3 u, u), \quad i = 2, 3, \qquad (4.2.6)$$

we get mappings $L = (L_1, L_2, L_3)$ and $N = (N_1, N_2, N_3)$ from Y to X. Moreover, $u \in Y$ is a solution of the BVP (4.2.1) if and only if it satisfies the operator equation $Lu = Nu$.

Proof. The assertions are obvious consequences of the given hypotheses and equations (4.2.1), (4.2.2), and (4.2.6). □

As an immediate consequence of Theorem 4.1.1 and Proposition 4.1.3 we obtain the following result.

Lemma 4.2.2. *The BVP*

$$\begin{cases} -\frac{d}{dt}(\mu(t) u'(t)) = \lambda\, g(t, u, u(t), u'(t)) + h_0(t) & \text{for a.e. } t \in J, \\ L_i u = c_i, \quad i = 2, 3, \end{cases} \qquad (4.2.7)$$

has for all fixed $h_0 \in L^1(J)$ and $c_2, c_3 \in \mathbb{R}$ extremal solutions, and they are increasing with respect to h_0, c_2, and c_3.

Next we define a subset V of Y which contains all the solutions of (4.2.1).

Lemma 4.2.3. *Define*

$$V = \{u \in Y \mid h_- \le Lu \le h_+\}, \quad h_\pm = \left(\frac{\pm M(\cdot)}{1 - \gamma(\cdot)}, \frac{\pm d_2}{1 - c_2}, \frac{\pm d_3}{1 - c_3} \right). \qquad (4.2.8)$$

If $u \in Y$ is a solution of the BVP (4.2.1), then $u \in V$.

Proof. Let $u \in Y$ be a solution of (4.2.1). Applying (f1) and (C1) we get

$$\begin{cases} |L_1 u(t))| = |f(t, u, L_1 u(t))| \le M(t) + \gamma(t)\,|L_1 u(t))| & \text{for a.e. } t \in J, \\ |L_i u| = |C_i(L_2 u, L_3 u, u)| \le c_i |L_i u| + d_i, \quad i = 2, 3, \end{cases}$$

so that

$$\begin{cases} |L_1 u(t)| \le \frac{M(t)}{1 - \gamma(t)} & \text{for a.e. } t \in J, \\ |L_i u| \le \frac{d_i}{1 - c_i}, \quad i = 2, 3. \end{cases}$$

These inequalities, (4.2.8), and (4.2.4) imply that $u \in V$. □

In the rest of this section we assume that V is equipped with the pointwise ordering. In the next Lemma we derive some properties to operators L and N which are needed in the proof of our existence and comparison results.

Lemma 4.2.4. *Let L, N, V, and $h_{\pm}$ be defined by (4.2.6) and (4.2.8).*
a) If $u \in V$, then $h_- \leq Nu \leq h_+$.
b) If u, $v \in V$, $u \leq v$ and $Lu \leq Lv$, then $Nu \leq Nv$.
c) monotone sequences of $N[V]$ converge in X with respect to the norm defined by (4.2.5).

Proof. a) Let $u \in V$ be given. Applying properties (f1) and (C1) and definitions (4.2.4), (4.2.6), and (4.2.8) we get

$$|N_1 u(t))| = |f(t, u, L_1 u(t))| \leq M(t) + \gamma(t) |L_1 u(t))|$$

$$\leq M(t) + \gamma(t) \frac{M(t)}{1 - \gamma(t)} = \frac{M(t)}{1 - \gamma(t)}$$

for a.e. $t \in J$, and

$$|N_i u| = |C_i(L_2 u, L_3 u, u)| \leq c_i |L_i u| + d_i \leq \frac{c_i d_i}{1 - c_i} + d_i = \frac{d_i}{1 - c_i},$$

for $i = 2, 3$. These inequalities and the definitions (4.2.4) and (4.2.8) imply that $h_- \leq Nu \leq h_+$.

b) Let u, $v \in V$ satisfy $u \leq v$ and $Lu \leq Lv$. These inequalities, (4.2.4), (4.2.6), and the hypotheses (f0) and (C0) imply that

$$\begin{cases} N_1 u(t) = f(t, u, L_1 u(t)) \leq f(t, v, L_1 v(t)) = N_1 v(t) \quad \text{a.e. in } J, \\ N_2 u = C_2(L_2 u, L_3 u, u) \leq C_2(L_2 v, L_3 v, v) = N_2 v, \\ N_3 u = C_3(L_2 u, L_3 u, u) \leq C_3(L_2 v, L_3 v, v) = N_3 v, \end{cases}$$

or equivalently, that $Nu \leq Nv$.

c) Assume that $(Nu_n)_{n=0}^{\infty}$ is a monotone sequence in $N[V]$. By the proof of a) we have

$$|N_1 u_n(t)| \leq \frac{M(t)}{1 - \gamma(t)} \quad \text{for a.e. } t \in J.$$

Thus by the monotone convergence theorem there is a function $h_0 \in L^1(J)$ such that

$$\|N_1 u_n - h_0\|_1 \to 0 \quad \text{as } n \to \infty.$$

Since the sequences $(N_i u_n)_{n=0}^{\infty}$, $i = 2, 3$, are bounded by the proof of a) and monotone, there exist $x_0, y_0 \in \mathbb{R}$ such that

$$|N_2 u_n - x_0| \to 0, \ |N_3 u_n - y_0| \to 0 \quad \text{as } n \to \infty.$$

These limes relations and (4.2.5) imply that

$$\|Nu_n - (h_0, x_0, y_0)\| \to 0 \quad \text{as } n \to \infty.$$

This proves the conclusion c). $\square$

4.2.2   Existence and comparison results. The results proved in the above subsection and Theorem 1.1.2 will now be applied to prove the following existence and comparison theorem for the BVP (4.2.1).

Theorem 4.2.1. *Assume that the hypotheses (f0), (f1), (g0)–(g3), (C0), and (C1) hold, and that $\lambda \in [0, \lambda_1)$, where λ_1 is the least eigenvalue of the operator $A \colon C(J) \to C(J)$, defined by (4.1.9). Then the BVP (4.2.1) has extremal solutions u_* and u^* in Y in the sense that if $u \in Y$ is a solution of (4.2.1), then $u_*(t) \le u(t) \le u^*(t)$ in J. Moreover, u_* and u^* are increasing with respect to f, C_2, and C_3.*

Proof. Let the space X, the operators L and N, the subset V of Y, and the elements $h_{\pm}$ of X be defined by (4.2.2), (4.2.3), (4.2.6), and (4.2.8), respectively. Lemma 4.2.2 and Lemma 4.2.4 imply that when the norm and the partial ordering of X are defined by (4.2.4) and (4.2.5), and when V is ordered pointwise, then the following properties hold.

   (I) If $u, v \in V$, $u \le v$ and $Lu \le Lv$, then $h_- \le Nu \le Nv \le h_+$.
  (II) Equation $Lu = h$ has for each $h \in [h_-, h_+]$ extremal solutions, and they are increasing with respect to h.
 (III) X is an ordered normed space and monotone sequences of $N[V]$ converge.

It then follows from Theorem 1.1.2 that equation $Lu = Nu$ has least and greatest solutions u_* and u^* in the set V. This and Lemma 4.2.1 imply that u_* and u^* are also extremal solutions of the BVP (4.2.1) in V. Since all the solutions of (4.2.1) belong to V by Lemma 4.2.3, then u_* and u^* are least and greatest ones of all the solutions of (4.2.1). Moreover, u_* and u^* are increasing with respect to N by Theorem 1.1.2. This property and the definition (4.2.6) of N imply that u_* and u^* are increasing with respect to f, C_2, and C_3. $\square$

The hypotheses of Theorem 4.2.1 can be relaxed as follows.

Proposition 4.2.1. *The results of Theorem 4.2.1 hold if (f0) and (C0) are replaced by the following hypotheses.*

   (f2) $f(\cdot, u, v(\cdot))$ *is measurable for all $u \in C(J)$ and $v \in L^1(J)$, and there exists an $\alpha \in L_+^{\infty}(J)$ such that $f(t, u, y) + \alpha(t)y$ is increasing in u and y for a.e. $t \in J$.*
   (C2) *There exist β_2, $\beta_3 \ge 0$ such that $C_i(x_2, x_3, u) + \beta_i x_i$ is increasing in x_2, x_3 and u for $i = 2, 3$.*

Proof. It is easy to see that problems (4.2.1) and

$$\begin{cases} L_1 u(t) = \hat{f}(t, u, L_1 u(t)) & \text{for a.e. } t \in J, \\ L_i u = \hat{C}_i(L_2 u, L_3 u, u), & i = 2, 3, \end{cases} \qquad (4.2.9)$$

where $\hat{f}: J \times C(J) \times \mathbb{R} \to \mathbb{R}$ and $\hat{C}_i: \mathbb{R} \times \mathbb{R} \times C(J) \to \mathbb{R}$ are defined by

$$\begin{cases} \hat{f}(t, u, y) = \frac{f(t,u,y)+\alpha(t)y}{1+\alpha(t)}, & t \in J, \, u \in C(J), \, y \in \mathbb{R}, \\ \hat{C}_i(x_2, x_3, u) = \frac{C_i(x_2,x_3,u)+\beta_i x_i}{1+\beta_i}, & u \in C(J), \quad i = 2, 3, \end{cases} \qquad (4.2.10)$$

have the same solutions. Moreover, the functions $\hat{f}$ and $\hat{C}_i$ satisfy the hypotheses (f0), (f1), (C0), and (C1) with γ replaced by $\frac{\gamma+\alpha}{1+\alpha}$ and c_i by $\frac{c_i+\beta_i}{1+\beta_i}$. Thus the BVP (4.2.9), with $\hat{f}$ and $\hat{C}_i$ defined by (4.2.10), has by Theorem 4.2.1 extremal solutions u_* and u^*, and they are increasing with respect to $\hat{f}$ and $\hat{C}_i$. In view of (4.2.10), u_* and u^* are then extremal solutions of (4.2.1), and they are increasing with respect to f, C_2, and C_3. $\square$

When the last two variables of f are dropped we get, as a consequence of Theorem 4.2.1, the following result concerning explicit BVP

$$\begin{cases} -\frac{d}{dt}(\mu(t)u'(t)) = \lambda \, g(t, u, u(t), u'(t)) + f(t) & \text{for a.e. } t \in J, \\ L_i u = C_i(L_2 u, L_3 u, u), \quad i = 2, 3. \end{cases} \qquad (4.2.11)$$

Proposition 4.2.2. *The BVP (4.2.11) has for each $f \in L^1(J)$ extremal solutions, and they are increasing with respect to f, C_2, and C_3, if g has properties (g0)–(g3).*

Example 4.2.1. Choose $J = [0, 1]$, and define

$$f(t, u, y) = \sum_{n=1}^{\infty} \frac{\arctan([n(u(1-t) + y - t)])}{n^2}, \quad t \in J, \, u \in C(J), \, y \in \mathbb{R},$$

$$C_i(x, y, u) = i + \sum_{n=1}^{\infty} \frac{\arctan([n(x + y + \int_0^1 u(t)dt)])}{n^2}, \quad u \in C(J), \, x, y \in \mathbb{R}.$$

It is easy to see that f, C_2, and C_3 have properties (f0), (f1), (C0), and (C1). Thus the BVP (4.1.1), with these f, C_2, and C_3, and with $g = q + h$ as in Example 4.1.1, has by Theorem 4.2.1 extremal solutions, and they are increasing with respect to f, C_2, and C_3.

Example 4.2.2. Consider the BVP

$$
\begin{cases}
L_1 u(t) = \frac{q_0(t)[u(t)]}{1+|[u(t)]|} + h_0(t) + \frac{[L_1 u(t)]}{2} & \text{a.e. in } J = [0,1], \\
u(0) = 1 + \frac{1}{2}[u(0)] + \frac{[u(1)]}{1+|[u(1)]|}, \\
u(1) = 1 + \frac{1}{2}[u(1)] + \frac{2}{\pi} \arctan(|[u(0)]|),
\end{cases}
\tag{4.2.12}
$$

where

$$
L_1 u(t) = -u''(t) + p_0(t)u(t) - p_2(t)u'(t) + p_3(t), \quad t \in J.
$$

Since $a_0 = a_1 = 1$ and $b_0 = b_1 = 0$, then

$$
\ell(s,t) = \begin{cases}
s, & 0 \le s \le t \le 1, \\
1-s, & 0 \le t \le s \le 1.
\end{cases}
$$

By assuming that $q_0, h_0, p_3 \in L^1(J)$, that $p_0 \in L_+^\infty(J)$, $p_2 \in L^\infty(J)$, and denoting $p_1(t) \equiv 1$ and $a = \|p_0\|_\infty + \|p_2\|_\infty$, it follows from Theorem 4.2.1 that the BVP (4.2.12) has extremal solutions if $a < \lambda_1$, where λ_1 is the least eigenvalue of the operator A, defined by

$$
Ax(t) = \int_0^t sx(s)\,ds + \int_t^1 (1-s)x(s)\,ds. \tag{4.2.13}
$$

To determine λ_1, consider equation $x = \lambda Ax$, i.e.,

$$
x(t) = \lambda \int_0^t sx(s)\,ds + \lambda \int_t^1 (1-s)x(s)\,ds. \tag{4.2.14}
$$

This integral equation can be converted to the initial value problem

$$
\begin{cases}
x'(t) = \lambda(2t-1)x(t), \quad t \in J, \\
x(0) = \lambda \int_0^1 (1-s)x(s)\,ds.
\end{cases}
\tag{4.2.15}
$$

The general solution of the differential equation of (4.2.15) is

$$
x(t) = Ce^{\lambda(t^2-t)}.
$$

This and the initial condition of (4.2.15) imply that

$$
x(0) = C = \lambda C \int_0^1 (1-s)e^{\lambda(s^2-s)}\,ds = \frac{\lambda C}{2} \int_0^1 e^{\lambda(s^2-s)}\,ds.
$$

This equation has a nonzero solution C, and hence equation (4.2.14) has a nonzero solution if and only if $\lambda = \lambda_1$, where

$$\lambda_1 \int_0^1 e^{\lambda_1(s^2 - s)} ds = 2, \quad \text{i.e.} \quad \lambda_1 \approx 3.416130626. \tag{4.2.16}$$

Thus problem (4.2.12) has extremal solutions if

$$\|p_0\|_\infty + \|p_2\|_\infty < 3.416130626.$$

In the special case when $\|p_i\|_\infty \leq 1$, $i = 1, 2, 3$, $\|q_0\|_\infty \leq 1$ and $\|h_0\| \leq 1$, the hypotheses (f1) and (C1) hold with $M(t) \equiv 2$, $\gamma(t) \equiv \frac{1}{2}$, $c_i = \frac{1}{2}$, $d_i = 2$. Thus $h_\pm = (\pm 4, \pm 4, \pm 4)$. Thus it follows from Lemma 4.2.3 and Remark 4.1.2 that all the solutions of (4.2.12) belong to the set $B = \{u \in C^1(J) \mid \max\{|u(t)|, |u'(t)|\} \leq b(t)\}$, $t \in J$, where b is the solution of the integral equation

$$b(t) = 8 + \int_0^t s(2b(s) + 4) \, ds + \int_t^1 (1 - s)(2b(s) + 4) \, ds, \quad t \in J. \tag{4.2.17}$$

This integral equation can be converted to the initial value problem

$$\begin{cases} b'(t) = (4t - 2)b(t) + 8t - 4, & t \in J, \\ b(0) = 8 + \int_0^1 (1 - s)(2b(s) + 4) \, ds. \end{cases} \tag{4.2.18}$$

The general solution of the differential equation of (4.2.18) is

$$b(t) = Ce^{2(t^2 - t)} - 2.$$

This and the initial condition of (4.2.18) implies that

$$b(0) = C - 2 = 8 + 2C \int_0^1 (1 - s)e^{2(s^2 - s)} ds = 8 + C \int_0^1 e^{2(s^2 - s)} ds.$$

Thus

$$b(t) = Ce^{2(t^2 - t)} - 2, \quad C = 10\left(1 - \int_0^1 e^{2(s^2 - s)} ds\right)^{-1} \approx 36.33436526.$$

4.3 Convergence of successive approximations

In this section we present conditions which ensure that the least or the greatest solution of an explicit or an implicit Sturm-Liouville boundary value problem can be obtained by the method of successive approximations.

4.3.1 Successive approximations for an explicit BVP. In this subsection we shall prove that by adding one-sided continuity of the function $g(t, \cdot, x, z)$ to the hypotheses of Theorem 4.1.1 we get extremal solutions of the BVP

$$\begin{cases} -\frac{d}{dt}(\mu(t)u'(t)) = \lambda g(t, u, u(t), u'(t)) \text{ a.e. in } J = [t_0, t_1], \\ a_0 u(t_0) - b_0 u'(t_0) = c_0, \quad a_1 u(t_1) + b_1 u'(t_1) = c_1, \end{cases} \quad (4.3.1)$$

by the method of successive approximations. As in section 4.1 we assume that λ, a_j, $b_j \in \mathbb{R}_+$, $a_0 a_1 + a_0 b_1 + a_1 b_0 > 0$, and $c_j \in \mathbb{R}$, $j = 0, 1$, and denote

$$y_0(t) = \int_{t_0}^t \frac{a_0}{\mu(s)} ds + \frac{b_0}{\mu(t_0)}, \quad y_1(t) = \int_t^{t_1} \frac{a_1}{\mu(s)} ds + \frac{b_1}{\mu(t_1)},$$

$$D = \int_{t_0}^{t_1} \frac{a_0 a_1}{\mu(s)} ds + \frac{a_0 b_1}{\mu(t_1)} + \frac{a_1 b_0}{\mu(t_0)}, \quad k(t, s) = \begin{cases} \frac{y_1(t) y_0(s)}{D}, & t_0 \le s \le t, \\ \frac{y_0(t) y_1(s)}{D}, & t \le s \le t_1. \end{cases}$$

If a sequence $(v_n)_{n=1}^\infty$ of $C(J)$ converges uniformly to v, denote $v_n \nearrow v$ if $(v_n)_{n=1}^\infty$ is increasing and $v_n \searrow v$ if $(v_n)_{n=1}^\infty$ is decreasing.

Proposition 4.3.1. *Assume that $g: J \times C(J) \times \mathbb{R} \times \mathbb{R} \to \mathbb{R}$ has properties (g0)–(g3), and that $\lambda \in [0, \lambda_1)$, where λ_1 is the least eigenvalue of the operator A, defined by (4.1.9). Let $b \in C(J)$ and the subset B of $C^1(J)$ be defined by (4.1.10) and (4.1.11). Then for each $v_0 \in [-b, b]$, the relations*

$$v_{n+1}(t) = \frac{c_0 y_1(t) + c_1 y_0(t)}{D} + \lambda \int_{t_0}^{t_1} k(t, s) g(s, v_n, v_{n+1}(s), v_{n+1}'(s)) \, ds$$

$$(4.3.2)$$

define a sequence $(v_n)_{n=1}^\infty$ in B, which converges uniformly on J to the
a) least solution of the BVP (4.3.1) if $v_0 = -b$, and if

   (g4) $\sup\{|g(t, u_n, x, y) - g(t, u, x, y)| \mid x, y \in [-b(t), b(t)]\} \to 0$ as $u_n \nearrow u$ *in $[-b, b]$ for a.e. $t \in J$,*

b) greatest solution of the BVP (4.3.1) if $v_0 = b$, and if

   (g5) $\sup\{|g(t, u_n, x, y) - g(t, u, x, y)| \mid x, y \in [-b(t), b(t)]\} \to 0$ as $u_n \searrow u$ *in $[-b, b]$ for a.e. $t \in J$,*

Proof. By the proof of Theorem 4.1.1 the equation (4.1.19), which can be rewritten as

$$Gv(t) = \frac{c_0 y_1(t) + c_1 y_0(t)}{D} + \lambda \int_{t_0}^{t_1} k(t, s) g(s, v, Gv(s), (Gv)'(s)) \, ds, \quad (4.3.3)$$

defines an increasing mapping $G \colon [-b, b] \to B \subset [-b, b]$. Thus the sequence $(v_n)_{n=1}^{\infty}$, given by (4.3.2), is equal to the iteration sequence $(G^n v_0)_{n=1}^{\infty}$. As in the proof of Proposition 4.1.1 it can be shown that the sequences (v_n) and (v_n') are uniformly bounded and equicontinuous.

a) Choose $v_0 = -b$. Since G is increasing, then $(G^n v_0)_{n=1}^{\infty} = (v_n)_{n=1}^{\infty}$ is an increasing sequence in B. It then follows from Arzela-Ascoli theorem that $(v_n)_{n=1}^{\infty}$ converges uniformly on J to a function $v \in C(J)$, and that the sequence $(v_{n+1}')_{n=1}^{\infty}$ has a subsequence $(v_{n_k+1}')_{k=1}^{\infty}$ which converges uniformly on J to a function $w \in C(J)$. In particular, $w = v'$, and the hypotheses (g0) and (g4) imply that

$$\lim_{k \to \infty} g(s, v_{n_k}, v_{n_k+1}(s), v_{n_k+1}'(s)) = g(s, v, v(s), v'(s)) \quad \text{for a.e.} \quad s \in J.$$

Hence, replacing n by n_k in (4.3.2) and applying the dominated convergence theorem we obtain when $k \to \infty$,

$$v(t) = \frac{c_0 y_1(t) + c_1 y_0(t)}{D} + \lambda \int_{t_0}^{t_1} k(t, s) g(s, v, v(s), v'(s)) \, ds, \quad t \in J.$$

This equation implies by Lemma 4.1.2 that $u = v$ is a solution of the BVP (4.3.1), whence $v = Gv$ by (4.1.19).

The above proof shows that $v_n = G^n v_0 \to v$ in $C(J)$, equipped with the topology of uniform convergence, and that $v = Gv$. Thus v is by Proposition 1.1.3 the least fixed point of G. This implies by the proof of Theorem 4.1.1 that v is the least solution of the BVP (4.3.1).

The case b) can be proved similarly. $\qquad\qquad\qquad\qquad\qquad\qquad\qquad\square$

Example 4.3.1. Consider the BVP

$$\begin{cases} -u''(t) = \frac{[u(-t) - t^2]}{1 + |[u(-t) - t^2]|} & \text{a.e. in } J = (-1, 1), \\ u(-1) - 2u'(-1) = 1, \quad u(1) + 2u'(1) = 1, \end{cases} \qquad (4.3.4)$$

where $[z]$ means the greatest integer $\leq z$. Problem (4.3.4) is of the form (4.2.1), where

$$g(t, v, x, y) = \frac{[v(-t) - t^2]}{1 + |[v(-t) - t^2]|}, \quad \lambda = 1. \qquad (4.3.5)$$

The hypotheses of Theorem 4.1.1 are satisfied, whence problem (4.3.4) has extremal solutions. Moreover, since $[\cdot]$ is continuous from the right, the hypotheses of Proposition 4.3.1.b are satisfied. Thus the greatest solution of

(4.3.4) can be obtained by the method of successive approximations (4.3.2). In this case they can be rewritten as

$$v_{n+1}(t) = 1 + \frac{3-t}{6} \int_{-1}^{t} \frac{(3+s)[v_n(-s) - s^2]}{1 + |[v_n(-s) - s^2]|} \, ds$$
$$+ \frac{3+t}{6} \int_{t}^{1} \frac{(3-s)[v_n(-s) - s^2]}{1 + |[v_n(-s) - s^2]|} \, ds. \tag{4.3.6}$$

By choosing

$$v_0(t) = 1 + \frac{3-t}{6} \int_{-1}^{t} (3+s) \, ds + \frac{3+t}{6} \int_{t}^{1} (3-s) \, ds = \frac{7}{2} - \frac{t^2}{2},$$

and calculating approximations (4.3.6) by a numerical integration method, we get the following estimate for the greatest solution of (4.3.4).

$$u^*(t) \approx \begin{cases} -.25t^2 + .105t + 2.565, & -1 \leq t < -.632, \\ -.33333t^2 + 2.532, & -.632 \leq t \leq .632, \\ -.25t^2 - .106t + 2.565, & .632 < t \leq 1. \end{cases} \tag{4.3.7}$$

In view of this one can infer that the greatest solution of (4.3.4) is of the form

$$u(t) = \begin{cases} -\frac{1}{4}t^2 + at + b, & -1 \leq t < -x, \\ -\frac{1}{3}t^2 + c, & -x \leq t \leq x, \\ -\frac{1}{4}t^2 - at + b, & x < t \leq 1. \end{cases} \tag{4.3.8}$$

It remains to determine a, b, c, and x. Because u' is continuous at x, we get the equation

$$u'(x) = -\frac{2}{3}x = -\frac{1}{2}x - a \;\Rightarrow\; x = 6a.$$

The boundary conditions imply that $b = \frac{9}{4} + 3a$. At a point $t = x$ the second derivative of u has a jump, and the approximation (4.3.7) implies that it is caused by the jump of $[u(-t) - t^2]$, from the value 2 to the value 1 at x, whence $u(-x) - x^2 = 2$. This and (4.3.8) yield $-\frac{4}{3}x^2 + c = 2 = -\frac{5}{4}x^2 - ax + b$. Recalling that $x = 6a$ and $b = \frac{9}{4} + 3a$ we see that $c = 2 + 48a^2$, and that a is the positive solution of equation $204a^2 + 12a + 1 = 0$. This and the above formulae for b, c, and x yield

$$\begin{cases} a = \frac{3+2\sqrt{15}}{102} \approx .1053526146, \\ b = \frac{9}{4} + 3a = \frac{159}{68} + \frac{\sqrt{15}}{17} \approx 2.566057844, \\ c = 2 + 48a^2 = \frac{670}{289} + \frac{16\sqrt{15}}{289} \approx 2.532760323, \\ x = 6a = \frac{3+2\sqrt{15}}{17} \approx .6321156876. \end{cases} \tag{4.3.9}$$

In this case also the least solution of (4.3.4) can be obtained in a similar way when $u_0 = -b$. A numerical integration method gives the following estimate for the least solution of (4.3.4).

$$u(t) \approx .25t^2 - .252, \quad -1 \le t \le 1.$$

In view of this one can infer that the least solution of (4.3.4) is

$$u(t) = \frac{1}{4}t^2 - \frac{1}{4}, \quad -1 \le t \le 1. \tag{4.3.10}$$

This shows that the one-sided continuity condition assumed in Proposition 4.3.1.a is not necessary.

The function $u(t) \equiv 1$ is also a solution of (4.3.4).

4.3.2 Successive approximations for an implicit BVP.

Consider next the following special case of the BVP (4.2.1):

$$\begin{cases} -\frac{d}{dt}(\mu(t)u'(t)) = q(t, u(t), u'(t)) \\ +f(t, u(t), -\frac{d}{dt}(\mu(t)u'(t)) - q(t, u(t), u'(t))) \quad \text{a.e. in } J, \\ L_i u = C_i(L_2 u, L_3 u, u), \quad i = 2, 3, \\ L_2 u = a_0 u(t_0) - b_0 u'(t_0), \ L_3 u = a_1 u(t_1) + b_1 u'(t_1), \end{cases} \tag{4.3.11}$$

where

$$\mu \in C(J, (0, \infty)), \ a_i, b_i \in \mathbb{R}_+, \ i = 0, 1, \text{ and } a_0 a_1 + a_0 b_1 + a_1 b_0 > 0.$$

Assume that the functions $q \colon J \times \mathbb{R} \times \mathbb{R} \to \mathbb{R}$, $f \colon J \times \mathbb{R} \times \mathbb{R} \to \mathbb{R}$ and $C_i \colon \mathbb{R} \times \mathbb{R} \times C(J) \to \mathbb{R}$, $i = 2, 3$, satisfy the following hypotheses when $C(J)$ is ordered pointwise.

(qa) $q(t, \cdot, y)$ is continuous and decreasing for a.e. $t \in J$ and for all $y \in \mathbb{R}$, and there exist a $p \in L^1_+(J)$ and an increasing function $\phi \colon \mathbb{R}_+ \to \mathbb{R}_+$ with $\int_{0+}^{1} \frac{dz}{\phi(z)} = \infty$ such that $|q(t, x, y) - q(t, x, z)| \le p(t)\phi(|y - z|)$ for a.e. $t \in J$ and all $x, y, z \in \mathbb{R}$.

(qb) $|q(t, x, y)| \le p_1(t) \max\{|x|, |y|\} + m(t)$ for all $x, y \in \mathbb{R}$ and for a.e. $t \in J$, where $p_1, m \in L^1_+(J)$, and $\|A^n\|^{\frac{1}{n}} < 1$ for some $n = 1, 2, \ldots$, with $A \colon C(J) \to C(J)$ given by (4.1.9).

(fa) f is sup-measurable, and $f(t, x, y)$ is increasing in x and in y for a.e. $t \in J$.

(fb) $|f(t, x, y)| \le M(t) + \gamma(t)|y|$ for a.e. $t \in J$ and for all $x, y \in \mathbb{R}$, where $M \in L^1_+(J)$, $\gamma \colon J \to [0, 1)$ and $\frac{1}{1-\gamma} \in L^\infty(J)$.

(C0) The functions C_2 and C_3 are increasing in all their variables.

(C1) $|C_i(x_2, x_3, u)| \le c_i|x_i| + d_i$ for all $x_2, x_3 \in \mathbb{R}$ and $u \in C(J)$, where $d_i \ge 0$, and $c_i \in [0, 1)$, $i = 2, 3$.

We are going to show that under these hypotheses the BVP (4.3.11) has extremal solutions. Moreover, under extra one-sided continuity hypotheses for the functions C_i, $i = 2, 3$, in all their arguments and for the function f in its last two arguments, we show that these extremal solutions can be obtained by the method of successive approximations.

To prove this we first reduce (4.3.11) to an equation $Lu = Nu$. Let L_2 and L_3 be as in (4.3.11). Denoting

$$\begin{cases} X = L^1(J) \times \mathbb{R} \times \mathbb{R}, \quad Y = \{u \in C^1(J) \mid \mu\, u' \in AC(J)\}, \\ L_1 u = t \mapsto -\frac{d}{dt}(\mu(t)u'(t)) - q(t, u(t), u'(t)), \\ N_1 u = f(\cdot, u(\cdot), L_1 u(\cdot)), \quad N_i u = C_i(L_2 u, L_3 u, u), \; i = 2, 3, \end{cases} \quad (4.3.12)$$

we get mappings $L = (L_1, L_2, L_3)$ and $N = (N_1, N_2, N_3)$ from Y to X. The given assumptions ensure that the following result holds.

Lemma 4.3.1. *$u \in Y$ is a solution of the BVP (4.3.11) if and only if $Lu = Nu$.*

Assuming that X is normed and ordered by (4.2.4) and (4.2.5), and that Y is ordered pointwise, the results of subsection 4.2 imply the following Lemma.

Lemma 4.3.2. *Assume that the hypotheses (qa), (qb), (fa), (fb), (C0), and (C1) are valid. Then the BVP (4.3.11) has extremal solutions. Moreover, denoting*

$$V = \{u \in Y \mid h_- \leq Lu \leq h_+\}, \; h_{\pm} = \left(\frac{\pm M(\cdot)}{1 - \gamma(\cdot)}, \frac{\pm d_2}{1 - c_2}, \frac{\pm d_3}{1 - c_3}\right), \quad (4.3.13)$$

the following properties hold.

(III) *Monotone sequences of $N[V]$ converge in X*
(IV) *If $u, v \in V$, and $Lu \leq Lv$, then $u \leq v$, and $h_- \leq Nu \leq Nv \leq h_+$.*

Proof. Properties (qa) and (qb) imply that the hypotheses (g0)–(g3) hold when

$$g(t, u, x, y) = q(t, x, y), \quad t \in J, \; x, y \in \mathbb{R}, \; u \in C(J),$$

and that $r(A)^{-1} > 1$. It then follows from the proof of Theorem 4.2.1 that the operators L and N given by (4.3.12) satisfy the hypotheses (I), (II), and (III) of Theorem 1.1.2, and that (4.3.11) has extremal solutions. In view of Proposition 4.1.2 we have $u \leq v$ whenever $u, v \in V$ and $Lu \leq Lv$. This property and (I) imply that (IV) is valid. □

Proposition 4.3.2. *Assume that* $q\colon J \times \mathbb{R} \times \mathbb{R} \to \mathbb{R}$, $f\colon J \times \mathbb{R} \times \mathbb{R} \to \mathbb{R}$ *and* $C_i\colon \mathbb{R} \times \mathbb{R} \times C(J) \to \mathbb{R}$, $i = 0, 1$, *have properties (qa), (qb), (fa), (fb), (C0), and (C1). Then there exist* $u_\pm \in Y$ *such that the sequence* $(u_n)_{n=0}^\infty$, *defined on* J *by*

$$u_{n+1}(t) = \begin{cases} \dfrac{C_2(L_2 u_n, L_3 u_n, u_n)y_1(t) + C_3(L_2 u_n, L_3 u_n, u_n)y_0(t)}{D} \\[2mm] + \int_{t_0}^{t_1} k(t,s)(q(s, u_{n+1}(s), u'_{n+1}(s)) \\[2mm] + f(s, u_n(s), -\frac{d}{ds}(\mu(s)u'_n(s)) - q(s, u_n(s), u'_n(s)))) \, ds, \end{cases}$$

$$(4.3.14)$$

where k, D, *and the functions* y_0 *and* y_1 *are given by (4.1.4), converge uniformly on* J *to the*
a) *least solution* u_* *of the BVP (4.3.11) if* $u_0 = u_-$, *and if*
$C_i(x_n, y_n, v_n) \to C_i(x, y, v)$ *and* $f(t, x_n, y_n) \to f(t, x, y)$ *for a.e.* $t \in J$
whenever $x_n \nearrow x$ *and* $y_n \nearrow y$ *in* $\mathbb{R}$, *and* $v_n \nearrow v \in C(J)$ *uniformly on* J;
b) *greatest solution* u^* *of the BVP (4.3.11) if* $u_0 = u_+$, *and if*
$C_i(x_n, y_n, v_n) \to C_i(x, y, v)$ *and* $f(t, x_n, y_n) \to f(t, x, y)$ *for a.e.* $t \in J$
whenever $x_n \searrow x$ *and* $y_n \searrow y$ *in* $\mathbb{R}$, *and* $v_n \searrow v \in C(J)$ *uniformly on* J.

Proof. In view of Lemma 4.1.2 the successive approximations (4.3.14) can be rewritten as

$$\begin{cases} L_1 u_{n+1}(t) = q(t, u_{n+1}(t), u'_{n+1}(t)) + f(t, u_n(t), L_1 u_n(t)), \\ L_i u_{n+1} = C_i(L_2 u_n, L_3 u_n, u_n), \quad i = 2, 3, \end{cases} \quad (4.3.15)$$

which in turn can be represented by (4.3.12) as

$$L u_{n+1} = N u_n, \quad n \in \mathbb{N}. \quad (4.3.16)$$

Denote $u_\pm = L^{-1} h_\pm$. It follows from (4.3.13), and (IV) that $u_- = \min V$ and $u_+ = \max V$.

a) Choose $u_0 = u_-$. Applying properties (III) and (IV) one can show by induction that

$$h_- \le L u_n \le N u_n = L u_{n+1} \le h_+, \quad u_- \le u_n \le u_{n+1} \le u_+, \quad n \in \mathbb{N}.$$

These relations imply the existence of the limits

$$u(t) := \lim_{n \to \infty} u_n(t), \quad t \in J,$$
$$c_i := \lim_{n \to \infty} L_i u_n = \lim_{n \to \infty} C_i(L_2 u_n, L_3 u_n, u_n), \quad i = 2, 3, \quad (4.3.17)$$
$$h_1(t) := \lim_{n \to \infty} L_1 u_n(t) = \lim_{n \to \infty} N_1 u_n(t) \quad \text{a.e. in } J.$$

As in the proof of Proposition 4.1.1 it can be shown that the sequences (u_n) and (u'_n) are equicontinuous. Hence, noticing that (u_n) is an increasing sequence Ascoli-Arzela theorem implies that the sequence $(u_n)_{n=1}^{\infty}$ converges uniformly on J to a function $u \in C(J)$, and that the sequence $(u'_{n+1})_{n=1}^{\infty}$ has a subsequence $(u'_{n_k+1})_{k=1}^{\infty}$ which converges uniformly on J to a function $w \in C(J)$. In particular, $w = u'$, and the hypotheses (qa) imply that

$$\lim_{k \to \infty} q(s, u_{n_k+1}(s), u'_{n_k+1}(s)) = q(s, u(s), u'(s)) \quad \text{for a.e. } t \in J.$$

Hence, replacing n by n_k in (4.3.14) and applying (4.3.17) and the dominated convergence theorem we obtain when $k \to \infty$,

$$u(t) = \frac{c_3 y_0(t) + c_2 y_1(t)}{D} + \int_{t_0}^{t_1} k(t, s)(q(s, u(s), u'(s)) + h_1(s)) \, ds, \quad t \in J.$$

This equation implies by Lemma 4.1.2 that u is a solution of the BVP

$$\begin{cases} -\frac{d}{dt}(\mu(t)u'(t)) = q(t, u(t), u'(t)) + h_1(t) \quad \text{a.e. in } J, \\ L_i u = c_i, \quad i = 2, 3. \end{cases}$$

In particular, $u \in Y$, $L_1 u = h_1$, $L_2 u = c_2$, and $L_3 u = c_3$. These results, (4.3.12), and (4.3.17) imply that

$$u(t) := \lim_{n \to \infty} u_n(t), \quad \text{uniformly on } J,$$

$$L_i u := \lim_{n \to \infty} L_i u_n = \lim_{n \to \infty} C_i(L_2 u_n, L_3 u_n, u_n), \quad i = 2, 3, \qquad (4.3.18)$$

$$L_1 u(t) := \lim_{n \to \infty} L_1 u_n(t) = \lim_{n \to \infty} N_1 u_n(t) \quad \text{a.e. in } J.$$

Since all the sequences in (4.3.18) are increasing, the one-sided continuity hypotheses given for f, C_0, and C_1 in a) imply that

$$\lim_{n \to \infty} C_i(L_2 u_n, L_3 u_n, u_n) = C_i(L_2 u, L_3 u, u), \quad i = 2, 3,$$

$$\lim_{n \to \infty} N_1 u_n = \lim_{n \to \infty} f(\cdot, u_n(\cdot), L_1 u_n(\cdot)) = f(\cdot, u(\cdot), L_1 u(\cdot)) = N_1 u. \qquad (4.3.19)$$

It then follows from (4.3.18) and (4.3.19) that $Lu = Nu$. This and Lemma 4.3.1 imply that u is a solution of the BVP (4.3.11).

To prove that u is the least solution of (4.3.11), let $v \in Y$ be a solution of (4.3.11). Then $v \in V$ by Lemma 4.3.2 and $Lv = Nv$ by Lemma 4.3.1, whence properties (IV) imply by induction that

$$Lu_{n+1} = Nu_n \leq Nv = Lv \qquad (4.3.20)$$

for each $n \in \mathbb{N}$. When $n \to \infty$ in (4.3.20) we get $Lu \leq Lv$, so that $u \leq v$ by (IV). Thus u is the least solution of the BVP (4.3.11).

The case b) can be proved similarly. $\qquad \square$

Example 4.3.2. Consider the BVP

$$\begin{cases} -u''(t) = \frac{[u(t)-t^2]}{1+|[u(t)-t^2]|} + \frac{[-u''(t)]}{2+2|[-u''(t)]|} & \text{a.e. in } J = [-1,1], \\ u(-1) - 4u'(-1) = 1 + [u(-1) - 4u'(-1)]/2, \\ u(1) + 4u'(1) = 1 + [u(1) + 4u'(1)]/2. \end{cases} \tag{4.3.21}$$

Problem (4.3.21) is of the form (4.3.11), where

$$\begin{cases} q(t,x,y) = 0, \quad f(t,x,y) = \frac{[x-t^2]}{1+|[x-t^2]|} + \frac{[y]}{2+2|[y]|}, \\ C_i(L_2u, L_3u, u) = 1 + [L_iu]/2, \quad i = 2, 3. \end{cases} \tag{4.3.22}$$

The hypotheses of Theorem 4.3.1 are satisfied, whence problem (4.3.21) has extremal solutions. Moreover, since the integer function $[\cdot]$ is continuous from the right, the hypotheses of Proposition 4.3.2 b) are satisfied. Thus the greatest solution of (4.3.21) can be obtained by the method of successive approximations (4.3.14). In this case they can be rewritten as

$$\begin{aligned} u_{n+1}(t) = & \frac{5-t}{10}(1 + [u_n(-1) - 4u'_n(-1)]/2) \\ &+ \frac{5+t}{10}(1 + [u_n(1) + 4u'_n(1)]/2) \\ &+ \frac{5-t}{10}\int_{-1}^{t}(5+s)\left(\frac{[u_n(s) - s^2]}{1 + |[u_n(s) - s^2]|} + \frac{[-u''_n(s)]}{2 + 2|[-u''_n(s)]|}\right)ds \\ &+ \frac{5+t}{10}\int_{t}^{1}(5-s)\left(\frac{[u_n(s) - s^2]}{1 + |[u_n(s) - s^2]|} + \frac{[-u''_n(s)]}{2 + 2|[-u''_n(s)]|}\right)ds. \end{aligned} \tag{4.3.23}$$

Let u_0 be the solution of the BVP

$$-u''(t) = 2,\ t \in [-1,1], \quad u(-1) - 4u'(-1) = u(1) + 4u'(1) = 2,$$

i.e.,

$$u_0(t) = 2 + \frac{5-t}{5}\int_{-1}^{t}(5+s)\,ds + \frac{5+t}{5}\int_{t}^{1}(5-s)\,ds = 11 - t^2.$$

Calculating the approximations (4.3.23) by a numerical integration method one obtains the following estimate for the greatest solution of the BVP (4.3.21).

$$u(t) \approx \begin{cases} -.5416667t^2 + .018t + 6.967, & -1 \le t < -.786, \\ -.5535714t^2 + 6.96, & -.786 \le t \le .786, \\ -.5416667t^2 - .019t + 6.968, & .786 < t \le 1. \end{cases} \tag{4.3.24}$$

In view of this one can infer that the greatest solution of (4.3.21) is of the form

$$u(t) = \begin{cases} -\frac{13}{24}t^2 + at + b, & -1 \leq t < -x, \\ -\frac{31}{56}t^2 + c, & -x \leq t \leq x, \\ -\frac{13}{24}t^2 - at + b, & x < t \leq 1. \end{cases} \qquad (4.3.25)$$

It remains to determine a, b, c, and x. The boundary conditions imply that $b = \frac{55}{8} + 5a$. Because u' is continuous at x, we get the equation

$$u'(x) = -\frac{31}{28}x = -\frac{13}{12}x - a \implies a = \frac{x}{42}.$$

At a point $t = x$ the second derivative of u has a jump, and the approximation (4.3.24) implies that it is caused by the jump of $[u(t) - t^2]$, from the value 6 to the value 5 at x, whence $u(x) - x^2 = 6$. This and (4.3.25) yield

$$-\frac{87}{56}x^2 + c = 6 = -\frac{37}{24}x^2 - ax + b.$$

Recalling that $a = \frac{x}{42}$ and $b = \frac{55}{8} + 5a$ we see that $c = 6 + \frac{87}{56}x^2$, and that x is the positive solution of equation

$$263x^2 - 20x - 147 = 0.$$

This and the above formulae for b, c, and x yield

$$\begin{cases} x = \frac{10 + 7\sqrt{789}}{263} \approx .7856426109, \\ a = \frac{x}{42} = \frac{5}{5523} + \frac{\sqrt{789}}{1578} \approx .01876426109, \\ b = \frac{55}{8} + 5a = \frac{303965}{44184} + \frac{5\sqrt{789}}{1578} \approx 6.968528882, \\ c = 6 + \frac{87}{56}x^2 = \frac{2612991}{3873464} + \frac{435\sqrt{789}}{138338} \approx 6.958917592. \end{cases} \qquad (4.3.26)$$

Next we shall study whether the least solution of (4.3.21) can be obtained by the method of successive approximations. By choosing $u_0 = -b$, and calculating numerically the successive approximations (4.3.23), we get the following estimate for a solution of (4.3.21).

$$u(t) \approx \frac{7}{12}t^2 - 4.251, \quad -1 \leq t \leq 1.$$

This suggests to study whether the function

$$\bar{u}(t) = \frac{7}{12}t^2 - \frac{17}{4}, \quad -1 \leq t \leq 1,$$

is a solution of (4.3.21). It satisfies the differential equation of (4.3.21), but not the boundary conditions. To proceed we choose $u_0 = \bar{u}$ and calculate again the successive approximations (4.3.23) by a numerical integration method. The resulting estimate is

$$u(t) \approx \begin{cases} .5833333t^2 + .031t - 3.592, & -1 \le t < -.954, \\ .5666667t^2 - 3.607 & -.954 \le t \le .954, \\ .5833333t^2 - .032t - 3.591, & .954 < t \le 1. \end{cases} \qquad (4.3.27)$$

In view of this one can infer that the least solution of (4.3.21) is of the form

$$u(t) = \begin{cases} \frac{7}{12}t^2 + at - b, & -1 \le t < -x, \\ \frac{17}{30}t^2 - c, & -x \le t \le x, \\ \frac{7}{12}t^2 - at - b, & x < t \le 1. \end{cases} \qquad (4.3.28)$$

The exact values of a, b, c, and x, can be calculated as above. The boundary conditions imply that $b = \frac{15}{4} - 5a$. Because u' is continuous at x, we get the equation

$$u'(x) = \frac{17}{15}x = -\frac{7}{6}x - a \;\Rightarrow\; a = \frac{x}{30}.$$

At a point $t = x$ the second derivative of u has a jump, and the approximation (4.3.27) implies that it is caused by the jump of $[u(t) - t^2]$, from the value -4 to the value -5 at x, whence $u(x) - x^2 = -4$. This and (4.3.28) yield

$$-\frac{13}{30}x^2 - c = -4 = -\frac{5}{12}x^2 - ax - b.$$

Recalling that $a = \frac{x}{30}$ and $b = \frac{15}{4} - 5a$ we see that $c = 4 - \frac{13}{30}x^2$, and that x is the positive solution of equation

$$27x^2 - 10x - 15 = 0.$$

This and the above formulae for a, b, and c yield

$$\begin{cases} x = \frac{5+\sqrt{430}}{27} \approx .9532015316, \\ a = \frac{x}{30} = \frac{5+\sqrt{430}}{810} \approx .03177338438, \\ b = \frac{15}{4} - 5a = \frac{1205-2\sqrt{430}}{324} \approx 3.591133078, \\ c = 4 - \frac{13}{30}x^2 = \frac{16313+26\sqrt{430}}{4317} \approx 3.606276297. \end{cases} \qquad (4.3.29)$$

4.4 Explicit phi-Laplacian problems

In this section we derive existence and comparison results for the problem

$$\begin{cases} \frac{d}{dt}\varphi(u'(t)) = g(t, u, u', u'(t)) & \text{a.e. in } J = [t_0, t_1], \\ u(t_0) = B_0(u(t_0), u, u'), \quad u'(t_0) = B_1(u'(t_0), u, u'), \end{cases} \tag{4.4.1}$$

where $g\colon J \times C(J) \times C(J) \times \mathbb{R} \to \mathbb{R}$, $\varphi\colon \mathbb{R} \to \mathbb{R}$ and $B_i\colon \mathbb{R} \times C(J) \times C(J) \to \mathbb{R}$, $i = 0, 1$. We assume that $C(J)$ is equipped with the pointwise ordering $\leq$.

Definition 4.4.1. We say that $u \in C^1(J)$ is a *lower solution* of (4.4.1) if $\varphi \circ u' \in AC(J)$, and

$$\begin{cases} \frac{d}{dt}\varphi(u'(t)) \leq g(t, u, u', u'(t)) & \text{a.e. in } J, \\ u(t_0) \leq B_0(u(t_0), u, u'), \quad u'(t_0) \leq B_1(u'(t_0), u, u'), \end{cases} \tag{4.4.2}$$

and an *upper solution* of (4.4.1) if the reversed inequalities hold. If equalities hold in (4.4.2), we say that u is a *solution* of (4.4.1).

4.4.1 Hypotheses and main results. The following hypotheses are imposed on the functions φ, g, B_0, and B_1.

  (φ0) The function φ is an increasing homeomorphism.
  (g0) $g(\cdot, u, v, x)$ is measurable for all $x \in \mathbb{R}$ and $u, v \in C(J)$, and
  $$\limsup_{y \to x-} g(t, u, v, y) \leq g(t, u, v, x) \leq \liminf_{y \to x+} g(t, u, v, y) \text{ for a.e. } t \in J.$$
  (g1) $g(t, u, v, x)$ is increasing in u and v for a.e. $t \in J$ and all $x \in \mathbb{R}$.
  (Bi0) For each $x \in \mathbb{R}$ the function $B_i(x, u, v)$ is increasing in u and v, and
  $$\limsup_{y \to x-} B_i(y, u, v) \leq B_i(x, u, v) \leq \liminf_{y \to x+} B_i(y, u, v) \text{ when } u, v \in C(J).$$

Moreover, if

  (A) (4.4.1) has lower and upper solutions $\underline{u}$, $\overline{u}$ with $\underline{u} \leq \overline{u}$, $\underline{u}' \leq \overline{u}'$, and g is L^1-bounded in the set $\Omega \subset J \times C(J) \times C(J) \times \mathbb{R}$, defined by
  $$\Omega = \{(t, u, v, x) \mid t \in J, \ \underline{u} \leq u \leq \overline{u}, \ \underline{u}' \leq v \leq \overline{u}', \ \underline{u}'(t) \leq x \leq \overline{u}'(t)\},$$

we show that (4.4.1) has the least and the greatest of those solutions u for which $\underline{u} \leq u \leq \overline{u}$ and $\underline{u}' \leq u' \leq \overline{u}'$.

If (A) is replaced by the following hypotheses:

  (g2) $|g(t, u, v, x)| \leq p_1(t)\psi(|\varphi(x)|)$ for all $x \in \mathbb{R}$ and $u, v \in C(J)$ and for a.e. $t \in J$, where $p_1 \in L^1_+(J)$, $\psi\colon \mathbb{R}_+ \to (0, \infty)$ is increasing and $\int_0^\infty \frac{dx}{\psi(x)} = \infty$,
  (Bi1) $|B_i(x, u, v)| \leq a_i|x| + b_i$ for all $x \in \mathbb{R}$ and $u, v \in C(J)$, where $a_i \in [0, 1)$ and $b_i \geq 0$, $i = 0, 1$,

we prove that problem (4.4.1) has extremal solutions u_* and u^* in the sense that if u is any solution of (4.4.1), then $u_* \leq u \leq u^*$ and $u_*' \leq u' \leq (u^*)'$. We also show that these extremal solutions and their first derivatives depend monotonically on the data. Finally, special cases of the above problems and examples are given.

4.4.2   Preliminaries. In this subsection we introduce auxiliary results which are used in the proof of our first existence result. First we convert problem (4.4.1) to a pair of first order problems. Denoting

$$Y = \{u \in C(J) \mid \varphi \circ u \in AC(J)\}, \quad Z = \{u \in C^1(J) \mid \varphi \circ u' \in AC(J)\},$$

it follows from Definition 4.4.1 that the solutions of (4.4.1) belong to Z.

Lemma 4.4.1. *If $u \in Z$ is a solution of problem (4.4.1), then $v = u' \in Y$, and u, v are solutions of problems*

$$u' = v, \quad u(t_0) = B_0(u(t_0), u, v), \tag{4.4.3}$$

$$\begin{cases} \frac{d}{dt}\varphi(v(t)) = g(t, u, v, v(t)) \quad a.e. \text{ in } J, \\ v(t_0) = B_1(v(t_0), u, v). \end{cases} \tag{4.4.4}$$

Conversely, if $u \in C(J)$ and $v \in Y$ satisfy (4.4.3) and (4.4.4), then $u \in Z$, $u' = v$, and u is a solution of problem (4.4.1).

Proof. Assume first that $u \in Z$ is a solution of problem (4.4.1). Denoting $v = u'$, then $v \in Y$, and u, v satisfy problems (4.4.3) and (4.4.4). Conversely, let $u \in C(J)$ and $v \in Y$ satisfy equations (4.4.3) and (4.4.4). Since $u' = v \in Y$, and since v is a solution of (4.4.4), then $u \in Z$ and u is a solution of problem (4.4.1).   □

In the following we denote $[v, w] = \{u \in C(J) \mid v \leq u \leq w\}$ when $v, w \in C(J)$. As a consequence of Theorem 2.3.1 we get the following result.

Lemma 4.4.2. *If the hypotheses $(\varphi 0)$, $(g0)$, $(g1)$, $(Bi0)$, and (A) are valid, then problem (4.4.4) has for each $u \in [\underline{u}, \overline{u}]$ extremal solutions v_* and v^* in the order interval $[\underline{u}', \overline{u}']$. Moreover,*

$$\begin{cases} v_* = \min\{v_+ \mid v_+ \text{ is an upper solution of (4.4.4) in } [\underline{u}', \overline{u}']\}, \\ v^* = \max\{v_- \mid v_- \text{ is a lower solution of (4.4.4) in } [\underline{u}', \overline{u}']\}. \end{cases} \tag{4.4.5}$$

Proof. The given hypotheses ensure that if $u \in [\underline{u}, \overline{u}]$, then the hypotheses of Theorem 2.3.1 are valid when $(t, x, v) \mapsto g(t, u, v, x)$ stands for g, and $(x, v) \mapsto B_1(x, u, v)$ stands for B_0, which implies the assertions.   □

4.4.3 An extremality result for problem (4.4.1). We shall first prove
the existence of the least solution of (4.4.1) between its lower and upper
solutions $\underline{u}$ and $\overline{u}$. This will be done by constructing an increasing mapping
$G\colon [\underline{u}, \overline{u}] \to [\underline{u}, \overline{u}]$ whose least fixed point exists and is the least solution of
(4.4.1) in $[\underline{u}, \overline{u}]$. The definition of G is given in the following Lemma.

Lemma 4.4.3. *Assume that the hypotheses ($\varphi 0$), ($g0$), ($g1$), ($Bi0$), and
(A) are valid. The relation which assigns to each $u \in [\underline{u}, \overline{u}]$ the least solution
$y = Gu$ of problem*

$$y' = v, \quad y(t_0) = B_0(y(t_0), u, v), \tag{4.4.6}$$

*in $[\underline{u}, \overline{u}]$, where v is the least solution of (4.4.4) in $[\underline{u}', \overline{u}']$, defines an in-
creasing mapping $G\colon [\underline{u}, \overline{u}] \to [\underline{u}, \overline{u}]$.*

Proof. Let $u \in [\underline{u}, \overline{u}]$, be given, and let v be the least solution of (4.4.4)
in $[\underline{u}', \overline{u}']$. The hypotheses of Theorem 2.2.1 are valid for problem (4.4.6)
when $(t, x) \mapsto x$ stands for g, and $(x, w) \mapsto x - B_0(x, u, v)$ stands for B.
Thus problem (4.4.6) has the least solution $y = Gu$ in $[\underline{u}, \overline{u}]$, so that (4.4.6)
defines a mapping $G\colon [\underline{u}, \overline{u}] \to [\underline{u}, \overline{u}]$.

To prove that G is increasing, let $u, u_+ \in [\underline{u}, \overline{u}]$, $u \le u_+$, be given. Let v
be the least solution of (4.4.4) in $[\underline{u}', \overline{u}']$, and let v_+ be the least solution of
problem

$$\frac{d}{dt}\varphi(v_+(t)) = g(t, u_+, v_+(t)) \quad \text{a.e. in } J,$$

$$v_+(t_0) = B_1(v_+(t_0), u_+, v_+),$$

in $[\underline{u}', \overline{u}']$. Because of ($g1$), and ($Bi0$), v_+ is an upper solution of (4.4.4)
in $[\underline{u}', \overline{u}']$. Since v is by Lemma 4.4.2 the least of such upper solutions of
(4.4.4), then $v \le v_+$. The inequalities $u \le u_+$ and $v \le v_+$, the definition
(4.4.6) of G and the hypothesis ($Bi0$) imply that $y_+ = Gu_+$ is an upper
solution of (4.4.6) in $[\underline{u}, \overline{u}]$. Since $y = Gu$ is by Theorem 2.2.1 the least
of such upper solutions of (4.4.6), it follows that $Gu \le Gu_+$, whence G is
increasing. $\square$

The following result is needed to ensure that the mapping G, defined in
Lemma 4.4.3, has the least fixed point.

Lemma 4.4.4. *Assume that the hypotheses ($\varphi 0$), ($g0$), ($g1$), ($Bi0$), and
(A) are valid. Then monotone sequences of $G[\underline{u}, \overline{u}]$ converge uniformly on
J.*

Proof. Given $u \in [\underline{u}, \overline{u}]$, let v denote the least solution of (4.4.4) in $[\underline{u}', \overline{u}']$.
(4.4.6) implies that

$$y'(t) = (Gu)'(t) = v(t), \qquad t \in J.$$

Since Gu belongs to $[\underline{u}, \overline{u}]$ and v belongs to $[\underline{u}', \overline{u}']$, it follows from the above equation the existence of a constant K, which does not depend on u or v, such that

$$|Gu(t) - Gu(s)| \le K|t - s|, \quad t, s \in J, \ u \in [\underline{u}, \overline{u}].$$

In particular, $G[\underline{u}, \overline{u}]$ is an equicontinuous and uniformly bounded subset of $C(J)$, which implies the assertion. $\qquad\qquad\qquad\qquad\qquad\qquad\square$

Now we are able to prove the existence of extremal solutions of (4.4.1) between its lower and upper solutions.

Theorem 4.4.1. *Assume that properties ($\varphi0$), ($g0$), ($g1$), ($Bi0$), and (A) hold. Then problem (4.4.1) has extremal solutions u_* and u^* in $[\underline{u}, \overline{u}]$ in the sense that if u is any solution of (4.4.1) in $[\underline{u}, \overline{u}]$ such that u' belongs to $[\underline{u}', \overline{u}']$, then $u_* \le u \le u^*$ and $u'_* \le u' \le (u^*)'$.*

Proof. It follows from Lemma 4.4.3 and Lemma 4.4.4 that the mapping $G \colon [\underline{u}, \overline{u}] \to [\underline{u}, \overline{u}]$, defined in Lemma 4.4.3, satisfies the hypotheses of Corollary 1.1.1 when $[a, b] = J$, whence G has the least fixed point u_*. This means that $u_* = Gu_*$, i.e. (4.4.6) holds when $y = u = u_*$, so that $u = u_*$ is the least solution of problem (4.4.3) in $[\underline{u}, \overline{u}]$, where $v = v_*$ is the least solution of equation (4.4.4) in $[\underline{u}', \overline{u}']$. These properties imply by Lemma 4.4.1 that u_* is a solution of problem (4.4.1) in $[\underline{u}, \overline{u}]$. If u is any solution of (4.4.1) in $[\underline{u}, \overline{u}]$, then u and $v = u'$ satisfy equations (4.4.3) and (4.4.4) by Lemma 4.4.1. Since $y = Gu$ is the least solution of (4.4.6) in $[\underline{u}, \overline{u}]$, where v is the least solution of (4.4.4) in $[\underline{u}', \overline{u}']$, it then follows that $Gu \le u$. Thus $u_* \le u$ by the first relation of (1.1.2). This inequality, (4.4.1), and the hypotheses ($g1$) and ($Bi0$) imply that $v = u'$ is an upper solution of problem

$$\frac{d}{dt}\varphi(v(t)) = g(t, u_*, v, v(t)) \text{ a.e. in } J, \quad v(t_0) = B_1(v(t_0), u_*, v), \quad (4.4.7)$$

in $[\underline{u}', \overline{u}']$. Since $v_* = u'_*$ is the least solution of problem (4.4.7) in $[\underline{u}', \overline{u}']$, it follows from Lemma 4.4.2 that $u'_* \le u'$. This proves that u_* is the least solution of (4.4.1) in $[\underline{u}, \overline{u}]$ in the asserted sense. The proof that problem (4.4.1) has the greatest solution in $[\underline{u}, \overline{u}]$ is similar. $\qquad\square$

4.4.4 Main existence and comparison results. In this subsection we prove that if the hypothesis (A) is replaced in Theorem 4.4.1 by properties ($g2$) and ($Bi1$), then problem (4.4.1) has extremal solutions. By Theorem 4.4.1 it suffices to construct such lower and upper solutions $\underline{u}$, $\overline{u}$ of (4.4.1) that (A) holds, and if u is any solution of (4.4.1), then $\underline{u} \le u \le \overline{u}$ and

$\underline{u}' \leq u' \leq \overline{u}'$. In the next three lemmas we shall construct these lower and upper solutions by assuming that the properties $(\varphi 0)$, (g2), and (Bi1) hold. Choose $w_0 \in \mathbb{R}$ so that

$$-w_0 \leq \varphi(-\frac{b_1}{1 - a_1}), \varphi(\frac{b_1}{1 - a_1}) \leq w_0, \qquad (4.4.8)$$

and let w be the solution of the IVP

$$w'(t) = p_1(t)\psi(w(t)) \quad \text{a.e. in} \quad J, \quad w(t_0) = w_0. \qquad (4.4.9)$$

Lemma 4.4.5. *Assume that properties $(\varphi 0)$, (g2), and (Bi1) hold. The functions $\underline{v}$, $\overline{v}$, defined by*

$$\underline{v}(t) = \varphi^{-1}(-w(t)), \ t \in J, \quad \text{and} \quad \overline{v}(t) = \varphi^{-1}(w(t)), \ t \in J, \qquad (4.4.10)$$

belong to Y. Moreover, if $u \in Z$ is a solution of (4.4.1), then $u' \in [\underline{v}, \overline{v}]$.

Proof. Since w, as a solution of (4.4.9), belongs to $AC(J)$, it follows from (4.4.10) that $\varphi \circ \underline{v} = -w$ and $\varphi \circ \overline{v} = w$ belong to $AC(J)$. Thus $\underline{v}$ and $\overline{v}$ belong to $C(J)$ by $(\varphi 0)$, and hence $\underline{v}, \overline{v} \in Y$.

To prove the second assertion, assume that $u \in Z$ is a solution of (4.4.1). Applying (4.4.1) and properties (Bi1) and (g2) we obtain

$$|u'(t_0)| = |B_1(u'(t_0), u, u')| \leq a_1|u'(t_0)| + b_1, \quad \text{i.e.} \quad |u'(t_0)| \leq \frac{b_1}{1 - a_1},$$

$$|\frac{d}{dt}\varphi(u'(t))| = |g(t, u, u', u'(t))| \leq p_1(t)\psi(|\varphi(u'(t))|) \quad \text{a.e. in} \quad J.$$

In view of these inequalities and the choice (4.4.8) of w_0 we see that

$$|\varphi(u'(t))| \leq |\varphi(u'(t_0))| + \int_{t_0}^t p_1(s)\psi(|\varphi(u'(s))|)\, ds$$

$$\leq w_0 + \int_{t_0}^t p_1(s)\psi(|\varphi(u'(s))|)\, ds$$

for all $t \in J$. This implies by Lemma B.7.1 that $|\varphi(u'(t))| \leq w(t)$ on J, where w is the solution of the IVP (4.4.9). It then follows from this inequality, (4.4.10) and from the monotonicity of φ^{-1} that $u' \in [\underline{v}, \overline{v}]$. $\square$

Next we shall construct such functions $\underline{u}, \overline{u} \in Z$ that $u \in [\underline{u}, \overline{u}]$ and $u' \in [\underline{u}', \overline{u}']$ whenever $u \in Z$ is a solution of problem (4.4.1).

Lemma 4.4.6. *Assume that properties $(\varphi 0)$, $(g2)$, and $(Bi1)$ hold. Let the functions $\underline{v}$, $\overline{v} \in Y$ be defined by $(4.4.10)$, and define $\underline{u}$, $\overline{u} \in Z$ by*

$$\underline{u}(t) = -\frac{b_0}{1 - a_0} + \int_{t_0}^t \underline{v}(s)\, ds, \quad \overline{u}(t) = \frac{b_0}{1 - a_0} + \int_{t_0}^t \overline{v}(s)\, ds, \quad t \in J. \quad (4.4.11)$$

If $u \in Z$ is a solution of $(4.4.1)$, then $u \in [\underline{u}, \overline{u}]$ and $u' \in [\underline{u}', \overline{u}']$.

Proof. Let $u \in Z$ be a solution of $(4.4.1)$. It follows from $(4.4.11)$ that $\underline{u}' = \underline{v}$ and $\overline{u}' = \overline{v}$, whence $u' \in [\underline{u}', \overline{u}']$ by Lemma 4.4.5. The first initial condition of $(4.4.1)$, and $(Bi1)$ imply that

$$|u(t_0)| = |B_0(u(t_0), u, u')| \leq a_0|u(t_0)| + b_0 \quad \text{i.e.} \quad |u(t_0)| \leq \frac{b_0}{1 - a_0}.$$

Thus

$$\begin{cases} u(t) = u(t_0) + \int_{t_0}^t u'(s),\, ds \leq \frac{b_0}{1-a_0} + \int_{t_0}^t \overline{u}'(s)\, ds = \overline{u}(t), \ t \in J, \\ u(t) = u(t_0) + \int_{t_0}^t u'(s),\, ds \geq -\frac{b_0}{1-a_0} + \int_{t_0}^t \underline{u}'(s)\, ds = \underline{u}(t), \ t \in J, \end{cases}$$

so that $u \in [\underline{u}, \overline{u}]$. $\square$

Next we show that the hypothesis (A) holds for problem $(4.4.1)$, when the functions $\underline{u}$ and $\overline{u}$ are those defined in Lemma 4.4.6.

Lemma 4.4.7. *Assume that properties $(\varphi 0)$, $(g2)$, and $(Bi1)$ hold. Then the functions $\underline{u}$ and $\overline{u}$, defined by $(4.4.11)$, where the functions $\underline{v}$ and $\overline{v}$ are given by $(4.4.10)$, are lower and upper solutions of problem $(4.4.1)$. Moreover, the function g is L^1-bounded in the set*
$$\Omega = \{(t, u, v, x) \mid t \in J, \ u \in [\underline{u}, \overline{u}], \ v \in [\underline{u}', \overline{u}'], \ x \in [\underline{u}'(t), \overline{u}'(t)]\}.$$

Proof. We shall first show that $\underline{u}$ and $\overline{u}$ are lower and upper solutions of $(4.4.1)$. Applying $(g2)$ and definitions $(4.4.10)$ and $(4.4.11)$, and recalling that $\underline{u}' = \underline{v}$ and $\overline{u}' = \overline{v}$, we obtain

$$\frac{d}{dt}\varphi(\overline{u}'(t)) = w'(t) = p_1(t)\psi(w(t)) = p_1(t)\psi(|\varphi(\overline{u}'(t))|) \geq g(t, \overline{u}, \overline{u}'(t)),$$

$$\frac{d}{dt}\varphi(\underline{u}'(t)) = -w'(t) = -p_1(t)\psi(w(t)) = -p_1(t)\psi(|\varphi(\underline{u}'(t))|) \leq g(t, \underline{u}, \underline{u}'(t))$$

for a.e. $t \in J$. The choice $(4.4.8)$ of w_0, monotonicity of φ^{-1} and $(4.4.11)$ imply that

$$\underline{u}'(t_0) \leq -\frac{b_1}{1 - a_1}, \quad \overline{u}'(t_0) \geq \frac{b_1}{1 - a_1}, \quad \underline{u}(t_0) = -\frac{b_0}{1 - a_0}, \quad \overline{u}(t_0) = \frac{b_0}{1 - a_0}.$$

By (Bi1) we have

$$|B_i(x, u, u')| \leq a_i|x| + b_i, \quad \text{for all } x \in \mathbb{R}, \ u \in Z, \ i = 0, 1.$$

In view of these inequalities we obtain

$$\begin{cases} B_1(\overline{u}'(t_0), \overline{u}, \overline{u}') \leq a_1|\overline{u}'(t_0)| + b_1 = a_1\overline{u}'(t_0) + b_1 \leq \overline{u}'(t_0), \\ B_1(\underline{u}'(t_0), \underline{u}, \underline{u}') \geq -a_1|\underline{u}'(t_0)| - b_1 = a_1\underline{u}'(t_0) - b_1 \geq \underline{u}'(t_0), \end{cases}$$

and

$$\begin{cases} B_0(\overline{u}(t_0), \overline{u}, \overline{u}') \leq a_0|\overline{u}(t_0)| + b_0 = a_0\overline{u}(t_0) + b_0 = \overline{u}(t_0), \\ B_0(\underline{u}(t_0), \underline{u}, \underline{u}') \geq -a_0|\underline{u}(t_0)| - b_0 = a_0\underline{u}(t_0) - b_0 = \underline{u}(t_0). \end{cases}$$

The above inequalities show that $\underline{u}$ and $\overline{u}$ are lower and upper solutions of (4.4.1). Moreover, applying (g2), (4.4.9), and (4.4.10) we see that

$$|g(t, u, x)| \leq p_1(t)\psi(|\varphi(x)|) \leq p_1(t)\psi(w(t)) = w'(t)$$

for a.e. $t \in J$, and for all $u \in [\underline{u}, \overline{u}]$ and $x \in [\underline{u}'(t), \overline{u}'(t)]$. This ensures that g is L^1-bounded by w' in the set Ω, which concludes the proof. $\quad \square$

Our main existence and comparison result for problem (4.4.1) reads as follows.

Theorem 4.4.2. *Assume that the functions* $g \colon J \times C(J) \times C(J) \times \mathbb{R} \to \mathbb{R}$, $\varphi \colon \mathbb{R} \to \mathbb{R}$ *and* $B_i \colon \mathbb{R} \times C(J) \times C(J) \to \mathbb{R}$ *have properties* $(\varphi 0)$, $(g0)$, $(g1)$, $(g2)$, $(Bi0)$, *and* $(Bi1)$, $i = 0, 1$. *Then problem (4.4.1) has extremal solutions* u_* *and* u^* *in* Z *in the sense that if* $u \in Z$ *is any solution of (4.4.1), then* $u_* \leq u \leq u^*$ *and* $u_*' \leq u' \leq (u^*)'$. *Moreover,*

$$\begin{cases} \text{If } u \text{ is an upper solution of (4.4.1), then } u_* \leq u \text{ and } u_*' \leq u'. \\ \text{If } u \text{ is a lower solution of (4.4.1), then } u \leq u^* \text{ and } u' \leq (u^*)'. \end{cases}$$
$$(4.4.12)$$

Proof. The given hypotheses imply by Lemma 4.4.7 that the hypotheses of Theorem 4.4.1 are satisfied when $\underline{u}$ and $\overline{u}$ are defined by (4.4.11). According to Lemma 4.4.6 every solution u of (4.4.1) satisfies $u \in [\underline{u}, \overline{u}]$ and $u' \in [\underline{u}', \overline{u}']$. These properties imply by Theorem 4.4.1 the existence of the least solution u_* and the greatest solution u^* of (4.4.1) in the asserted sense. To prove the first assertion of (4.4.12), let u_+ be an upper solution of (4.4.1). Choose the constants b_i in (Bi1) so that

$$-\frac{b_1}{1 - a_1} \leq \varphi(u_+'(t)) \leq \frac{b_1}{1 - a_1} \quad \text{on } J, \text{ and} \quad -\frac{b_0}{1 - a_0} \leq u_+(t_0) \leq \frac{b_0}{1 - a_0}.$$

Then $-w \leq \varphi \circ u'_+ \leq w$, whence $\underline{v} = \underline{u}' \leq u'_+ \leq \overline{v} = \overline{u}'$. These relations and (4.4.11) imply that $\underline{u} \leq u_+ \leq \overline{u}$. Thus problem (4.4.1) has by Theorem 4.4.1 such a solution $u \in [\underline{u}, u_+] \subseteq [\underline{u}, \overline{u}]$ that $u' \in [\underline{u}', u'_+] \subseteq [\underline{u}', \overline{u}']$. These results and the above properties of u_* imply that $u_* \leq u_+$ and $u'_* \leq u'_+$. Similarly, it can be shown that if u_- is a lower solution of (4.4.1), then $u_- \leq u^*$ and $u'_- \leq (u^*)'$. □

As a consequence of Theorem 4.4.2 we obtain the following existence and comparison results for problem

$$\begin{cases} \frac{d}{dt}\varphi(u'(t)) = g(t, u, u', u'(t)) + h(t) & \text{for a.e. } t \in J, \\ u(t_0) = B_0(u(t_0), u, u') + x_0, \quad u'(t_0) = B_1(u'(t_0), u, u') + x_1. \end{cases} \tag{4.4.13}$$

Proposition 4.4.1. *If the hypotheses of Theorem 4.4.2 hold, then problem (4.4.13) has for all $h \in L^1(J)$ and x_0, $x_1 \in \mathbb{R}$ extremal solutions. Moreover, they and their first derivatives are increasing with respect to h, x_0, and x_1.*

Proof. Given h, $\hat{h} \in L^1(J)$, x_i, $\hat{x}_i \in \mathbb{R}$, $i = 0, 1$, assume that

$$h \leq \hat{h}, \quad \text{and} \quad x_i \leq \hat{x}_i, \ i = 0, 1.$$

The functions $(t, x) \mapsto g(t, u, x) + h(t)$ and $(t, x) \mapsto g(t, u, x) + \hat{h}(t)$ have properties (g0) and (g1), and also (g2) when p_1 and ψ are replaced by $t \mapsto p_1(t) + |h(t)| + |\hat{h}(t)|$ and $z \mapsto \psi(z) + 1$, respectively. Similarly, the functions $B_i(\cdot, \cdot, \cdot) + x_i$ and $B_i(\cdot, \cdot, \cdot) + \hat{x}_i$ have properties (Bi0), and also (Bi1) when b_0 and b_1 are replaced by $b_0 + |x_0| + |\hat{x}_0|$ and $b_1 + |x_1| + |\hat{x}_1|$. Denoting by $\hat{u}$ the least solution of problem

$$\begin{cases} \frac{d}{dt}\varphi(u'(t)) = g(t, u, u', u'(t)) + \hat{h}(t) & \text{for a.e. } t \in J, \\ u(t_0) = B_0(u(t_0), u, u') + \hat{x}_0, \quad u'(t_0) = B_1(u'(t_0), u, u') + \hat{x}_1, \end{cases}$$

then $\hat{u}$ is an upper solution of (4.4.13). This and (4.4.12) imply that $u_* \leq \hat{u}$ and $u'_* \leq \hat{u}'$. The assertions for the greatest solution of (4.4.13) can be proved similarly. □

Example 4.4.1. Choose $J = [0, 1]$ and consider the problem

$$\begin{cases} u''(t) = H(u'(1-t) - 1) + \frac{[u'(1-t)-t]}{1+|[u'(1-t)-t]|} + \frac{[\int_0^1 u(s)ds]}{1+|[\int_0^1 u(s)ds]|}, & \text{a.e. in } J, \\ u(0) = \frac{2[u(0)]}{1+|[u(0)]|}, \quad u'(0) = \frac{[u(1)-u(0)]}{2+2|[u(1)-u(0)]|}, \end{cases}$$

$$\tag{4.4.14}$$

where H is the Heaviside function, and $[x]$ denotes the greatest integer less than or equal to x. Problem (4.4.14) is of the form (4.4.1) with

$$\begin{cases} g(t,u,v,x) = H(v(1-t)-1) + \frac{[v(1-t)-t]}{1+|[v(1-t)-t]|} + \frac{[\int_0^1 u(s)ds]}{1+|[\int_0^1 u(s)ds]|}, \\ B_0(x,u,v) = \frac{2[x]}{1+|[x]|}, \quad B_1(x,u,v) = \frac{[u(1)-u(0)]}{2+2|[u(1)-u(0)]|}. \end{cases}$$

It is easy to see that the hypotheses of Theorem 4.4.2 hold, whence problem (4.4.14) has the least solution u_* and the greatest solution u^*. Problem (4.4.4) can be written in this case in the form

$$\begin{cases} v'(t) = H(v(1-t)-1) + \frac{[v(1-t)-t]}{1+|[v(1-t)-t]|} \\ \qquad + \frac{[u(0)+\int_0^1 \int_0^t v(s)ds\, dt]}{1+|[u(0)+\int_0^1 \int_0^t v(s)ds\, dt]|} \quad \text{a.e. in } J, \\ v(0) = \frac{[\int_0^1 v(s)ds]}{2+2|[\int_0^1 v(s)ds]|}. \end{cases} \qquad (4.4.15)$$

Thus it is of the form

$$v'(t) = f(t,v,u(0)) \text{ a.e. in } J, \quad v(0) = B(v),$$

where B is increasing, bounded, and right-continuous, and $f(t,v,x)$ is increasing and right-continuous in v and x. Since the least and greatest solutions of equation $x = \frac{2[x]}{1+|[x]|}$ are $-\frac{4}{3}$ and 1, these solutions are the least and the greatest values of $u(0)$. Then the greatest solution of (4.4.14) is $u^*(t) = 1 + \int_0^t v^*(s)\,ds$, where v^* is the greatest solution of problem (4.4.15) with $u(0) = 1$. Because of the above-mentioned monotonicity and right-continuity properties, v^* can be obtained by Proposition 1.1.3 as the limit of the successive approximations

$$\begin{cases} v'_{n+1}(t) = H(v_n(1-t)-1) + \frac{[v_n(1-t)-t]}{1+|[v_n(1-t)-t]|} \\ \qquad + \frac{[u(0)+\int_0^1 \int_0^t v_n(s)ds\, dt]}{1+|[u(0)+\int_0^1 \int_0^t v_n(s)ds\, dt]|} \quad \text{a.e. in } J, \\ v_{n+1}(0) = \frac{[\int_0^1 v_n(s)ds]}{2+2|[\int_0^1 v_n(s)ds]|}, \end{cases} \qquad (4.4.16)$$

by choosing $u(0) = 1$ and $v_0(t) = \frac{1}{2} + 3t$, $t \in J$. These approximations can be estimated numerically, and the so obtained estimations can be used to infer the exact formula for v^*. Denoting by χ_W the characteristic function of $W \subset \mathbb{R}$, we get the following representation for v^*.

$$v^*(t) = (\frac{1}{4}+2t)\chi_{[0,\frac{3}{8}]}(t) + (\frac{7}{16}+\frac{3}{2}t)\chi_{[\frac{3}{8},\frac{5}{8}]}(t) + (\frac{17}{16}+\frac{t}{2})\chi_{[\frac{5}{8},\frac{3}{4}]}(t) + \frac{23}{16}\chi_{[\frac{3}{4},1]}(t).$$

Thus

$$u^*(t) = 1 + \int_0^t v^*(s)\, ds = \begin{cases} 1 + \frac{t}{4} + t^2, & 0 \le t \le \frac{3}{8}, \\ \frac{247}{256} + \frac{7}{16}t + \frac{3}{4}t^2, & \frac{3}{8} \le t \le \frac{5}{8}, \\ \frac{197}{256} + \frac{17}{16}t + \frac{1}{4}t^2, & \frac{5}{8} \le t \le \frac{3}{4}, \\ \frac{161}{256} + \frac{23}{16}t, & \frac{3}{4} \le t \le 1. \end{cases}$$

By choosing $u(0) = -\frac{4}{3}$ and $v_0(t) = -\frac{1}{2} - 3t$, $t \in J$, in (4.4.16), the numerical estimations show that the so-obtained approximations converge to the function

$$v_*(t) = -\frac{1}{3} - \frac{4}{3}t, \quad t \in J.$$

Since v_* is a solution of (4.4.15), then it is also the least one by Proposition 1.1.3. Consequently, the least solution of (4.4.14) is

$$u_*(t) = -\frac{4}{3} + \int_0^t v_*(s)\, ds = -\frac{4}{3} - \frac{t}{3} - \frac{2}{3}t^2, \quad t \in J.$$

4.4.5 Special cases. In this subsection we shall consider the solvability of the problem

$$\begin{cases} u''(t) = q(u'(t))g(t, u, u', u'(t)) \quad \text{a.e. in } J, \\ u(t_0) = B_0(u(t_0), u, u'), \quad u'(t_0) = B_1(u'(t_0), u, u'), \end{cases} \tag{4.4.17}$$

where $g\colon J \times C(J) \times C(J) \times \mathbb{R} \to \mathbb{R}$, $B_i\colon \mathbb{R} \times C(J) \times C(J) \to \mathbb{R}$, $i = 0, 1$, and $q\colon \mathbb{R} \to (0, \infty)$. A function $u \in AC^1(J)$ is said to be a *lower solution* of problem (4.4.17) if

$$\begin{cases} u''(t) \le q(u'(t))g(t, u, u', u'(t)) \quad \text{a.e. in } J, \\ u(t_0) \le B_0(u(t_0), u, u'), \quad u'(t_0) \le B_1(u'(t_0), u, u'). \end{cases}$$

and an *upper solution* if the reversed inequalities hold. If equalities hold, we say that u is a *solution* of problem (4.4.17).

Lemma 2.1.2 and the proof of Lemma 2.1.3 imply the following result.

Lemma 4.4.8. *Assume that $q\colon \mathbb{R} \to (0, \infty)$ has the following property.*

(q0) *q and $\frac{1}{q}$ belong to $L_{loc}^\infty(\mathbb{R})$, and $\int_0^{\pm\infty} \frac{dz}{q(z)} = \pm\infty$.*

Then $u \in AC^1(J)$ is a lower solution, an upper solution or a solution of (4.4.17) if and only if u is a lower solution, an upper solution or a solution of problem (4.4.1), respectively, where $\varphi\colon \mathbb{R} \to \mathbb{R}$ is defined by $\varphi(x) = \int_0^x \frac{dz}{q(z)}$. Moreover, φ has property $(\varphi 0)$.

In view of Lemma 4.4.8 we obtain the following result.

Proposition 4.4.2. *The results of Theorem 4.4.1 and Theorem 4.4.2 hold for problem (4.4.17) if the hypothesis ($\varphi 0$) is replaced by property (q0).*

Remark 4.4.1. By Remarks 2.1.1 we can replace $\psi(|\varphi(x)|)$ by $\psi(|x|)$ in (g2) if φ is Lipschitz-continuous. The function φ, defined in Lemma 4.4.8, is Lipschitz-continuous if $\frac{1}{q}$ is essentially bounded.

4.5 Implicit phi-Laplacian problems

In this section we consider first an implicit initial value problem of the form

$$\begin{cases} \frac{d}{dt}\varphi(u'(t)) = g(t, u, u', u'(t)) \\ \quad + f(t, u, u', u'(t), \frac{d}{dt}\varphi(u'(t)) - g(t, u, u', u'(t))) \quad \text{a.e. in } J, \\ u(t_0) = B_0(u(t_0), u, u') + C_0(u, u(t_0) - B_0(u(t_0), u, u')), \\ u'(t_0) = B_1(u'(t_0), u, u') + C_1(u, u'(t_0) - B_1(u'(t_0), u, u')), \end{cases} \qquad (4.5.1)$$

where $g \colon J \times C(J) \times C(J) \times \mathbb{R} \to \mathbb{R}$, $f \colon J \times C(J) \times C(J) \times \mathbb{R} \times \mathbb{R} \to \mathbb{R}$, $\varphi \colon \mathbb{R} \to \mathbb{R}$, $B_i \colon \mathbb{R} \times C(J) \times C(J) \to \mathbb{R}$ and $C_i \colon C(J) \times \mathbb{R} \to \mathbb{R}$, $i = 0, 1$. Results derived for the explicit problem

$$\begin{cases} \frac{d}{dt}\varphi(u'(t)) = g(t, u, u', u'(t)) + h_1(t) \quad \text{for a.e. } t \in J, \\ u(t_0) = B_0(u(t_0), u, u') + x_0, \quad u'(t_0) = B_1(u'(t_0), u, u') + x_1, \end{cases} \qquad (4.5.2)$$

in section 4.4 will be used in the sequel.

Assuming that $C(J)$ is equipped with pointwise ordering, we are going to prove that problem (4.5.1) has extremal solutions if φ, g, B_0, and B_1 have properties ($\varphi 0$), (g0), (g1), (g2), (Bi0), and (Bi1) given in subsection 4.4.1, and if the following hypotheses hold for the functions f, C_0, and C_1.

(f0) $f(\cdot, u, v, v(\cdot), w(\cdot))$ is measurable for all u, $v \in C(J)$ and $w \in L^1(J)$, and $f(t, u, v, x, y)$ is increasing in u, v, x and y for a.e. $t \in J$.

(f1) $|f(t, u, v, x, y)| \le p_2(t)\psi(|\varphi(x)|) + \lambda(t)|y|$ for a.e. $t \in J$ and all $x, y \in \mathbb{R}$ and u, $v \in C(J)$, where $\lambda \colon J \to [0, 1)$, $\frac{p_2}{1-\lambda} \in L^1_+(J)$, and the function ψ is as in (g2).

(C01) C_i is increasing in both arguments, and $|C_i(u, x)| \le c_i|x| + d_i$ for all $x \in \mathbb{R}$ and $u \in C(J)$, where $c_i \in [0, 1)$ and $d_i \in \mathbb{R}_+$, $i = 0, 1$.

4.5.1 Preliminaries. In this subsection we reduce problem (4.5.1) to an operator equation of the form $Lu = Nu$, by assuming that the hypotheses ($\varphi 0$), (g0), (g1), (g2), (f0), (f1), (Bi0), (Bi1), and (C01) hold. The domain Z of L and N is defined and ordered by

$$\begin{cases} Z = \{u \in C^1(J) | \varphi \circ u' \in AC(J) \text{ and } g(\cdot, u, u', u'(\cdot)) \text{ is measurable}\}, \\ u \le v \text{ iff } u(t) \le v(t) \text{ and } u'(t) \le v'(t) \text{ for all } t \in J, \end{cases}$$

$$(4.5.3)$$

and assume that the range X of L and N is defined, ordered, and normed as follows.

$$\begin{cases} X = L^1(J) \times \mathbb{R} \times \mathbb{R}, \\ (h_1, x_0, x_1) \le (h_2, y_0, y_1) \text{ iff } h_1 \le h_2, \ x_0 \le y_0 \text{ and } \ x_1 \le y_1, \quad (4.5.4) \\ \|(h, x, y)\| = \|h\|_1 + |x| + |y|. \end{cases}$$

Denoting for each $u \in Z$,

$$\begin{cases} L_1 u(\cdot) = (\varphi \circ u')'(\cdot) - g(\cdot, u, u', u'(\cdot)), \\ N_1 u(\cdot) = f(\cdot, u, u', u'(\cdot), L_1 u(\cdot)), \\ L_2 u = u(t_0) - B_0(u(t_0), u, u'), \quad N_2 u = C_0(u, L_2 u), \\ L_3 u = u'(t_0) - B_1(u'(t_0), u, u'), \quad N_3 u = C_1(u, L_3 u), \end{cases} \quad (4.5.5)$$

we obtain mappings $L = (L_1, L_2, L_2)$ and $N = (N_1, N_2, N_3)$ from Z to X. Moreover, the following result holds.

Lemma 4.5.1. *$u \in Z$ is a solution of (4.5.1) if and only if $Lu = Nu$.*

Proof. The assertions are direct consequences of (4.5.1) and (4.5.5) and the given hypotheses. $\quad\square$

In the next Lemma we define a subset V of Z which contains all the possible solutions of (4.5.1) in Z.

Lemma 4.5.2. *Choose $w_0 \in \mathbb{R}$ such that*

$$\varphi^{-1}(-w_0) \le \pm \frac{d_1 + (1 - c_1)b_1}{(1 - a_1)(1 - c_1)} \le \varphi^{-1}(w_0). \quad (4.5.6)$$

Let $z \in AC(J)$ be the solution of the IVP

$$z'(t) = \left(p_1(t) + \frac{p_2(t)}{1 - \lambda(t)} \right) \psi(z(t)) \quad a.e. \text{ in } J, \quad z(t_0) = w_0. \quad (4.5.7)$$

If $u \in Z$ is a solution of problem (4.5.1), then u belongs to the set V given by

$$V = \{u \in Z | h_- \le Lu \le h_+\}, \quad h_\pm = \left(\frac{\pm p_2 \psi(z)}{1 - \lambda}, \frac{\pm d_0}{1 - c_0}, \frac{\pm d_1}{1 - c_1} \right). \quad (4.5.8)$$

Proof. The hypotheses (f1) and (g2) imply by Lemma 4.4.5 that the IVP (4.5.7) has a unique solution $z \in AC(J)$. Let $u \in Z$ be a solution of (4.5.1). Applying (f1) we get for a.e. $t \in J$,

$$|(\varphi \circ u')'(t) - g(t, u, u', u'(t))|$$
$$= |f(t, u, u', u'(t), (\varphi \circ u')'(t) - g(t, u, u', u'(t)))|$$
$$\leq p_2(t)\psi(|\varphi(u'(t))|) + \lambda(t)|(\varphi \circ u')'(t) - g(t, u, u', u'(t))|,$$

so that

$$|(\varphi \circ u')'(t) - g(t, u, u', u'(t))| \leq \frac{p_2(t)}{1 - \lambda(t)}\psi(|\varphi(u'(t))|) \text{ for a.e. } t \in J. \quad (4.5.9)$$

This inequality and (g2) imply that

$$|(\varphi \circ u')'(t)| \leq |(\varphi \circ u')'(t) - g(t, u, u', u'(t))| + |g(t, u, u', u'(t))|)$$
$$\leq \frac{p_2(t)\psi(|\varphi(u'(t))|)}{1 - \lambda(t)} + p_1(t)\psi(|\varphi(u'(t))|)$$
$$= \left(p_1(t) + \frac{p_2(t)}{1 - \lambda(t)}\right)\psi(|\varphi(u'(t))|)$$

for a.e. $t \in J$. In view of (C01) and the second initial condition of (4.5.1) we obtain

$$|L_3u| = |C_1(u, L_3u)| \leq c_1|L_3u| + d_1, \text{ i.e. } |L_3u| \leq \frac{d_1}{1 - c_1}. \quad (4.5.10)$$

Since $B_1(u'(t_0), u, u) \leq a_1|u'(t_0)| + b_1$ by (Bi1), we get

$$|u'(t_0)| = |L_3u - B_1(u'(t_0), u, u')| \leq |L_3u| + |B_1(u'(t_0), u, u')|$$
$$\leq \frac{d_1}{1 - c_1} + a_1|u(t_0)| + b_1, \text{ i.e. } |u'(t_0)| \leq \frac{d_1 + (1 - c_1)b_1}{(1 - a_1)(1 - c_1)} \leq \varphi^{-1}(w_0).$$

Thus

$$|\varphi(u'(t))| \leq |\varphi(u'(t_0))| + \int_{t_0}^{t} |(\varphi \circ u')'(s)|ds$$
$$\leq w_0 + \int_{t_0}^{t} \left(p_1(s) + \frac{p_2(s)}{1 - \lambda(s)}\right)\psi(|\varphi(u'(s))|)ds, \quad t \in J.$$

Noticing that z is a solution of the IVP (4.5.7), this implies by Lemma B.7.1 that $|\varphi(u'(t))| \leq z(t)$ on J. In view of this property, (4.5.9), and the monotonicity of ψ we obtain

$$|L_1 u(t)| = |(\varphi \circ u')'(t) - g(t, u, u', u'(t))| \leq \frac{p_2(t)\psi(z(t))}{1 - \lambda(t)} \quad \text{a.e. in } J. \quad (4.5.11)$$

The hypothesis (C01) and the first initial condition of (4.5.1) imply that

$$|L_2 u| = |C_0(u, L_2 u)| \leq c_0|L_2 u| + d_0 \quad \text{i.e.} \quad |L_2 u| \leq \frac{d_0}{1 - c_0}. \quad (4.5.12)$$

The inequalities (4.5.10), (4.5.11), and (4.5.12), and the definition (4.5.8) of V imply that $u \in V$. $\qquad\qquad\square$

Lemma 4.5.3. *Let V be given by (4.5.8). If $u \in V$, then $h_- \leq Nu \leq h_+$.*

Proof. It follows from (4.5.5) and (4.5.8) that if $u \in V$, then

$$|L_1 u(t)| = |(\varphi \circ u')'(t) - g(t, u, u', u'(t))| \leq \frac{p_2(t)}{1 - \lambda(t)}\psi(z(t)) \quad (4.5.13)$$

for a.e. $t \in J$. This and (g2) imply that

$$|(\varphi \circ u')'(t)| \leq |(\varphi \circ u')'(t) - g(t, u, u', u'(t))| + |g(t, u, u', u'(t))|$$
$$\leq \frac{p_2(t)\psi(z(t))}{1 - \lambda(t)} + p_1(t)\psi(|\varphi(u'(t))|) \quad \text{for a.e. } t \in J.$$

Denoting $v(t) = |\varphi(u'(t))|$, $t \in J$, then $v(t_0) = |\varphi(u'(t_0))| \leq w_0$ by the choice of w_0 in Lemma 4.5.2. The above inequalities imply that

$$v'(t) \leq \frac{p_2(t)\psi(z(t))}{1 - \lambda(t)} + p_1(t)\psi(v(t)) \quad \text{for a.e. } t \in J, \quad v(t_0) \leq w_0. \quad (4.5.14)$$

Because z is the solution of the IVP (4.5.7), then it is a solution of the IVP

$$z'(t) = \frac{p_2(t)\psi(z(t))}{1 - \lambda(t)} + p_1(t)\psi(z(t)) \quad \text{for a.e. } t \in J, \quad z(t_0) = w_0. \quad (4.5.15)$$

Denoting $y = \max\{v, z\}$, it follows from (4.5.14) and (4.5.15) by the monotonicity of ψ that

$$y'(t) \leq \left(\frac{p_2(t)}{1 - \lambda(t)} + p_1(t) \right) \psi(y(t)) \quad \text{for a.e. } t \in J, \quad y(t_0) = w_0.$$

In view of (4.5.7), the above properties of y and Lemma B.7.1 we see that $y(t) \leq z(t)$ on J. Thus

$$|\varphi(u'(t))| = v(t) \leq z(t), \quad t \in J.$$

Applying this result, (4.5.5), (4.5.13), and (f1) we get

$$|N_1 u(t)| = |f(t, u, u', u'(t), L_1 u(t))| \leq p_2(t)\psi(|\varphi(u'(t))|) + \lambda(t)|L_1 u(t)|$$

$$\leq p_2(t)\psi(z(t)) + \frac{\lambda(t)p_2(t)\psi(z(t))}{1 - \lambda(t)} = \frac{p_2(t)\psi(z(t))}{1 - \lambda(t)} \quad \text{a.e. in } J.$$

Since $u \in V$, then $|L_{2+i}u| \leq \frac{d_i}{1-c_i}$, $i = 0, 1$. Applying these inequalities and (C01) we obtain

$$|N_{2+i}u| = |C_i(u, L_{2+i}u)| \leq c_i|L_{2+i}u| + d_i \leq c_i\frac{d_i}{1-c_i} + d_i = \frac{d_i}{1-c_i}, \ i = 0, 1.$$

The inequalities derived above for $|N_1 u(t)|$, $|N_2 u|$, and $|N_3 u|$ and the definitions (4.5.5) and (4.5.8) of N and $h_\pm$ imply $h_- \leq Nu \leq h_-$. □

Lemma 4.5.4. *Let L, N, and V be defined by (4.5.5) and (4.5.8).*
a) If u, $v \in V$, $u \leq v$, and $Lu \leq Lv$, then $Nu \leq Nv$.
b) Monotone sequences of $N[V]$ converge in X with respect to the norm given by (4.5.4).

Proof. a) Let u, $v \in V$ satisfy $u \leq v$ and $Lu \leq Lv$. It follows from (4.5.3), (4.5.4), (4.5.5), (4.5.6) and from the hypotheses (f0) and (C01) that

$$\begin{cases} N_1 u(t) = f(t, u, u', u'(t), L_1 u(t)) \leq f(t, v, v', v'(t), L_1 v(t)) = N_1 v(t), \\ N_{2+i}u = C_i(u, L_{2+i}u) \leq C_i(v, L_{2+i}v) = N_{2+i}v, \ i = 0, 1. \end{cases}$$

These inequalities imply that $Nu \leq Nv$.

b) Assume that $(Nu_n)_{n=0}^{\infty}$ is a monotone sequence in $N[V]$. In view of Lemma 4.5.3 we have $-h \leq Nu_n \leq h$ for each $n \in \mathbb{N}$, so that the sequence $(N_1 u_n)$ is monotone and

$$|N_1 u_n(t)| \leq \frac{p_2(t)\psi(z(t))}{1 - \lambda(t)} \quad \text{for a.e. } t \in J.$$

Then there is by monotone convergence theorem a function $h_2 \in L^1(J)$ such that

$$\int_J |N_1 u_n(t) - h_2(t)|dt \to 0 \quad \text{as } n \to \infty.$$

Since $|N_{i+2}u_n| \le \frac{d_i}{1-c_i}$, $i = 0, 1$, $n \in \mathbb{N}$, and since the sequences $(N_{2+i}u_n)$ are monotone, they converge, say, to $x_i \in \mathbb{R}$. These convergence relations and (4.5.4) imply that

$$\|Nu_n - (h_2, x_0, x_1)\| \to 0 \quad \text{as} \quad n \to \infty.$$

Since $(h_2, x_0, x_1) \in X = L^1(J) \times \mathbb{R} \times \mathbb{R}$, this proves b).   □

4.5.2   Existence and comparison results for problem (4.5.1). As a consequence of the results of subsection 4.5.1 we now prove our main existence and comparison result.

Theorem 4.5.1. *Assume that the hypotheses (φ), (f0), (f1), (g0), (g1), (g2), (Bi0), (Bi1), and (C01) are satisfied. Then problem (4.5.1) has extremal solutions u_* and u^* in the sense that if $u \in Z$ is any solution of (4.5.1), then $u_*(t) \le u(t) \le u^*(t)$ and $u'_*(t) \le u'(t) \le (u^*)'(t)$ for all $t \in J$. Moreover, u_* and u^* are increasing with respect to f, C_0, and C_1.*

Proof. Let the operators L and N, the subset V of Z and the elements $h_\pm$ of X be defined by (4.5.5) and (4.5.8). It follows from Lemma 4.5.3, Proposition 4.4.1, and Lemma 4.5.4 that the following properties hold.

(I) If $u, v \in V$, $u \le v$ and $Lu \le Lv$, then $h_- \le Nu \le Nv \le h_+$.

(II) Equation $Lu = h$ has for each $h \in [h_-, h_+]$ extremal solutions, and they are increasing with respect to h.

(III) Monotone sequences of $N[V]$ converge in X.

Thus equation $Lu = Nu$ has by Theorem 1.1.2 extremal solutions u_* and u^* in the set V. In view of this result, Lemma 4.5.1, Lemma 4.5.2, and the definition (4.5.3) of the partial ordering of Z, we conclude that u_* and u^* are extremal solutions of problem (4.5.1) in the asserted sense. Moreover, u_* and u^* are increasing with respect to N by Theorem 1.1.2, which implies by the definitions (4.5.3), (4.5.4), and (4.5.5) that they are increasing with respect to f, C_0, and C_1.   □

The hypotheses of Theorem 4.5.1 can be relaxed as follows.

Proposition 4.5.1. *The result of Theorem 4.5.1 holds if (f0) and (C01) are replaced by*

(f0') *The function $t \mapsto f(t, u, v, v(t), w(t))$ is measurable for all $u, v \in C(J)$ and $w \in L^1(J)$, and there is a function $\alpha \in L^\infty_+(J)$ such that $f(t, u, v, x, y) + \alpha(t)y$ is increasing in x, u, v and y for a.e. $t \in J$.*

(C01') $C_i(u, x) + \alpha_i x$ *is increasing in x and u for some $\alpha_i \ge 0$, $i = 0, 1$, and $|C_i(u, x)| \le c_i|x| + d_i$ for all $x \in \mathbb{R}$ and $u, v \in C(J)$, where $c_i \in [0, 1)$ and $d_i \in \mathbb{R}_+$, $i = 0, 1$.*

Proof. It is easy to see that problems (4.5.1) and

$$\begin{cases} \frac{d}{dt}\varphi(u'(t)) = g(t, u, u', u'(t)) \\ \quad + \hat{f}(t, u, u', u'(t), \frac{d}{dt}\varphi(u'(t)) - g(t, u, u', u'(t))) \quad \text{a.e. in } J, \\ u(t_0) = B_0(u(t_0), u, u') + \hat{C}_0(u, u(t_0) - B_0(u(t_0), u, u')), \\ u'(t_0) = B_1(u'(t_0), u, u') + \hat{C}_1(u, u'(t_0) - B_1(u'(t_0), u, u')), \end{cases} \quad (4.5.16)$$

where $\hat{f} \colon J \times C(J) \times \mathbb{R} \to \mathbb{R}$ and $\hat{C}_i \colon C(J) \times \mathbb{R} \to \mathbb{R}$, are defined by

$$\begin{cases} \hat{f}(t, u, v, x, y) = \frac{f(t, u, v, x, y) + \alpha(t)y}{1 + \alpha(t)}, \quad t \in J, \ x, y \in \mathbb{R}, \ u, v \in C(J), \\ \hat{C}_i(u, x) = \frac{C_i(u, x) + \alpha_i x}{1 + \alpha_i}, \quad x \in \mathbb{R}, \ u \in C(J), \ i = 0, 1, \end{cases} \quad (4.5.17)$$

have the same solutions. Moreover, the functions $\hat{f}$, and $\hat{C}_i$, $i = 0, 1$, satisfy the hypotheses (f0), (f1), and (C01) with λ replaced by $\frac{\lambda+\alpha}{1+\alpha}$ and c_i by $\frac{c_i+\alpha_i}{1+\alpha_i}$. Thus problem (4.5.16), with $\hat{f}$ and $\hat{C}_i$ defined by (4.5.17), has by Theorem 4.5.1 extremal solutions u_* and u^*, and they are increasing with respect to $\hat{f}$ and $\hat{C}_i$. In view of (4.5.17), u_* and u^* are then extremal solutions of (4.5.1), and they are increasing with respect to f and C_i. $\qquad\square$

Remarks 4.5.1. In view of Proposition 1.1.2 the extremal solutions of (4.5.1) are obtained sometimes by the following method of successive approximations. Denote by u_0 the greatest solution of $Lu = h_+$, i.e., the greatest solution of problem (4.5.2) with $(h_1, x_0, x_1) = h_+$, defined by (4.5.8), and define a sequence $(u_n)_{n=0}^\infty$ recursively by choosing u_{n+1}, $n \in \mathbb{N}$, as the greatest solution of equation $Lu = Nu_n$, i.e., the greatest solution of problem

$$\begin{cases} \frac{d}{dt}\varphi(u'(t)) = g(t, u, u', u'(t)) \\ \quad + f(t, u_n, u'_n, u'_n(t), \frac{d}{dt}\varphi(u_n(t)) - g(t, u_n, u'_n, u'_n(t))) \quad \text{a.e. in } J, \\ u(t_0) = B_0(u(t_0), u, u') + C_1(u_n, u_n(t_0) - B_0(u(t_0), u_n, u'_n)), \\ u'(t_0) = B_1(u'(t_0), u, u') + C_1(t, u_n, u'_n(t_0) - B_1(u'_n(t_0), u_n, u'_n)). \end{cases} \quad (4.5.18)$$

The hypotheses of Theorem 4.5.1 ensure that (I) and (II) hold, whence the sequence (u_n) is decreasing. If $u_{n+1} = u_n$ for some $n \in \mathbb{N}$, it follows from Proposition 1.1.2 that u_n is the greatest solution of equation $Lu = Nu$, and hence the greatest solution of problem (4.5.1). The next possible candidate for the greatest solution is $\lim_{n\to\infty} u_n$. This is the case, e.g., under suitable right-continuity hypotheses for the functions f, g, B_i, and C_i, $i = 0, 1$.

If u_0 is the least solution of (4.5.2) with $h_- = -(h_1, x_0, x)$, and u_{n+1}, $n \in \mathbb{N}$, is the least solution of (4.5.18), then the above comments hold for the least solution of (4.5.1) when right continuity is replaced by left continuity.

Example 4.5.1. Choose $J = [0, 1]$ and consider the problem

$$\begin{cases} u''(t) = H(u'(1-t)-1) + \frac{[u'(1-t)-t]}{1+|[u'(1-t)-t]|} \\ + \frac{[\int_0^1 u(s)ds]}{1+|[\int_0^1 u(s)ds]|} + \frac{[u''(t)-H(u'(1-t)-1)]}{1+|[u''(t)-H(u'(1-t)-1)]|} \quad \text{a.e. in } J, \qquad (4.5.19) \\ u(0) = \frac{2[u(0)]}{1+|[u(0)]|}, \quad u'(0) = \frac{[u(1)-u(0)]}{2+2|[u(1)-u(0)]|}, \end{cases}$$

where H is the Heaviside function. and $[x]$ denotes the greatest integer less than or equal to x. Problem (4.5.19) is of the form (4.5.1) with

$$\begin{cases} g(t, u, v, x) = H(v(1-t)-1), \\ f(t, u, v, x, y) = \frac{[v(1-t)-t]}{1+|[v(1-t)-t]|} + \frac{[\int_0^1 u(s)ds]}{1+|[\int_0^1 u(s)ds]|} + \frac{[y]}{1+|[y]|}, \\ B_0(x, u, v) = \frac{2[x]}{1+|[x]|}, \quad B_1(x, u, v) = \frac{[u(1)-u(0)]}{2+2|[u(1)-u(0)]|}, \end{cases}$$

and where C_0 and C_1 are zero-functions.

It is easy to see that the hypotheses of Theorem 4.5.1 hold, whence problem (4.5.19) has the least solution u_* and the greatest solution u^*. Denoting $v = u'$, problem (4.5.19) can be written in the form

$$\begin{cases} v'(t) = H(v(1-t)-1) + \frac{[v(1-t)-t]}{1+|[v(1-t)-t]|} \\ + \frac{[u(0)+\int_0^1 \int_0^t v(s)ds\, dt]}{1+|[u(0)+\int_0^1 \int_0^t v(s)ds\, dt]|} + \frac{[v'(t)-H(v(1-t)-1)]}{1+|[v'(t)-H(v(1-t)-1)]|} \quad \text{a.e. in } J, \qquad (4.5.20) \\ v(0) = \frac{[\int_0^1 v(s)ds]}{2+2|[\int_0^1 v(s)ds]|}, \quad u(0) = \frac{2[u(0)]}{1+|[u(0)]|}. \end{cases}$$

Thus it is of the form

$$\begin{cases} v'(t) = g_1(v) + f_1(t, v, u(0), v'(t)-g_1(v)) \quad \text{a.e. in } J, \\ v(0) = B(v), \quad u(0) = \frac{2[u(0)]}{1+|[u(0)]|}, \end{cases}$$

where the functions B and g_1 are increasing, bounded, and right-continuous, and the function $f_1(t, v, x, y)$ is increasing and right-continuous in v, x, and y. Since the least and greatest solutions of equation $x = \frac{2[x]}{1+|[x]|}$ are $-\frac{4}{3}$ and 1, these solutions are the least and the greatest values of $u(0)$. Then the greatest solution of (4.5.19) is $u^*(t) = 1 + \int_0^t v^*(s)\, ds$, where v^* is the greatest solution of problem (4.5.20) with $u(0) = 1$. Because of the above-mentioned monotonicity and right-continuity properties, v^* can be obtained as the limit of the successive approximations

$$
\begin{cases}
v'_{n+1}(t) = H(v_{n+1}(1-t) - 1) + \dfrac{[v_n(1-t)-t]}{1+|[v_n(1-t)-t]|} \\[2mm]
\quad + \dfrac{[u(0)+\int_0^1 \int_0^t v_n(s)ds\,dt]}{1+|[u(0)+\int_0^1 \int_0^t v_n(s)ds\,dt]|} + \dfrac{[v'_n(t)-H(v_n(1-t)-1)]}{1+|[v'_n(t)-H(v_n(1-t)-1)]|} \quad \text{a.e. in } J, \quad (4.5.21)\\[2mm]
v_{n+1}(0) = \dfrac{[\int_0^1 v_n(s)ds]}{2+2|[\int_0^1 v_n(s)ds]|},
\end{cases}
$$

by choosing $u(0) = 1$ and $v_0(t) = \frac{1}{2} + 4t$, $t \in J$. These approximations can be estimated numerically, and these estimations can be used to infer that

$$
v^*(t) = (\frac{1}{4}+\frac{5t}{2})\chi_{[0,\frac{1}{2}]}(t) + (\frac{1}{2}+2t)\chi_{[\frac{1}{2},\frac{7}{10}]}(t) + (\frac{6}{5}+t)\chi_{[\frac{7}{10},\frac{11}{14}]}(t) + \frac{139}{70}\chi_{[\frac{11}{14},1]}(t).
$$

Thus

$$
u^*(t) = 1 + \int_0^t v^*(s)\,ds =
\begin{cases}
1 + \frac{t}{4} + \frac{5}{4}t^2, & 0 \le t \le \frac{1}{2}, \\[1mm]
\frac{15}{6} + \frac{t}{2} + t^2, & \frac{1}{2} \le t \le \frac{7}{10}, \\[1mm]
\frac{277}{400} + \frac{6}{5}t + \frac{1}{2}t^2, & \frac{7}{10} \le t \le \frac{11}{14}, \\[1mm]
\frac{7523}{19600} + \frac{139}{70}t, & \frac{11}{14} \le t \le 1.
\end{cases}
$$

By choosing $u(0) = -\frac{4}{3}$ and $v_0(t) = -\frac{1}{2} - 4t$, $t \in J$ in (4.5.21), the numerical estimations show that the so-obtained approximations converge to the function

$$
v_*(t) = -(\frac{1}{3} + \frac{13}{6}t)\chi_{[0,\frac{3}{7}]}(t) - (\frac{31}{84} + \frac{25}{12}t)\chi_{[\frac{3}{7},1]}(t).
$$

v_* is the least solution of (4.5.21) by Proposition 1.1.2. Thus the least solution of (4.5.19) is

$$
u_*(t) = -\frac{4}{3} + \int_0^t v_*(s)\,ds =
\begin{cases}
-\frac{4}{3} - \frac{t}{3} - \frac{13}{12}t^2, & t \in [0, \frac{3}{7}], \\[1mm]
-\frac{41}{24} - \frac{31}{84}t - \frac{25}{24}t^2, & t \in [\frac{3}{7}, 1].
\end{cases}
$$

4.5.3 Special cases. Applying Lemma 2.1.3 it can be shown that problem

$$
\begin{cases}
\dfrac{u'(t)}{q(u'(t))} = g(t,u,u',u'(t)) + f(t,u,\dfrac{u'(t)}{q(u'(t))} - g(t,u,u',u'(t))), \\[2mm]
u(t_0) = B_0(u(t_0),u,u') + C_0(u,u(t_0) - B_0(u(t_0),u,u')), \qquad (4.5.22)\\[2mm]
u'(t_0) = B_1(u'(t_0),u,u') + C_1(u,u'(t_0) - B_1(u'(t_0),u,u')),
\end{cases}
$$

where $q: \mathbb{R} \to (0,\infty)$ has property

(q0) q and $\frac{1}{q}$ belong to $L^\infty_{loc}(\mathbb{R})$, and $\int_0^{\pm\infty} \frac{dz}{q(z)} = \pm\infty$, has same solutions as (4.5.1), where $\varphi \colon \mathbb{R} \to \mathbb{R}$ is defined by $\varphi(x) = \int_0^x \frac{dz}{q(z)}$. Moreover, this function φ has property $(\varphi 0)$. Thus the results of Theorem 4.5.1 and Proposition 4.5.1 are valid for problem (4.5.22) if, instead of $(\varphi 0)$ we assume that (q0) holds.

The function $\varphi \colon \mathbb{R} \to \mathbb{R}$, defined by $\varphi(x) = |x|^{p-2}x$, $x \in \mathbb{R}$, has property $(\varphi 0)$ for each $p > 1$. Thus the results of Theorem 4.5.1 and Proposition 4.5.1 hold for the p-Laplacian problem

$$
\begin{cases}
\frac{d}{dt}(|u'(t)|^{p-2}u'(t)) = g(t, u, u', u'(t)) \\
\quad + f(t, u, \frac{d}{dt}(|u'(t)|^{p-2}u'(t)) - g(t, u, u', u'(t))) \quad \text{a.e. in } J, \\
u(t_0) = B_0(u(t_0), u, u') + C_0(u, u(t_0) - B_0(u(t_0), u, u')), \\
u'(t_0) = B_1(u'(t_0), u, u') + C_1(u, u'(t_0) - B_1(u'(t_0), u, u')).
\end{cases}
\tag{4.5.23}
$$

When the function f is dropped from problems (4.5.1), (4.5.22), and (4.5.23) we get results for initial value problems of explicit differential equations. Problem

$$
\begin{cases}
F(t, u, u', u'(t), \frac{d}{dt}\varphi(u'(t)) - g(t, u, u', u'(t))) = 0 \quad \text{a.e. in } J, \\
A_0(u, u(t_0) - B_0(u(t_0), u, u')) = 0, \\
A_1(u, u'(t_0) - B_1(u'(t_0), u, u')) = 0,
\end{cases}
\tag{4.5.24}
$$

has the same solutions as (4.5.1) if $f \colon J \times C(J) \times C(J) \times \mathbb{R} \times \mathbb{R} \to \mathbb{R}$ and $C_i \colon C(J) \times \mathbb{R} \to \mathbb{R}$, $i = 0, 1$, are defined by

$$
\begin{cases}
f(t, u, v, x, y) = y - \mu(t, u, v, x, y)F(t, u, v, x, y), \\
C_i(u, x) = x - \nu_i(u, x)A_i(u, x), \quad x \in \mathbb{R}, \; u \in C(J), \; i = 0, 1,
\end{cases}
\tag{4.5.25}
$$

where $\mu \colon J \times C(J) \times C(J) \times \mathbb{R} \times \mathbb{R} \to (0, \infty)$ and $\nu_i \colon C(J) \times \mathbb{R} \to (0, \infty)$. Hence, if μ and ν_i can be chosen so that the hypotheses of Theorem 4.5.1 hold when f and C_i are defined by (4.5.25), then problem (4.5.24) has extremal solutions.

Remarks 4.5.2. We have assumed that the hypotheses (g2) and (f1) hold with the same ψ. One can replace ψ by $\hat{\psi}$ in (f1), if $\int_0^\infty \frac{dx}{\max\{\psi(x), \hat{\psi}(x)\}} = \infty$. This and all the other properties given for ψ:s in (g2) and (f1) hold when ψ:s are any of the functions $\psi_0(x) = ax + b$, $x \ge 0$, $a \ge 0$, $b > 0$ and

$$
\psi_n(x) = (x + 1)\ln(x + e) \cdots \ln_n(x + exp_n(1)), \quad x \ge 0, \; n = 1, 2, \dots.
$$

By Remarks 2.1.1 we can replace $\psi(|\varphi(x)|)$ by $\psi(|x|)$ in (g2) and (f1) if φ is Lipschitz continuous.

4.6 Notes and comments

The extremality results derived in sections 4.1, 4.2, and 4.3 for discontinuous
explicit and implicit Sturm-Liouville boundary value problems are taken
from [5]. The special case when $\mu(t) \equiv 1$ is considered, e.g., in [71], [74],
[78], [111], [140], [141], [153], [176], [192], [209].

The extremality results presented in sections 4.4 and 4.5 for explicit and
implicit discontinuous phi-Laplacian differential equations are taken from
[85]. Recently, the phi-Laplacian and the p-Laplacian differential equations

$$\frac{d}{dt}\varphi(u'(t)) = f(t, u(t), u'(t)) \quad \text{and} \quad \frac{d}{dt}(|u'(t)|^{p-2}u'(t)) = f(t, u(t), u'(t)),$$

equipped with different kinds of boundary conditions, have been studied
quite intensively (see, e.g., [1], [33], [34], [35], [36], [37], [38], [40], [93],
[113], [115], [116], [124], [151], [186], [213], [214], [215]). In these references
the function f is assumed to be continuous, or to satisfy Carathéodory
conditions, except in [34] and [35], where also the case when f may depend
discontinuously on its second argument is considered. The functions in
boundary conditions are assumed to be continuous.

In [41] existence results are provided for the quasilinear differential equation

$$\frac{d}{dt}\varphi(t, u, u(t), u'(t)) = f(t, u(t), u'(t)),$$

equipped with functional boundary conditions.

Chapter 5

Extremality results for quasilinear PDE

The upper and lower solution method combined with the monotone iteration technique has been proved to be a powerful tool to obtain the existence of extremal solutions of elliptic and parabolic boundary value problems within an order interval $[\underline{u}, \bar{u}]$ formed by an ordered pair of upper and lower solutions $\bar{u}$ and $\underline{u}$, respectively. However, in order to apply the monotone iteration technique to such type of problems given in an abstract form by $\mathcal{A}u = \mathcal{F}u$, the operator $\mathcal{A}$ which stands for an elliptic or parabolic operator related with some boundary and initial conditions has to have an increasing inverse, and the operator $\mathcal{F}$ which stands for the Nemytskij operator of the lower order terms has to be increasing with respect to the underlying natural partial ordering of functions. For example, semilinear elliptic and parabolic problems with a nonlinear lower order term depending only on u which is either an increasing or a Lipschitz continuous function of u can always be transformed by the maximum principle into an abstract setting given above. Thus the above problem may be rewritten as a fixed point equation

$$u = \mathcal{A}^{-1} \circ \mathcal{F}u$$

within some function space governed by an increasing fixed point operator $\mathcal{A}^{-1} \circ \mathcal{F} : [\underline{u}, \bar{u}] \to [\underline{u}, \bar{u}]$ of the order interval $[\underline{u}, \bar{u}]$ to itself which allows one to obtain the extremal solutions within $[\underline{u}, \bar{u}]$ by successive approximation.

In this chapter we provide existence and extremality results for general quasilinear elliptic and parabolic problems. For such problems none of the above key properties of $\mathcal{A}$ and $\mathcal{F}$, respectively, is satisfied so that, in general, the monotone iteration technique cannot be applied. The crucial point in our approach toward the existence of extremal solutions within an order interval is to prove that the solution set $\mathcal{S} \subseteq [\underline{u}, \bar{u}]$ of the problem under consideration is a *directed* set. Moreover, we shall show that $\mathcal{S}$ is compact. On the basis of these extremality results we are then able to treat quasilinear problems involving discontinuous lower order terms.

153

5.1 Quasilinear elliptic boundary value problems

5.1.1 Notations, hypotheses, and the main result. Let $\Omega \subset \mathbb{R}^N$ be a bounded domain with Lipschitz boundary $\partial\Omega$. In this section we consider the following quasilinear Dirichlet boundary value problem (BVP for short)

$$Au = Fu \quad \text{in } \Omega, \quad u = 0 \quad \text{on } \partial\Omega, \tag{5.1.1}$$

where A is a second order quasilinear differential operator in divergence form given by

$$Au(x) = -\sum_{i=1}^{N} \frac{\partial}{\partial x_i} a_i(x, u(x), \nabla u(x)), \quad \text{with} \quad \nabla u = \left(\frac{\partial u}{\partial x_1}, \dots, \frac{\partial u}{\partial x_N} \right),$$

and F denotes the Nemytskij operator of the lower order terms generated by a function $f : \Omega \times \mathbb{R} \times \mathbb{R}^N \to \mathbb{R}$ and defined by

$$Fu(x) = f(x, u(x), \nabla u(x)).$$

Let $V = W^{1,p}(\Omega)$ and $V_0 = W_0^{1,p}(\Omega)$ denote the usual Sobolev spaces (see section C.1) with $1 < p < \infty$, and V^* and $V_0^* = W^{-1,q}(\Omega)$, $1/p + 1/q = 1$, their corresponding dual spaces, respectively, and $\langle \cdot, \cdot \rangle$ the duality pairing between them. We introduce the natural partial ordering in $L^p(\Omega)$, that is, $u \leq w$ if and only if $w - u$ belongs to the cone $L_+^p(\Omega)$ of all nonnegative elements of $L^p(\Omega)$, which induces also a partial ordering in the Sobolev space V. If $u, w \in V$, and $u \leq w$, then

$$[u, w] := \{v \in V \mid u \leq v \leq w\}$$

denotes the order interval formed by u and w. Moreover, for functions $u, w \in L^p(\Omega)$ we also use the notation

$$\{u \leq w\} := \{x \in \Omega \mid u(x) \leq w(x)\}.$$

We impose the following conditions of Leray-Lions type on the coefficient functions a_i, $i = 1, \dots, N$:

(A1) Each $a_i : \Omega \times \mathbb{R} \times \mathbb{R}^N \to \mathbb{R}$ satisfies Carathéodory conditions, i.e., $a_i(x, s, \xi)$ is measurable in $x \in \Omega$ for all $(s, \xi) \in \mathbb{R} \times \mathbb{R}^N$ and continuous in (s, ξ) for a.e. $x \in \Omega$. There exist a constant $c_0 > 0$ and a function $k_0 \in L^q(\Omega)$, $1/p + 1/q = 1$, such that

$|a_i(x, s, \xi)| \le k_0(x) + c_0(|s|^{p-1} + |\xi|^{p-1})$ for a.e. $x \in \Omega$ and for all $(s, \xi) \in \mathbb{R} \times \mathbb{R}^N$.

(A2) $\sum_{i=1}^N (a_i(x, s, \xi) - a_i(x, s, \xi'))(\xi_i - \xi_i') \ge \mu|\xi - \xi'|^p$

for a.e. $x \in \Omega$, for all $s \in \mathbb{R}$, and for all $\xi, \xi' \in \mathbb{R}^N$ with μ being some positive constant.

(A3) $|a_i(x, s, \xi) - a_i(x, s', \xi)|$

$\le [k_1(x) + |s|^{p-1} + |s'|^{p-1} + |\xi|^{p-1}] \omega(|s - s'|)$

for some function $k_1 \in L^q(\Omega)$, for a.e. $x \in \Omega$, for all $s, s' \in \mathbb{R}$ and for all $\xi \in \mathbb{R}^N$, where $\omega : \mathbb{R}_+ \to \mathbb{R}_+$ is the *modulus of continuity* satisfying

$$\int_{0+} \frac{dr}{\omega^q(r)} = +\infty, \qquad (5.1.2)$$

which means that for any $\varepsilon > 0$ the integral taken over $[0, \varepsilon]$ is divergent, i.e., we have $\int_0^\varepsilon \frac{dr}{\omega^q(r)} = +\infty$.

Remark 5.1.1. As will be seen the proof of the extremality result requires a strong monotonicity condition (A2) with respect to ξ (or p-ellipticity) which is related with the q-modulus of continuity condition (A3). There is an interplay between p-ellipticity and the q-modulus of continuity. Hypothesis (A3) is satisfied for example in case that $\omega(|s - s'|) = c\,|s - s'|^{1/q}$ with some positive constant c, i.e., the coefficients $a_i(x, s, \xi)$ satisfy a Hölder condition with respect to s. However, if we impose instead of (5.1.2) the more restrictive condition

$$\int_{0+} \frac{dr}{\omega(r)} = +\infty, \qquad (5.1.3)$$

which includes for example $\omega(|s - s'|) = c\,|s - s'|$, i.e., a Lipschitz condition with respect to s, then one can replace condition (A2) by a strict monotonicity condition (A2)$_1$ and a coercivity condition (A2)$_2$ with respect to ξ given as follows:

(A2)$_1$ $\sum_{i=1}^N (a_i(x, s, \xi) - a_i(x, s, \xi'))(\xi_i - \xi_i') > 0$

for a.e. $x \in \Omega$, for all $s \in \mathbb{R}$, and for all $\xi, \xi' \in \mathbb{R}^N$ with $\xi \ne \xi'$.

(A2)$_2$ $\sum_{i=1}^N a_i(x, s, \xi)\xi_i \ge \nu|\xi|^p - k(x)$

for a.e. $x \in \Omega$, for all $s \in \mathbb{R}$, and for all $\xi \in \mathbb{R}^N$ with some constant $\nu > 0$ and some function $k \in L^1(\Omega)$.

In particular, (A2) may be replaced by the weaker conditions (A2)$_1$ and (A2)$_2$ if the coefficients a_i do not depend on s.

As a consequence of (A1) the semilinear form a associated with the operator A by

$$\langle Au, \varphi \rangle = a(u, \varphi) = \sum_{i=1}^{N} \int_{\Omega} a_i(x, u, \nabla u) \frac{\partial \varphi}{\partial x_i} \, dx \quad \text{for all} \ \ \varphi \in V_0$$

is well defined for any $u \in V$ and the operator $A : V \rightarrow V^* \subset V_0^*$ is continuous and bounded. We denote the norm (strong) convergence by $\rightarrow$, and the weak convergence by $\rightharpoonup$. Further, we assume that the function $f : \Omega \times \mathbb{R} \times \mathbb{R}^N \rightarrow \mathbb{R}$ associated with the Nemytskij operator F satisfies the Carathéodory conditions.

Let us introduce the notion of a (weak) solution of the BVP (5.1.1).

Definition 5.1.1. A function $u \in V_0$ is called a *solution* of problem (5.1.1) if $Fu \in L^q(\Omega)$ and

$$a(u, \varphi) = \int_{\Omega} Fu \, \varphi \, dx \quad \text{for all} \ \varphi \in V_0.$$

An upper solution for (5.1.1) is defined as follows.

Definition 5.1.2. A function $\bar{u} \in V$ is called an *upper solution* of the BVP (5.1.1) if $F\bar{u} \in L^q(\Omega)$ and

  (i) $\bar{u} \geq 0$ on $\partial\Omega$,
  (ii) $a(\bar{u}, \varphi) \geq \int_{\Omega} F\bar{u} \, \varphi \, dx$ for all $\varphi \in V_0 \cap L_+^p(\Omega)$.

Similarly a function $\underline{u} \in V$ is a *lower solution* of (5.1.1) if the reversed inequalities hold in (i) and (ii) of Definition 5.1.2 with $\bar{u}$ replaced by $\underline{u}$. Further, we shall make the following hypotheses.

  (H1) The BVP (5.1.1) has an upper solution $\bar{u}$ and a lower solution $\underline{u}$ such that $\underline{u} \leq \bar{u}$.
  (H2) There exist a function $k_2 \in L_+^q(\Omega)$ and a constant $c_1 \geq 0$ such that

$$|f(x, s, \xi)| \leq k_2(x) + c_1 |\xi|^{p-1}$$

for a.e. $x \in \Omega$, for all $\xi \in \mathbb{R}^N$, and for all $s \in [\underline{u}(x), \bar{u}(x)]$.

Note that by hypotheses (H2) the Nemytskij operator F is continuous and bounded from $[\underline{u}, \bar{u}] \subset V$ into $L^q(\Omega)$; see D.2.

Definition 5.1.3. A solution u^* is the *greatest solution* within $[\underline{u}, \bar{u}]$ if for any solution $u \in [\underline{u}, \bar{u}]$ we have $u \leq u^*$. Similarly, u_* is the *least solution* in $[\underline{u}, \bar{u}]$ if for any solution $u \in [\underline{u}, \bar{u}]$ it holds $u_* \leq u$. The least and greatest solutions are called the *extremal* ones.

The main result of this section is the following existence and extremality theorem.

Theorem 5.1.1. *Let hypotheses (A1)–(A3) and (H1), (H2) be satisfied. Then the BVP (5.1.1) possesses extremal solutions within the sector $[\underline{u}, \bar{u}]$ formed by the lower and upper solution $\underline{u}$ and $\bar{u}$, respectively.*

In the proof of Theorem 5.1.1 which will be given in subsection 5.1.3 we focus on the existence of the greatest solution u^* only, since the existence of the least solution u_* can be shown analogously.

5.1.2 Preparatory results. Let us assume throughout this subsection that the hypotheses (A1)–(A3) and (H1), (H2) are satisfied. The following lemma generalizes the well-known enclosure result by Deuel and Hess [102] and provides the main tool for the proof of Theorem 5.1.1.

Lemma 5.1.1. *Let $\underline{u}_i \in [\underline{u}, \bar{u}]$, $i = 1, ..., m$, be lower solutions and let $\bar{u}_j \in [\underline{u}, \bar{u}]$, $j = 1, ..., m'$, be upper solutions of the BVP (5.1.1) such that $u_0 := \max(\underline{u}_1, ..., \underline{u}_m) \leq u^0 := \min(\bar{u}_1, ..., \bar{u}_{m'})$. Then there exists a solution u of the BVP (5.1.1) satisfying $u_0 \leq u \leq u^0$.*

Proof. The proof of the lemma will be done in several steps.

(a) Truncation and cut-off mappings.

We introduce truncation operators T_i^0, T_0^j, T_0^0 related with the functions $\underline{u}_i$, $\bar{u}_j$, u_0, and u^0, and defined as follows:

$$
T_i^0 u(x) = \begin{cases} u^0(x) & \text{if} \quad u(x) > u^0(x), \\ u(x) & \text{if} \quad \underline{u}_i(x) \leq u(x) \leq u^0(x), \\ \underline{u}_i(x) & \text{if} \quad u(x) < \underline{u}_i(x), \end{cases}
$$

$$
T_0^j u(x) = \begin{cases} \bar{u}^j(x) & \text{if} \quad u(x) > \bar{u}^j(x), \\ u(x) & \text{if} \quad u_0(x) \leq u(x) \leq \bar{u}^j(x), \\ u_0(x) & \text{if} \quad u(x) < u_0(x), \end{cases}
$$

$$
T_0^0 u(x) = \begin{cases} u^0(x) & \text{if} \quad u(x) > u^0(x), \\ u(x) & \text{if} \quad u_0(x) \leq u(x) \leq u^0(x), \\ u_0(x) & \text{if} \quad u(x) < u_0(x). \end{cases}
$$

The operators $T_0^0, T_i^0, T_0^j : V \to V$ are bounded and continuous; see C.4. Furthermore, we introduce the following cut-off function $b : \Omega \times \mathbb{R} \to \mathbb{R}$ by

$$
b(x, s) = \begin{cases} (s - u^0(x))^{p-1} & \text{if} \quad s > u^0(x), \\ 0 & \text{if} \quad u_0(x) \le s \le u^0(x), \\ -(u_0(x) - s)^{p-1} & \text{if} \quad s < u_0(x). \end{cases}
$$

Obviously, b is a Carathéodory function satisfying a growth condition of the form

$$
|b(x, s)| \le (|s| + |u^0(x)| + |u_0(x)|)^{p-1} \le k_3(x) + c_2 |s|^{p-1} \tag{5.1.4}
$$

for some positive constant c_2 and some function $k_3 \in L^q(\Omega)$. This shows that the Nemytskij operator B associated with the function b is bounded and continuous from $L^p(\Omega)$ into $L^q(\Omega)$, and moreover, satisfies the following estimate

$$
\int_\Omega (Bu)u \, dx = \int_\Omega b(x, u(x))u(x) \, dx \ge c_3 \|u\|_{L^p(\Omega)}^p - c_4 \tag{5.1.5}
$$

for some positive constants c_3, c_4. To verify the estimate (5.1.5) we use inequality B.3 to obtain by elementary calculations the following estimates:

$$
\text{If } u > u^0 \text{ then } (u - u^0)^{p-1}u \ge a_1 |u|^p - a_2 |u^0|^{p-1}|u| \tag{5.1.6}
$$

and

$$
\text{if } u < u_0 \text{ then } -(u_0 - u)^{p-1}u \ge a_3 |u|^p - a_4 |u_0|^{p-1}|u| \tag{5.1.7}
$$

for some positive constants a_i, $i = 1, 2, 3, 4$. By means of (5.1.6), (5.1.7), and the definition of the cut-off function b we are able to prove (5.1.5) as follows

$$
\int_\Omega b(\cdot, u)u \, dx = \int_{\{u > u^0\}} (u - u^0)^{p-1}u \, dx + \int_{\{u < u_0\}} -(u_0 - u)^{p-1}u \, dx
$$

$$
\ge \int_{\{u > u^0\}} (a_1 |u|^p - a_2 |u^0|^{p-1}|u|) \, dx + \int_{\{u < u_0\}} (a_3 |u|^p - a_4 |u_0|^{p-1}|u|) \, dx
$$

$$
\ge \int_{\{u > u^0\}} (a_5 |u|^p - a_6 |u^0|^p) \, dx + \int_{\{u < u_0\}} (a_7 |u|^p - a_8 |u_0|^p) \, dx
$$

$$
\ge \int_\Omega a_9 |u|^p \, dx - a_9 \int_{\{u_0 \le u \le u^0\}} |u|^p \, dx
$$

$$
- \int_{\{u > u^0\}} a_6 |u^0|^p \, dx - \int_{\{u < u_0\}} a_8 |u_0|^p \, dx
$$

$$
\ge c_3 \int_\Omega |u|^p \, dx - c_4.
$$

(b) Truncated auxiliary BVP.

Our approach is heavily based on existence and comparison results of the following auxiliary BVP. Find $u \in V_0$ such that

$$Au + \lambda Bu = Pu, \tag{5.1.8}$$

where the constant $\lambda \geq 0$ will be specified later. The operator P on the right-hand side of (5.1.8) is defined by

$$Pu := F \circ T_0^0 u + \sum_{i=1}^{m} |F \circ T_i^0 u - F \circ T_0^0 u| - \sum_{j=1}^{m'} |F \circ T_0^j u - F \circ T_0^0 u|,$$

where $F \circ T_i^0$, $F \circ T_0^j$, and $F \circ T_0^0$ stand for the compositions of the Nemytskij operator F and the truncation operators T_i^0, T_0^j, and T_0^0, respectively. Thus the BVP (5.1.8) can be rewritten as the following operator equation in V_0^*:

$$\text{Find } u \in V_0: \quad (A - P + \lambda B)u = 0. \tag{5.1.9}$$

Since the truncations T_i^0, T_0^j, $T_0^0 : V \to [\underline{u}, \bar{u}] \subset V$ are bounded and continuous, it follows by hypothesis (H2) that also the composed operators $F \circ T_0^0$, $F \circ T_i^0$, $F \circ T_0^j : V \to L^q(\Omega) \subset V_0^*$ are bounded and continuous, which implies that $P : V_0 \to L^q(\Omega) \subset V_0^*$ given by

$$\langle Pu, \varphi \rangle = \int_{\Omega} (Pu)\,\varphi\,dx, \quad \varphi \in V_0,$$

is bounded and continuous too. By applying Young's inequality (see B.2) and taking (H2) into account we obtain for any $\varepsilon > 0$ an estimate of the form

$$|\langle Pu, u \rangle| \leq \varepsilon \|\nabla u\|_{L^p(\Omega)}^p + C(\varepsilon)\|u\|_{L^p(\Omega)}^p + C\|u\|_{L^p(\Omega)}, \tag{5.1.10}$$

where $C(\varepsilon)$ and C are some generic positive constants not depending on u. By hypotheses (A1) and (A2) for any $\eta > 0$ we have an estimate below

$$\langle Au, u \rangle \geq \mu \|\nabla u\|_{L^p(\Omega)}^p - \eta \|\nabla u\|_{L^p(\Omega)}^p - C(\eta)(\|k_0\|_{L^q(\Omega)}^q + \|u\|_{L^p(\Omega)}^p). \tag{5.1.11}$$

The Leray-Lions conditions (A1) and (A2) along with the properties of the operators B and P imply that the operator

$$\mathcal{A} := A - P + \lambda B$$

gives rise to a continuous and bounded mapping from V_0 into its dual V_0^* which is, in addition, pseudomonotone, see section D. Thus by the main theorem on pseudomonotone operators (Theorem D.1.1) the mapping $\mathcal{A}$: $V_0 \to V_0^*$ is surjective provided that $\mathcal{A}$ is coercive, i.e., that the following condition holds:

$$\frac{\langle \mathcal{A}u, u \rangle}{\|u\|_{V_0}} \to \infty \quad \text{as } \|u\|_{V_0} \to \infty.$$

The coercivity of $\mathcal{A}$, however, immediately follows from (5.1.5), (5.1.10), and (5.1.11) by taking ε and η sufficiently small such that $\mu > \varepsilon + \eta$ and by choosing λ sufficiently large. Hence, the main theorem on pseudomonotone operators (Theorem D.1.1) implies the existence of at least one solution of the auxiliary BVP (5.1.8) (respectively, (5.1.9)).

(c) Completion of the proof by comparison.

To complete the proof we only need to show that any solution u of the auxiliary problem (5.1.8) satisfies

$$\bar{u}_j \geq u \geq \underline{u}_i, \quad 1 \leq i \leq m, \ 1 \leq j \leq m'. \tag{5.1.12}$$

This is because (5.1.12) implies that also $u^0 \geq u \geq u_0$ is fulfilled, and thus for any solution u of (5.1.8) it follows $T_0^0 u = u$, and $T_i^0 u = u$ for $i = 1, ..., m$, and $T_0^j u = u$ for $j = 1, ..., m'$. Consequently we get $Pu = Fu$ and $Bu = 0$, and thus u must be a solution of the original problem (5.1.1) satisfying $u_0 \leq u \leq u^0$, which proves the lemma.

Proof of $\underline{u}_k \leq u$ for any $k = 1, ..., m$.

By definition the lower solution $\underline{u}_k \in V$ satisfies $\underline{u}_k \leq 0$ on $\partial \Omega$ and the inequality

$$\langle A\underline{u}_k, \varphi \rangle \leq \int_\Omega (F\underline{u}_k) \, \varphi \, dx \quad \text{for all } \varphi \in V_0 \cap L_+^p(\Omega). \tag{5.1.13}$$

Subtracting the equation (5.1.8) from (5.1.13) yields for any nonnegative test function $\varphi \in V_0 \cap L_+^p(\Omega)$ the inequality

$$\langle A\underline{u}_k - Au, \varphi \rangle \leq \int_\Omega (F\underline{u}_k - Pu) \, \varphi \, dx + \lambda \int_\Omega Bu \, \varphi \, dx. \tag{5.1.14}$$

The special test function technique will be used in the comparison of u and $\underline{u}_k$. By using hypothesis (A3) it follows that for any $\varepsilon > 0$ there exists a $\delta(\varepsilon) \in (0, \varepsilon)$ such that

$$\int\limits_{\delta(\varepsilon)}^{\varepsilon} \frac{dr}{\omega^q(r)} = 1.$$

Define the function $\theta_\varepsilon : \mathbb{R} \to [0, \infty)$ by

$$
\theta_\varepsilon(t) = \begin{cases} 0 & \text{if} \quad t < \delta(\varepsilon), \\[2mm] \displaystyle\int\limits_{\delta(\varepsilon)}^{t} \frac{dr}{\omega^q(r)} & \text{if} \quad \delta(\varepsilon) \le t \le \varepsilon, \\[2mm] 1 & \text{if} \quad t > \varepsilon, \end{cases}
$$

then for any $\varepsilon > 0$ the function θ_ε is Lipschitz continuous, increasing, and satisfies

$$
\theta_\varepsilon(t) \to \chi_{\{t>0\}} \quad \text{as} \quad \varepsilon \to 0,
$$

where $\chi_{\{t>0\}}$ denotes the characteristic function of the set $\{t > 0\}$, as well as

$$
0 \le \theta'_\varepsilon(t) = \begin{cases} \dfrac{1}{\omega^q(t)} & \text{for} \quad \delta(\varepsilon) \le t \le \varepsilon, \\[2mm] 0 & \text{otherwise}. \end{cases}
$$

Since $\underline{u}_k - u \le 0$ on $\partial\Omega$, the composition of θ_ε with $\underline{u}_k - u$ yields an admissible nonnegative test function $\varphi = \theta_\varepsilon(\underline{u}_k - u) \in V_0$. Using this special test function in (5.1.14) we get the inequality

$$
\langle A\underline{u}_k - Au, \theta_\varepsilon(\underline{u}_k - u)\rangle \le \int_\Omega (F\underline{u}_k - Pu + \lambda Bu)\theta_\varepsilon(\underline{u}_k - u)\, dx. \quad (5.1.15)
$$

By means of (A2) and (A3) the left-hand side of the last inequality can be estimated in the following way:

$$
\langle A\underline{u}_k - Au, \theta_\varepsilon(\underline{u}_k - u)\rangle
$$

$$
= \sum_{i=1}^{N} \int_\Omega (a_i(x, \underline{u}_k, \nabla\underline{u}_k) - a_i(x, u, \nabla u))\frac{\partial}{\partial x_i}\theta_\varepsilon(\underline{u}_k - u)\, dx
$$

$$
\ge \mu \int_\Omega |\nabla(\underline{u}_k - u)|^p\, \theta'_\varepsilon(\underline{u}_k - u)\, dx
$$

$$
- N \int_\Omega [|k_1| + |\underline{u}_k|^{p-1} + |u|^{p-1} + |\nabla u|^{p-1}]\,\omega(|\underline{u}_k - u|)
$$

$$
\times \theta'_\varepsilon(\underline{u}_k - u)\, |\nabla(\underline{u}_k - u)|\, dx
$$

$$
\ge \frac{\mu}{2} \int_\Omega |\nabla(\underline{u}_k - u)|^p\, \theta'_\varepsilon(\underline{u}_k - u)\, dx
$$

$$
- c(\mu) \int_\Omega g^q \omega^q(|\underline{u}_k - u|)\, \theta'_\varepsilon(\underline{u}_k - u)\, dx, \quad (5.1.16)
$$

where $g = |k_1| + |\underline{u}_k|^{p-1} + |u|^{p-1} + |\nabla u|^{p-1} \in L^q(\Omega)$. By definition of θ_ε we obtain from (5.1.16)

$$\langle A\underline{u}_k - Au, \theta_\varepsilon(\underline{u}_k - u)\rangle \geq -c(\mu) \int_{\{\delta(\varepsilon)<\underline{u}_k-u<\varepsilon\}} g^q \, dx, \qquad (5.1.17)$$

where the integral on the right-hand side of (5.1.17) tends to zero as $\varepsilon \to 0$. Applying Lebesgue's dominated convergence theorem to the right-hand side of (5.1.15) as $\varepsilon \to 0$ we obtain

$$\lim_{\varepsilon \to 0}\left[\int_\Omega (F\underline{u}_k - Pu)\,\theta_\varepsilon(\underline{u}_k - u)\,dx + \lambda\int_\Omega Bu\,\theta_\varepsilon(\underline{u}_k - u)\,dx\right]$$

$$= \int_\Omega (F\underline{u}_k - Pu)\,\chi_{\{\underline{u}_k-u>0\}}\,dx + \lambda\int_\Omega Bu\,\chi_{\{\underline{u}_k-u>0\}}\,dx$$

$$\leq \lambda\int_\Omega Bu\,\chi_{\{\underline{u}_k-u>0\}}\,dx = -\lambda\int_{\{\underline{u}_k-u>0\}}(u_0 - u)^{p-1}dx$$

$$\leq -\lambda\int_\Omega [(\underline{u}_k - u)^+]^{p-1}\,dx \leq 0. \qquad (5.1.18)$$

In the estimate (5.1.18) we have used the definitions of the truncation operators which imply that for $\underline{u}_k(x) - u(x) > 0$ we have $T_0^0 u(x) = u_0(x)$, $T_k^0 u(x) = \underline{u}_k(x)$, and $T_0^j u(x) = u_0(x)$ $(j = 1, ..., m')$, and hence

$$(F\underline{u}_k - Pu)\,\chi_{\{\underline{u}_k-u>0\}}$$

$$= \left(F\underline{u}_k - F \circ T_0^0 u - \sum_{i=1}^m |F \circ T_i^0 u - F \circ T_0^0 u|\right.$$

$$\left. + \sum_{j=1}^{m'} |F \circ T_0^j u - F \circ T_0^0 u|\right)\chi_{\{\underline{u}_k-u>0\}}$$

$$\leq \left(F\underline{u}_k - F \circ T_0^0 u - |F \circ T_k^0 u - F \circ T_0^0 u|\right.$$

$$\left. + \sum_{j=1}^{m'} |F \circ T_0^j u - F \circ T_0^0 u|\right)\chi_{\{\underline{u}_k-u>0\}}$$

$$= \left(F\underline{u}_k - Fu_0 - |F\underline{u}_k - Fu_0| + \sum_{j=1}^{m'} |Fu_0 - Fu_0|\right)\chi_{\{\underline{u}_k-u>0\}} \leq 0,$$

which implies that

$$\int_\Omega (F\underline{u}_k - Pu)\,\chi_{\{\underline{u}_k-u>0\}}\,dx \leq 0.$$

Passing to the limit as $\varepsilon \to 0$ from (5.1.15), (5.1.17), and (5.1.18) we get

$$0 \leq \int_{\Omega} [(\underline{u}_k - u)^+]^{p-1} \, dx \leq 0,$$

which proves that $\underline{u}_k \leq u$ for any $k = 1, ..., m$.

Proof of $\bar{u}_k \geq u$, for any $k = 1, ..., m'$.

By definition the upper solution $\bar{u}_k \in V$ satisfies $\bar{u}_k \geq 0$ on $\partial\Omega$ and the inequality

$$A\bar{u}_k \geq F\bar{u}_k. \tag{5.1.19}$$

Subtracting the inequality (5.1.19) from (5.1.8) we get for any nonnegative test function $\varphi \in V_0 \cap L_+^p(\Omega)$

$$\langle Au - A\bar{u}_k, \varphi \rangle \leq \int_{\Omega} (Pu - F\bar{u}_k) \, \varphi \, dx - \lambda \int_{\Omega} Bu \, \varphi \, dx. \tag{5.1.20}$$

Substituting φ in (5.1.20) by the special test function $\varphi = \theta_\varepsilon(u - \bar{u}_k) \in V_0 \cap L_+^p(\Omega)$ yields in a similar way

$$\langle Au - A\bar{u}_k, \theta_\varepsilon(u - \bar{u}_k) \rangle \geq -c(\mu) \int_{\{\delta(\varepsilon) < u - \bar{u}_k < \varepsilon\}} g^q \, dx \to 0 \text{ as } \varepsilon \to 0 \tag{5.1.21}$$

with some $g \in L^q(\Omega)$, as well as

$$\lim_{\varepsilon \to 0} \left[\int_{\Omega} (Pu - F\bar{u}_k) \, \theta_\varepsilon(u - \bar{u}_k) \, dx - \lambda \int_{\Omega} Bu \, \theta_\varepsilon(u - \bar{u}_k) \, dx \right]$$

$$= \int_{\Omega} (Pu - F\bar{u}_k) \, \chi_{\{u - \bar{u}_k > 0\}} \, dx - \lambda \int_{\Omega} Bu \, \chi_{\{u - \bar{u}_k > 0\}} \, dx$$

$$\leq -\lambda \int_{\Omega} Bu \, \chi_{\{u - \bar{u}_k > 0\}} \, dx = -\lambda \int_{\{u - \bar{u}_k > 0\}} (u - u^0)^{p-1} dx$$

$$\leq -\lambda \int_{\Omega} [(u - \bar{u}_k)^+]^{p-1} \, dx \leq 0, \tag{5.1.22}$$

In (5.1.22) we have used the fact that the inequality

$$\int_{\Omega} (Pu - F\bar{u}_k) \, \chi_{\{u - \bar{u}_k > 0\}} \, dx \leq 0$$

holds, which can be seen as follows: For $u(x) - \bar{u}_k(x) > 0$ one readily verifies that $T_0^0 u(x) = u^0(x)$, $T_0^k u(x) = \bar{u}_k(x)$, and $T_i^0 u(x) = u^0(x)$, $i = 1, ..., m$. Thus we have

$$(Pu - F\bar{u}_k)\, \chi_{\{u - \bar{u}_k > 0\}}$$

$$= \Big(F \circ T_0^0 u - F\bar{u}_k + \sum_{i=1}^{m} |F \circ T_i^0 u - F \circ T_0^0 u|$$

$$- \sum_{j=1}^{m'} |F \circ T_0^j u - F \circ T_0^0 u| \Big) \chi_{\{u - \bar{u}_k > 0\}}$$

$$\leq \Big(F \circ T_0^0 u - F\bar{u}_k + \sum_{i=1}^{m} |F \circ T_i^0 u - F \circ T_0^0 u|$$

$$- |F \circ T_0^k u - F \circ T_0^0 u| \Big) \chi_{\{u - \bar{u}_k > 0\}}$$

$$= \Big(Fu^0 - F\bar{u}_k - |F\bar{u}_k - Fu^0| + \sum_{i=1}^{m} |Fu^0 - Fu^0| \Big) \chi_{\{u - \bar{u}_k > 0\}} \leq 0.$$

From (5.1.20) and (5.1.21), (5.1.22) we finally get

$$0 \leq \int_{\Omega} [(u - \bar{u}_k)^+]^{p-1}\, dx \leq 0,$$

which proves $u \leq \bar{u}_k$ for any $k = 1, ..., m'$. This completes the proof of Lemma 5.1.1. □

Remark 5.1.2. An immediate consequence of Lemma 5.1.1 is the existence of a solution u of the BVP 5.1.1 within the order interval $[\underline{u}, \bar{u}]$ formed by an ordered pair of upper and lower solutions $\bar{u}$ and $\underline{u}$, respectively. Denote the set of all solutions of (5.1.1) contained in $[\underline{u}, \bar{u}]$ by $\mathcal{S}$, i.e.,

$$\mathcal{S} = \{u \in V_0 \mid \underline{u} \leq u \leq \bar{u} \text{ and } u \text{ solves the BVP (5.1.1)}\}.$$

Then $\mathcal{S} \neq \emptyset$ has the following property.

Corollary 5.1.1. *The solution set $\mathcal{S}$ enclosed by the upper and lower solutions $\bar{u}$ and $\underline{u}$ is a directed set which means that for any $u_1, u_2 \in \mathcal{S}$ there exists an element $u_3 \in \mathcal{S}$ such that $u_1 \leq u_3$ and $u_2 \leq u_3$, and there is an element $u_4 \in \mathcal{S}$ such that $u_1 \geq u_4$ and $u_2 \geq u_4$.*

Proof. Given solutions $u_1, u_2 \in \mathcal{S}$. Then these solutions are, in particular, lower solutions of the BVP (5.1.1) and by Lemma 5.1.1 there exists a solution u_3 of (5.1.1) within the order interval $[u_0, \bar{u}]$ with $u_0 = \max(u_1, u_2)$, which shows $u_3 \in \mathcal{S}$ and $u_1 \leq u_3$, $u_2 \leq u_3$. Since u_1, u_2 are also upper solutions of the BVP (5.1.1), then by Lemma 5.1.1 there exists a solution of (5.1.1) within the interval $[\underline{u}, u^0]$ with $u^0 = \min(u_1, u_2)$, which proves the assertion of the corollary.   □

Lemma 5.1.2. *The solution set $\mathcal{S}$ of the BVP (5.1.1) has the following properties:*

  (i) *$\mathcal{S}$ is bounded in V_0.*
  (ii) *If (u_n) is an increasing sequence in $\mathcal{S}$, it converges weakly in V_0 and strongly in $L^p(\Omega)$ and its limit belongs to $\mathcal{S}$.*

Proof. Let us first prove the boundedness of $\mathcal{S}$. To this end let $u \in \mathcal{S}$ be arbitrarily given. With the special test function $\varphi = u$ we get

$$\langle Au - Fu, u \rangle = 0. \tag{5.1.23}$$

Recalling the estimate (5.1.10) of part (b) of the proof of Lemma 5.1.1, which holds likewise also for F, since $u \in [\underline{u}, \bar{u}]$, we have the following estimate

$$|\langle Fu, u \rangle| \leq \varepsilon \|\nabla u\|_{L^p(\Omega)}^p + C(\varepsilon)\|u\|_{L^p(\Omega)}^p + C\|u\|_{L^p(\Omega)}$$

for any $\varepsilon > 0$. Estimate (5.1.11) yields

$$\langle Au, u \rangle \geq \mu \|\nabla u\|_{L^p(\Omega)}^p - \eta \|\nabla u\|_{L^p(\Omega)}^p - C(\eta)(\|k_0\|_{L^q(\Omega)}^q + \|u\|_{L^p(\Omega)}^p),$$

for any $\eta > 0$. By selecting ε and η sufficiently small and taking into account that $\mathcal{S}$ is $L^p(\Omega)$-bounded from (5.1.23) we obtain

$$\|u\|_V \leq C \quad \text{for all } u \in \mathcal{S}. \tag{5.1.24}$$

Let $(u_n) \subseteq \mathcal{S}$ be an increasing sequence. Then by (5.1.24) and applying Lebesgue's dominated convergence theorem there is a $u \in L^p(\Omega)$ such that $u_n \to u$ in $L^p(\Omega)$. Since by (5.1.24) the sequence (u_n) is relatively weakly compact in V_0 and since due to the compact embedding $V_0 \subset L^p(\Omega)$ each weakly convergent subsequence of (u_n) has the same limit u, it follows that $u_n \rightharpoonup u$ in V_0. To complete the proof we need to show that the limit u belongs to $\mathcal{S}$. By definition u_n satisfy

$$\langle Au_n - Fu_n, \varphi \rangle = 0 \quad \text{for all } \varphi \in V_0. \tag{5.1.25}$$

Since $u_n \rightharpoonup u$ in V_0 and $\varphi = u_n - u$ is an admissible test function we get, in particular,

$$\lim_{n \to \infty} \langle (A - F)u_n, u_n - u \rangle = 0.$$

The pseudomonotonicity of the continuous and bounded operator $A - F$: $V_0 \to V_0^*$ implies (see D.1, D.2)

$$(A - F)u_n \rightharpoonup (A - F)u \quad \text{in } V_0^*,$$

which allows to pass to the limit as $n \to \infty$ in the equation (5.1.25) showing that u is a solution of (5.1.1), and hence $u \in S$, since obviously $u \in [\underline{u}, \bar{u}]$. $\square$

5.1.3 Proof of Theorem 5.1.1.

In this subsection we are going to prove our main result. In the proof we focus on the existence of the greatest solution u^* of (5.1.1) within $[\underline{u}, \bar{u}]$, which corresponds with the greatest element of S, since the existence of the least solution u_* can be shown similarly.

Since the Sobolev space V is separable and $S \subset V_0 \subset V$, it follows that S is separable too. Let $U = \{v_n \mid n \in \mathbb{N}\}$ be a countable dense subset of S. By means of Lemma 5.1.1 we can construct a sequence $(u_n) \subseteq S$ in the following way:

Let $u_1 := v_1$. Let u_n be chosen, then select $u_{n+1} \in S$ so that

$$\max(v_n, u_n) \leq u_{n+1} \leq \bar{u}.$$

The existence of such an element $u_{n+1} \in S$ follows from Lemma 5.1.1. In this way we obtain an increasing sequence $(u_n) \subseteq S$, which due to Lemma 5.1.2 is weakly convergent in V_0 and strongly convergent in $L^p(\Omega)$ to $u = \sup_n u_n \in S$. Since by construction we have $\max(v_1, ..., v_n) \leq u_{n+1} \leq u$ for all $n = 1, 2, ...$, it follows that u is an upper bound of $U = \{v_n \mid n \in \mathbb{N}\}$. Thus U is contained in the order interval $[\underline{u}, u]$ which is a closed subset of V. Since U was a dense subset of S we obtain

$$S \subseteq \overline{U} \subseteq \overline{[\underline{u}, u]} = [\underline{u}, u].$$

Hence it follows that the limit $u \in S$ is an upper bound of S which shows that u must be the greatest element of S. This completes the proof of Theorem 5.1.1. $\square$

5.1.4 Compactness of the solution set.

The solution set S of the BVP (5.1.1) enclosed by the upper and lower solutions $\bar{u}$ and $\underline{u}$, respectively, which due to Theorem 5.1.1 possesses extremal elements is, in addition, also compact.

Theorem 5.1.2. *Let the hypotheses of Theorem 5.1.1 be satisfied. Then the solution set $S \subset V_0$ is compact.*

Proof. Given any sequence $(u_n) \subseteq S$. This sequence is norm-bounded in V_0 by Lemma 5.1.2, and thus there exists a subsequence (u_{n_k}) of (u_n) converging weakly to $u \in V_0$, i.e., $u_{n_k} \rightharpoonup u$. We are going to prove that (u_{n_k}) is strongly convergent in V_0 to u and $u \in S$. Denote for simplicity the subsequence by (u_k). Then by the pseudomonotonicity of $A - F$ one can show in just the same way as in the proof of Lemma 5.1.2 that the weak limit u belongs to S. To show the strong convergence $u_k \to u$ in V_0 let T be the truncation mapping defined by

$$ Tu = \begin{cases} \bar{u}(x) & \text{if} & u > \bar{u}(x), \\ u & \text{if} & \underline{u}(x) \le u \le \bar{u}(x), \\ \underline{u}(x) & \text{if} & u < \underline{u}(x), \end{cases} $$

and consider the operator $A_T : V_0 \to V_0^*$ defined by the semilinear form a_T as follows

$$ \langle A_T u, \varphi \rangle = a_T(u, \varphi) := \sum_{i=1}^N \int_\Omega a_i(x, Tu, \nabla u) \frac{\partial \varphi}{\partial x_i} \, dx. $$

Then any $v \in S$ is a solution of the modified operator equation

$$ v \in V_0 : \quad (A_T - F)v = 0 \quad \text{in } V_0^*, \tag{5.1.26} $$

such that, in particular, also u_k and u satisfy (5.1.26). The strong convergence of (u_k) now follows from the (S_+)-property (see D.2) of the operator $A_T - F$ which can be ensured provided the defining coefficients $a_i(x, Ts, \xi)$ of A_T admit the following additional estimate

$$ \sum_{i=1}^N a_i(x, Ts, \xi)\xi_i \ge c \, |\xi|^p - k(x) \tag{5.1.27} $$

for a.e. $x \in \Omega$, for all $s \in \mathbb{R}$ and for all $\xi \in \mathbb{R}^N$ with some constant $c > 0$ and some $k \in L^1(\Omega)$. Assumption (A2) yields with $\xi' = 0$

$$ \sum_{i=1}^N (a_i(x, Ts, \xi) - a_i(x, Ts, 0))\xi_i \ge \mu|\xi|^p. \tag{5.1.28} $$

By using Young's inequality $\sum_{i=1}^{N} a_i(x, Ts, 0)\xi_i$ can be estimated in the following way

$$|\sum_{i=1}^{N} a_i(x, Ts, 0)\xi_i| \leq \sum_{i=1}^{N}(|k_0(x)| + c_0|Ts|^{p-1})|\xi|$$

$$\leq \sum_{i=1}^{N}(|k_0(x)| + c_0(|\bar{u}(x)| + |\underline{u}(x)|)^{p-1})|\xi|$$

$$\leq C(\varepsilon)(|k_0(x)|^q + (|\bar{u}(x)| + |\underline{u}(x)|)^p + \varepsilon \, |\xi|^p \qquad (5.1.29)$$

for any $\varepsilon > 0$. Thus condition (5.1.27) follows from (5.1.28) and (5.1.29) with ε sufficiently small ($\varepsilon < \mu$) showing the (S_+)-property of $A_T - F$ (see D.2), which implies due to

$$\langle(A_T - F)u_k, u_k - u\rangle = 0 \quad \text{for all } k = 1, 2, \dots$$

that $u_k \to u$ in V_0, completing the proof. □

In Remark 5.1.1 we already mentioned that the extremality result of Theorem 5.1.1 remains valid if hypothesis (A2) is replaced by the weaker ones (A2)$_1$ and (A2)$_2$, and condition (5.1.2) of hypothesis (A3) is replaced by the stronger condition (5.1.3), so that there is an interplay between the p-ellipticity in (A2) and the q-modulus of continuity condition in (A3). Let (A3') denote the modified hypothesis (A3) with condition (5.1.2) replaced by (5.1.3), i.e.,

(A3') $|a_i(x, s, \xi) - a_i(x, s', \xi)|$
$\leq [k_1(x) + |s|^{p-1} + |s'|^{p-1} + |\xi|^{p-1}]\omega(|s - s'|),$
for some function $k_1 \in L^q(\Omega)$, for a.e. $x \in \Omega$, for all $s, s' \in \mathbb{R}$ and for all $\xi \in \mathbb{R}^N$, where the modulus of continuity ω satisfies $\int_{0+} \frac{dr}{\omega(r)} = +\infty$.

Then we have the following result.

Corollary 5.1.2. *Let hypotheses (A1), (A2)$_1$, (A2)$_2$, (A3'), and (H1), (H2) be satisfied. Then the BVP (5.1.1) possesses extremal solutions within the sector $[\underline{u}, \bar{u}]$ formed by the lower and upper solution $\underline{u}$ and $\bar{u}$, respectively.*

Proof. For the proof we only need to verify that the crucial Lemma 5.1.1 remains valid under the hypotheses of the corollary. The Leray-Lions conditions (A1), (A2)$_1$, and (A2)$_2$ along with hypotheses (H1) and (H2) imply that the operator $A - P + \lambda B : V_0 \to V_0^*$ of the truncated auxiliary BVP is continuous, bounded, pseudomonotone, and coercive which implies the existence of a solution of the auxiliary BVP. The only place where the modulus of continuity comes into picture and where the interplay with the p-ellipticity appears is in part (c) of the proof of Lemma 5.1.1, which deals with the comparison of lower (respectively, upper) solutions $\underline{u}_k$ (respectively, $\bar{u}_k$) of the BVP (5.1.1) with any solution u of the associated auxiliary truncated BVP. In order to prove that $\underline{u}_k \leq u$, one crucial step is to show that

$$\langle A\underline{u}_k - Au, \theta_\varepsilon(\underline{u}_k - u)\rangle \geq l(\varepsilon) \to 0 \text{ as } \varepsilon \to 0 \qquad (5.1.30)$$

holds, where the function $\theta_\varepsilon : \mathbb{R} \to [0, \infty)$ is related with the modulus of continuity ω by

$$\theta_\varepsilon(t) = \begin{cases} 0 & \text{if} \quad t < \delta(\varepsilon), \\ \displaystyle\int_{\delta(\varepsilon)}^{t} \frac{dr}{\omega(r)} & \text{if} \quad \delta(\varepsilon) \leq t \leq \varepsilon, \\ 1 & \text{if} \quad t > \varepsilon, \end{cases}$$

which makes sense due to condition (5.1.3). Again we have that for any $\varepsilon > 0$ the function θ_ε is Lipschitz continuous, increasing, and satisfies

$$\theta_\varepsilon(t) \to \chi_{\{t>0\}} \quad \text{as} \quad \varepsilon \to 0,$$

where $\chi_{\{t>0\}}$ denotes the characteristic function of the set $\{t > 0\}$, as well as

$$0 \leq \theta'_\varepsilon(t) = \begin{cases} \dfrac{1}{\omega(t)} & \text{for} \quad \delta(\varepsilon) \leq t \leq \varepsilon, \\ 0 & \text{otherwise}. \end{cases}$$

In a similar way as in part (c) of the proof of Lemma 5.1.1 we get

$$\langle A\underline{u}_k - Au, \theta_\varepsilon(\underline{u}_k - u)\rangle$$

$$= \sum_{i=1}^{N} \int_\Omega (a_i(x, \underline{u}_k, \nabla\underline{u}_k) - a_i(x, u, \nabla u)) \frac{\partial}{\partial x_i} \theta_\varepsilon(\underline{u}_k - u)\, dx$$

$$\geq \sum_{i=1}^{N} \int_\Omega (a_i(x, \underline{u}_k, \nabla\underline{u}_k) - a_i(x, \underline{u}_k, \nabla u)) \frac{\partial(\underline{u}_k - u)}{\partial x_i} \theta'_\varepsilon(\underline{u}_k - u)\, dx$$

$$- N \int_\Omega [|k_1| + |\underline{u}_k|^{p-1} + |u|^{p-1} + |\nabla u|^{p-1}]\, \omega(|\underline{u}_k - u|)$$

$$\times \theta'_\varepsilon(\underline{u}_k - u)\, |\nabla(\underline{u}_k - u)|\, dx$$

$$\geq -N \int_{\{\delta(\varepsilon) < \underline{u}_k - u < \varepsilon\}} g\, |\nabla(\underline{u}_k - u)|\, dx \qquad (5.1.31)$$

where $g = |k_1| + |\underline{u}_k|^{p-1} + |u|^{p-1} + |\nabla u|^{p-1} \in L^q(\Omega)$. Since the term on the right-hand side of (5.1.31) tends to zero as $\varepsilon \to 0$, we have an estimate of the form (5.1.30) which proves the corollary. $\square$

Remark 5.1.3. If the coefficients $a_i(x, s, \xi)$ of the operator A are independent on s, then in view of Corollary 5.1.2 the usual Leray-Lions conditions (A1), (A2)$_1$, (A2)$_2$, and the growth condition (H2) on the lower order terms are sufficient to ensure the existence of extremal solutions of the BVP (5.1.1) within the order interval $[\underline{u}, \bar{u}]$ of the given upper and lower solutions $\bar{u}$ and $\underline{u}$, respectively, satisfying $\underline{u} \leq \bar{u}$. Thus, for example, if $A = \Delta_p u = -\sum_{i=1}^N \frac{\partial}{\partial x_i}(|\nabla u|^{p-2}\frac{\partial u}{\partial x_i})$, $p > 1$, which is the p-Laplacian, then $a_i(x, s, \xi) = |\xi|^{p-2}\xi_i$ and conditions (A1), (A2)$_1$, (A2)$_2$ are satisfied.

Remark 5.1.4. Let γv denote the trace of $v \in V$ (see C.3) and $h \in V_0^*$, then Theorem 5.1.1 holds likewise also for the inhomogeneous Dirichlet BVP

$$Au = Fu + h \quad \text{in } \Omega, \quad u = \gamma v \text{ on } \partial\Omega.$$

Moreover, nonlinear boundary conditions of the form

$$\frac{\partial u}{\partial \nu} = g(x, u) \quad \text{on } \partial\Omega$$

can also be treated by the method developed in this section. Here $\partial/\partial\nu$ denotes the outer conormal derivative on $\partial\Omega$ related with A, and given by

$$\frac{\partial u}{\partial \nu} = \sum_{i=1}^N a_i(x, u, \nabla u)\nu_i,$$

where $\nu = (\nu_1, ..., \nu_N)$ is the outer unit normal vector on $\partial\Omega$, and $g :$ $\partial\Omega \times \mathbb{R} \to \mathbb{R}$ is supposed to be a Carathéodory function satisfying a growth condition of the form

$$|g(x, s)| \leq k(x), \quad x \in \partial\Omega,$$

with $k \in L^q(\partial\Omega)$ and for all $s \in [\gamma\underline{u}, \gamma\bar{u}]$. Thus the last condition holds, for example, if g satisfies for all $s \in \mathbb{R}$ and some $k \in L^q(\partial\Omega)$ the growth condition

$$|g(x, s)| \leq k(x) + c|s|^{p-1}.$$

Remark 5.1.5. There is an alternative approach to prove the extremality result of Theorem 5.1.1. After having shown the existence of solutions within the interval $[\underline{u}, \bar{u}]$, the directedness of the solution set $\mathcal{S}$, which is a consequence of Lemma 5.1.1, is the crucial property to ensure extremality. In our treatment both the existence of solutions within $[\underline{u}, \bar{u}]$ and the directedness of the solution set $\mathcal{S}$ follow from Lemma 5.1.1. Alternatively one can prove first the existence of solutions between upper and lower solutions which can be done in a much simpler way and which is a known result (cf. [102]), and after that prove the directedness of the solution set $\mathcal{S}$ by showing that $\max(u_1, u_2)$ is a lower solution and $\min(u_1, u_2)$ is an upper solution for any solutions $u_1, u_2 \in \mathcal{S}$. However, this alternative method can not be applied, in general, in the same way to treat the corresponding parabolic initial boundary value problem without imposing additional regularity properties on the data. One of the main reasons for the approach given in this section is that we can use the technique developed here in a similar way to study the existence of extremal weak solutions of quasilinear parabolic initial boundary value problems which is the aim of the next section. The alternative method described above will be used in chapter 6 where we provide extremality results of elliptic differential inclusions.

5.2 Quasilinear parabolic problems

5.2.1 Notations, hypotheses, and the main result.
Let $\Omega \subset \mathbb{R}^N$ be a bounded domain with Lipschitz boundary $\partial\Omega$, and let $Q = \Omega \times (0, \tau)$ be a cylindrical domain with the lateral boundary $\Gamma = \partial\Omega \times (0, \tau)$, $\tau > 0$. Using the notation of section 5.1 we denote by $V = W^{1,p}(\Omega)$ and $V_0 = W_0^{1,p}(\Omega)$ the usual Sobolev spaces, and assume for the sake of simplicity that $2 \le p < \infty$. Let q be the dual real satisfying $1/p + 1/q = 1$. Identifying $L^2(\Omega)$ with its dual, then $V \subset L^2(\Omega) \subset V^*$ forms an evolution triple with all the embeddings being continuous, dense, and compact; see E.2.

We introduce the space $\mathcal{V} = L^p(0, \tau; V)$ of vector-valued functions; see E.1. Then its dual space is given by $\mathcal{V}^* = L^q(0, \tau; V^*)$; see E.1. Define the function space $\mathcal{W}$ by

$$\mathcal{W} = \{ w \in \mathcal{V} \mid \frac{\partial w}{\partial t} \in \mathcal{V}^* \},$$

where the derivative $\partial/\partial t$ is understood in the sense of vector-valued distributions; see E.2. The space $\mathcal{W}$ endowed with the norm

$$\|w\|_{\mathcal{W}} = \|w\|_{\mathcal{V}} + \|\partial w/\partial t\|_{\mathcal{V}^*}$$

is a Banach space which is separable and reflexive due to the separability and reflexivity of $\mathcal{V}$ and $\mathcal{V}^*$, respectively. Furthermore, it is well known that the embedding $\mathcal{W} \subset C([0, \tau]; L^2(\Omega))$ is continuous; see Theorem E.2.3. Finally, because $V \subset L^p(\Omega)$ is compactly embedded, we have by Aubin's lemma (Theorem E.2.2) a compact embedding of $\mathcal{W} \subset L^p(Q)$.

Replacing V in the above definitions of $\mathcal{V}$, $\mathcal{V}^*$, and $\mathcal{W}$ by V_0 we denote the corresponding spaces by $\mathcal{V}_0$, $\mathcal{V}_0^*$, and $\mathcal{W}_0$, respectively. Obviously $V_0 \subset L^2(\Omega) \subset V_0^*$ forms an evolution triple with all the embeddings being continuous, dense, and compact such that all statements made above remain true also for $\mathcal{V}_0$, $\mathcal{V}_0^*$, and $\mathcal{W}_0$.

Let $h \in \mathcal{V}_0^*$ be given. This section deals with weak solutions of the following quasilinear initial-Dirichlet boundary value problem (IBVP for short)

$$\left. \begin{array}{c} \dfrac{\partial u}{\partial t} + Au = Fu + h \quad \text{in } Q, \\[2mm] u = 0 \text{ in } \Omega \times \{0\} \quad \text{and} \quad u = 0 \text{ on } \Gamma, \end{array} \right\} \tag{5.2.1}$$

where A is a second order quasilinear differential operator in divergence form given by

$$Au(x, t) = -\sum_{i=1}^{N} \frac{\partial}{\partial x_i} a_i(x, t, u(x, t), \nabla u(x, t)),$$

and F denotes the Nemytskij operator associated with a Carathéodory function $f : Q \times \mathbb{R} \times \mathbb{R}^N \to \mathbb{R}$ and given by

$$Fu(x, t) = f(x, t, u(x, t), \nabla u(x, t)).$$

Remark 5.2.1. Without loss of generality and only for the sake of simplifying the presentation homogeneous initial and boundary conditions have been assumed. The case of inhomogeneous initial and boundary conditions of the form

$$u = \psi \text{ in } \Omega \times \{0\}, \quad \text{and} \quad u = \gamma g \text{ on } \Gamma,$$

where $\psi \in L^2(\Omega)$ and γg is the trace on Γ of some function $g \in \mathcal{W}$, can always be reduced to a homogeneous IBVP (5.2.1) by a simple translation $u \to u + w$, where the translation function w may be any $w \in \mathcal{W}$ satisfying the given initial and boundary values ψ and γg, respectively. Such a function always exists and may be given, for instance, by the unique solution of the IBVP

$$\frac{\partial w}{\partial t} - \Delta_p w = 0, \quad w = \gamma g \quad \text{on } \Gamma \quad \text{and } w = \psi \quad \text{in } \Omega \times \{0\},$$

where Δ_p is the p-Laplacian. This last IBVP can be transformed by translation $w \to w + g$ to an IBVP of the form

$$\frac{\partial \hat{w}}{\partial t} - \Delta_p \hat{w} = \hat{h}, \quad \hat{w} = 0 \quad \text{on } \Gamma \quad \text{and } w = \hat{\psi} \quad \text{in } \Omega \times \{0\},$$

whose unique solvability follows immediately by applying, e.g., [218, Theorem 30.A], because $\Delta_p : V_0 \to V_0^*$ is strictly monotone and coercive. Moreover, the translation preserves all the structure conditions to be imposed on the coefficients a_i and the right-hand side f of the operator A and F, respectively.

We assume the following conditions of Leray-Lions type on the coefficient functions a_i, $i = 1, ..., N$.

(A1) Each $a_i : Q \times \mathbb{R} \times \mathbb{R}^N \to \mathbb{R}$ satisfies Carathéodory conditions, i.e., $a_i(x, t, s, \xi)$ is measurable in $(x, t) \in Q$ for all $(s, \xi) \in \mathbb{R} \times \mathbb{R}^N$ and continuous in (s, ξ) for a.e. $(x, t) \in Q$. There exist a constant $c_0 > 0$ and a function $k_0 \in L^q(Q)$ such that

$$|a_i(x, t, s, \xi)| \leq k_0(x, t) + c_0(|s|^{p-1} + |\xi|^{p-1})$$

for a.e. $(x, t) \in Q$ and for all $(s, \xi) \in \mathbb{R} \times \mathbb{R}^N$.

(A2) $\sum_{i=1}^{N} (a_i(x, t, s, \xi) - a_i(x, t, s, \xi'))(\xi_i - \xi_i') \geq \mu |\xi - \xi'|^p$
for a.e. $(x, t) \in Q$, for all $s \in \mathbb{R}$, and for all $\xi, \xi' \in \mathbb{R}^N$ with μ being some positive constant.

(A3) $|a_i(x, t, s, \xi) - a_i(x, t, s', \xi)|$
$\leq [k_1(x, t) + |s|^{p-1} + |s'|^{p-1} + |\xi|^{p-1}] \omega(|s - s'|),$
for some function $k_1 \in L^q(Q)$, for a.e. $(x, t) \in Q$, for all $s, s' \in \mathbb{R}$
and for all $\xi \in \mathbb{R}^N$, where $\omega : [0, \infty) \to [0, \infty)$ is the *modulus of continuity* satisfying $\int_{0+} \frac{dr}{\omega^q(r)} = +\infty$.

Let us denote by $\langle \cdot, \cdot \rangle$ the duality pairing between V_0^* and V_0. Then as a consequence of (A1) the semilinear form a associated with the operator A by

$$\langle Au, \varphi \rangle = a(u, \varphi) = \sum_{i=1}^{N} \int_Q a_i(x, t, u, \nabla u) \frac{\partial \varphi}{\partial x_i} \, dx dt, \quad \text{for all } \varphi \in V_0,$$

is well defined for any $u \in V$ and gives rise to a continuous and bounded operator $A : V \to V_0^*$.

We introduce again the natural partial ordering in $L^p(Q)$ defined by the order cone $L_+^p(Q)$ of all nonnegative elements of $L^p(Q)$ which induces a corresponding partial ordering also in the subset $\mathcal{W}$ of $L^p(Q)$. If $\underline{u}, \bar{u} \in \mathcal{W}$ with $\underline{u} \le \bar{u}$, then

$$[\underline{u}, \bar{u}] = \{u \in \mathcal{W} \mid \underline{u} \le u \le \bar{u}\}$$

denotes the order interval formed by $\underline{u}$ and $\bar{u}$.

Let us define the notion of (weak) solution of the IBVP (5.2.1).

Definition 5.2.1. A function $u \in \mathcal{W}_0$ is called a *solution* of problem (5.2.1) if $Fu \in L^q(Q)$ such that

(i) $u(x, 0) = 0$ in Ω,
(ii) $\langle \frac{\partial u}{\partial t}, \varphi \rangle + a(u, \varphi) = \int_Q Fu \, \varphi \, dxdt + \langle h, \varphi \rangle$ for all $\varphi \in \mathcal{V}_0$.

The upper and lower solutions for (5.2.1) are defined as follows.

Definition 5.2.2. A function $\bar{u} \in \mathcal{W}$ is called an *upper solution* of the IBVP (5.2.1) if $F\bar{u} \in L^q(Q)$ and

(i) $\bar{u}(x, 0) \ge 0$ on Ω and $\bar{u} \ge 0$ on Γ,
(ii) $\langle \frac{\partial \bar{u}}{\partial t}, \varphi \rangle + a(\bar{u}, \varphi) \ge \int_Q F\bar{u} \, \varphi \, dxdt + \langle h, \varphi \rangle$ for all $\varphi \in \mathcal{V}_0 \cap L_+^p(Q)$.

Similarly a function $\underline{u} \in \mathcal{W}$ is a *lower solution* of (5.2.1) if the reversed inequalities hold in Definition 5.2.2 with $\bar{u}$ replaced by $\underline{u}$. Further we assume the following hypotheses:

(H1) Suppose the IBVP (5.2.1) has an upper solution $\bar{u}$ and a lower solution $\underline{u}$ such that $\underline{u} \le \bar{u}$.

(H2) There exist a function $k_2 \in L_+^q(Q)$ and a constant $c_1 \ge 0$ such that

$$|f(x, t, s, \xi)| \le k_2(x, t) + c_1 |\xi|^{p-1}$$

for a.e. $(x, t) \in Q$ and for all $\xi \in \mathbb{R}^N$ and $s \in [\underline{u}(x, t), \bar{u}(x, t)]$.

The notion of extremal solutions within the order interval $[\underline{u}, \bar{u}]$ is the same as in section 5.1.

The main result of this section is the following existence and extremality theorem.

Theorem 5.2.1. *Let hypotheses (A1)–(A3) and (H1), (H2) be satisfied. Then the IBVP (5.2.1) possesses extremal solutions within the interval $[\underline{u}, \bar{u}]$ formed by the ordered pair of lower and upper solutions $\underline{u}$ and $\bar{u}$, respectively.*

The proof of Theorem 5.2.1 which will be given in subsection 5.2.3 follows the same idea as in the elliptic case. However, it requires new techniques and tools from the theory of nonlinear evolution equations; see E.3. For this purpose we provide in the next subsection some preliminary results needed in the proof.

Remark 5.2.2. As in the elliptic case the proof of the extremality result, in particular the proof of the directedness of the solution set, requires a strong monotonicity condition (A2) with respect to ξ which is related with the modulus of continuity condition (A3). One can weaken condition (A2) on the expense of condition (A3) by assuming instead of (A2) the assumptions (A2)$_1$ and (A2)$_2$, and instead of (A3) the condition (A3') given as follows:

(A2)$_1$ $\sum_{i=1}^{N}(a_i(x,t,s,\xi) - a_i(x,t,s,\xi'))(\xi_i - \xi_i') > 0$
   for a.e. $(x,t) \in Q$, for all $s \in \mathbb{R}$, and for all $\xi, \xi' \in \mathbb{R}^N$ with $\xi \neq \xi'$.

(A2)$_2$ $\sum_{i=1}^{N} a_i(x,t,s,\xi)\xi_i \geq \nu|\xi|^p - k(x,t)$
   for a.e. $(x,t) \in Q$, for all $s \in \mathbb{R}$, and for all $\xi \in \mathbb{R}^N$ with some constant $\nu > 0$ and some function $k \in L^1(Q)$.

(A3') $|a_i(x,t,s,\xi) - a_i(x,t,s',\xi)|$
   $\leq [k_1(x,t) + |s|^{p-1} + |s'|^{p-1} + |\xi|^{p-1}]\,\omega(|s - s'|)$,
   for some function $k_1 \in L^q(Q)$, for a.e. $(x,t) \in Q$, for all $s, s' \in \mathbb{R}$, $\xi \in \mathbb{R}^N$, and the modulus of continuity ω satisfying $\int_{0+} \frac{dr}{\omega(r)} = +\infty$.

In particular (A2) may be replaced by the weaker conditions (A2)$_1$ and (A2)$_2$ if the coefficients a_i do not depend on s.

5.2.2 Preparatory results. Let us assume throughout this subsection that hypotheses (A1)-(A3) and (H1), (H2) are satisfied. First we prove the following parabolic version of Lemma 5.1.1 which yields both the existence of solutions within the interval $[\underline{u}, \bar{u}]$ and the directedness of the solution set enclosed by the given upper and lower solutions $\bar{u}$ and $\underline{u}$, respectively.

Lemma 5.2.1. *Let $\underline{u}_i \in [\underline{u}, \bar{u}]$, $i = 1,...,m$, be lower solutions and let $\bar{u}_j \in [\underline{u}, \bar{u}]$, $j = 1,...,m'$, be upper solutions of the IBVP (5.2.1) such that $u_0 := \max(\underline{u}_1,...,\underline{u}_m) \leq u^0 := \min(\bar{u}_1,...,\bar{u}_{m'})$. Then there exists a solution u of the IBVP (5.2.1) satisfying $u_0 \leq u \leq u^0$.*

Proof. The proof will be done in several steps and follows essentially the idea of the proof of Lemma 5.1.1.

   (a) Truncation and cut-off mappings.

We introduce truncation operators T_i^0, T_0^j, T_0^0 related with the functions $\underline{u}_i, \bar{u}_j, u_0, u^0$ and defined by

$$T_i^0 u(x,t) = \begin{cases} u^0(x,t) & \text{if} \quad u(x,t) > u^0(x,t), \\ u(x,t) & \text{if} \quad \underline{u}_i(x,t) \leq u(x,t) \leq u^0(x,t), \\ \underline{u}_i(x,t) & \text{if} \quad u(x,t) < \underline{u}_i(x,t), \end{cases}$$

and analogously the definition of the truncations T_0^j is obtained from that of T_i^0 replacing u^0 by $\bar{u}_j$ and $\underline{u}_i$ by u_0, whereas the definition of T_0^0 is obtained from T_i^0 replacing $\underline{u}_i$ by u_0.

The operators $T_0^0, T_i^0, T_0^j : \mathcal{V} \to [\underline{u}, \bar{u}] \subset \mathcal{V}$ are bounded and continuous (cf., e.g., [103]) such that in view of (H2) the composed operators $F \circ T_0^0$, $F \circ T_i^0$, $F \circ T_0^j : \mathcal{V} \to L^q(\Omega) \subset V_0^*$ are bounded and continuous. Further we introduce the following cut-off function $b : Q \times \mathbb{R} \to \mathbb{R}$ by

$$b(x,t,s) = \begin{cases} (s - u^0(x,t))^{p-1} & \text{if} \quad s > u^0(x,t), \\ 0 & \text{if} \quad u_0(x,t) \le s \le u^0(x,t), \\ -(u_0(x,t) - s)^{p-1} & \text{if} \quad s < u_0(x,t), \end{cases}$$

which is a Carathéodory function satisfying

$$|b(x,t,s)| \le k_3(x,t) + c_2|s|^{p-1} \tag{5.2.2}$$

for some positive constant c_2 and some function $k_3 \in L^q(Q)$. Thus the Nemytskij operator B associated with the function b is bounded and continuous from $L^p(Q)$ into $L^q(Q)$, and satisfies the estimate

$$\int_Q (Bu)u \, dxdt = \int_Q b(x,t,u(x,t))u(x,t) \, dxdt \ge c_3\|u\|_{L^p(Q)}^p - c_4 \tag{5.2.3}$$

for some positive constants c_3, c_4. The proof of (5.2.2) and (5.2.3) is similar as in Lemma 5.1.1.

(b) Truncated auxiliary IBVP.

The basic auxiliary problem is now the following one: Find $u \in \mathcal{W}_0$ such that

$$\begin{aligned} \frac{\partial u}{\partial t} + Au + \lambda Bu &= F \circ T_0^0 u + \sum_{i=1}^m |F \circ T_i^0 u - F \circ T_0^0 u| \\ &\quad - \sum_{j=1}^{m'} |F \circ T_0^j u - F \circ T_0^0 u| + h \quad \text{in } Q, \\ u &= 0 \quad \text{in } \Omega \times \{0\}. \end{aligned} \right\} \tag{5.2.4}$$

Let $L = \partial/\partial t$ and its domain $D(L) \subset \mathcal{V}_0$ be given by

$$D(L) = \{u \in \mathcal{W}_0 \mid u(\cdot, 0) = 0 \quad \text{in } \Omega\},$$

where $L : D(L) \subset \mathcal{V}_0 \to \mathcal{V}_0^*$ is defined by

$$\langle Lu, \varphi \rangle = \int_0^\tau < \frac{\partial u}{\partial t}(t), \varphi(t) > dt \quad \text{for all } \varphi \in \mathcal{V}_0,$$

and where $< \cdot, \cdot >$ denotes the duality pairing between V_0^* and V_0. The linear operator $L : D(L) \subset \mathcal{V}_0 \to \mathcal{V}_0^*$ can be shown to be closed, densely defined, and maximal monotone, cf. [218, Chapter 32]; see also E.3. Defining the operator P by

$$Pu := F \circ T_0^0 u + \sum_{i=1}^m |F \circ T_i^0 u - F \circ T_0^0 u| - \sum_{j=1}^{m'} |F \circ T_0^j u - F \circ T_0^0 u|,$$

then the IBVP (5.2.4) may be rewritten as the following problem in $\mathcal{V}_0^*$:
Find $u \in D(L) \subset \mathcal{V}_0$ such that

$$(L + A - P + \lambda B)u = h. \tag{5.2.5}$$

By means of (A1) and (A2) we have for any $\eta > 0$ an estimate

$$\langle Au, u \rangle \geq \mu \|\nabla u\|_{L^p(Q)}^p - \eta \|\nabla u\|_{L^p(Q)}^p - C(\eta)(\|k_0\|_{L^q(Q)}^q + \|u\|_{L^p(Q)}^p). \tag{5.2.6}$$

Since the truncation operators introduced above are bounded and continuous from $\mathcal{V}$ to itself and have their images in the interval $[\underline{u}, \bar{u}]$, it follows from hypothesis (H2) that the composed operator $P : \mathcal{V}_0 \to L^q(Q) \subset \mathcal{V}_0^*$ is continuous and bounded, and for any $\varepsilon > 0$ the following estimate holds

$$|\langle Pu, u \rangle| \leq \varepsilon \|\nabla u\|_{L^p(Q)}^p + C(\varepsilon)\|u\|_{L^p(Q)}^p + C\|u\|_{L^p(Q)}. \tag{5.2.7}$$

Obviously the operator $\mathcal{A} := A - P + \lambda B : \mathcal{V}_0 \to \mathcal{V}_0^*$ is continuous and bounded. Moreover, hypotheses (A1), (A2), and (H2) imply that $\mathcal{A} : \mathcal{V}_0 \to \mathcal{V}_0^*$ is pseudomonotone with respect to the graph norm topology of $D(L)$, which means that for any sequence (u_n) in $D(L)$ with $u_n \rightharpoonup u$ in $\mathcal{V}_0$, $Lu_n \rightharpoonup Lu$ in $\mathcal{V}_0^*$ and $\limsup_n \langle \mathcal{A}u_n, u_n - u \rangle \leq 0$ it follows that $\mathcal{A}u_n \rightharpoonup \mathcal{A}u$ in $\mathcal{V}_0^*$ and $\langle \mathcal{A}u_n, u_n \rangle \to \langle \mathcal{A}u, u \rangle$; see E.3. Applying the existence result on nonlinear evolution equations given by Theorem E.3.1, the mapping $L + \mathcal{A} : D(L) \to \mathcal{V}_0^*$ is surjective provided that $\mathcal{A} : \mathcal{V}_0 \to \mathcal{V}_0^*$ is coercive, i.e.,

$$\frac{\langle \mathcal{A}u, u \rangle}{\|u\|_{\mathcal{V}_0}} \to \infty \quad \text{as } \|u\|_{\mathcal{V}_0} \to \infty. \tag{5.2.8}$$

The coercivity (5.2.8) of $\mathcal{A}$ readily follows from (5.2.3), (5.2.6), and (5.2.7) by choosing ε and η sufficiently small such that $\mu > \varepsilon + \eta$, and by choosing λ sufficiently large, which implies the existence of a solution of the operator equation (5.2.5), and thus of the auxiliary IBVP (5.2.4).

(c) Completion of the proof by comparison.

The assertion of the lemma is proved if any solution u of the auxiliary IBVP (5.2.4) can be shown to satisfy

$$\bar{u}_j \geq u \geq \underline{u}_i, \quad 1 \leq i \leq m, \ 1 \leq j \leq m', \qquad (5.2.9)$$

since (5.2.9) implies that also $u^0 \geq u \geq u_0$ is fulfilled, and thus for any solution u of (5.2.4) we have $T_0^0 u = u$, and $T_i^0 u = u$ for $i = 1, ..., m$, and $T_0^j u = u$ for $j = 1, ..., m'$. Consequently we get $Pu = Fu$ and $Bu = 0$, and thus u must be a solution of the original IBVP (5.2.1) which satisfies $u_0 \leq u \leq u^0$, and thus the assertion of the lemma.

Proof of $\underline{u}_k \leq u$ for any $k = 1, ..., m$.

By definition the lower solution $\underline{u}_k \in \mathcal{W}$ satisfies the inequality

$$\frac{\partial \underline{u}_k}{\partial t} + A\underline{u}_k \leq F\underline{u}_k + h, \qquad (5.2.10)$$

as well as

$$\underline{u}_k \leq 0 \text{ in } \Omega \times \{0\}, \quad \text{and } \underline{u}_k \leq 0 \text{ on } \Gamma.$$

Subtracting the equation (5.2.5) from (5.2.10) we obtain for any nonnegative test function $\varphi \in \mathcal{V}_0 \cap L_+^p(Q)$ the inequality

$$\langle \frac{\partial(\underline{u}_k - u)}{\partial t}, \varphi \rangle + \langle A\underline{u}_k - Au, \varphi \rangle \leq \int_Q (F\underline{u}_k - Pu)\,\varphi\,dx dt + \lambda \int_Q Bu\,\varphi\,dx dt, \qquad (5.2.11)$$

and

$$\underline{u}_k - u \leq 0 \text{ in } \Omega \times \{0\}, \quad \text{and } \underline{u}_k - u \leq 0 \text{ on } \Gamma. \qquad (5.2.12)$$

Let $\theta_\varepsilon : \mathbb{R} \to [0, \infty)$ be the function introduced in the proof of Lemma 5.1.1 which is related with the modulus of continuity. Then in view of (5.2.12), $\varphi = \theta_\varepsilon(\underline{u}_k - u) \in \mathcal{V}_0$ is an admissible nonnegative test function which will be used in inequality (5.2.11). Defining the primitive Θ_ε of the nonnegative Lipschitz function θ_ε by $\Theta_\varepsilon(r) = \int_0^r \theta_\varepsilon(s)\,ds$, we get by means of [94, Lemma 1.11] for the first term on the left-hand side of (5.2.11) the following estimate

$$\langle \frac{\partial(\underline{u}_k - u)}{\partial t}, \theta_\varepsilon(\underline{u}_k - u)\rangle = \int_\Omega \Theta_\varepsilon(\underline{u}_k - u)(x, \tau)\, dx - \int_\Omega \Theta_\varepsilon(\underline{u}_k - u)(x, 0)\, dx \geq 0,$$

$$(5.2.13)$$

because $\Theta_\varepsilon(\underline{u}_k - u)(x, 0) = 0$ a.e. in Ω in view of $(\underline{u}_k - u)(x, 0) \leq 0$. By means of (A2) and (A3) the second term on the left-hand side of (5.2.11) can be estimated below in just the same way as in the elliptic case (cf. (5.1.16)), and yields

$$\langle A\underline{u}_k - Au, \theta_\varepsilon(\underline{u}_k - u)\rangle \geq -c(\mu)\int_{\{\delta(\varepsilon)<\underline{u}_k - u<\varepsilon\}} g^q\, dxdt \to 0 \text{ as } \varepsilon \to 0,$$

$$(5.2.14)$$

where g is some element of $L^q(Q)$. Furthermore, for the right-hand side of (5.2.11) we have

$$\lim_{\varepsilon\to 0}\left[\int_Q (F\underline{u}_k - Pu)\,\theta_\varepsilon(\underline{u}_k - u)\, dxdt + \lambda\int_Q Bu\,\theta_\varepsilon(\underline{u}_k - u)\, dxdt\right]$$

$$= \int_Q (F\underline{u}_k - Pu)\,\chi_{\{\underline{u}_k - u>0\}}\, dxdt + \lambda\int_Q Bu\,\chi_{\{\underline{u}_k - u>0\}}\, dxdt$$

$$\leq \lambda\int_Q Bu\,\chi_{\{\underline{u}_k - u>0\}}\, dxdt = -\lambda\int_{\{\underline{u}_k - u>0\}} (u_0 - u)^{p-1}\, dxdt$$

$$\leq -\lambda\int_Q [(\underline{u}_k - u)^+]^{p-1}\, dxdt \leq 0,$$

$$(5.2.15)$$

where χ is the characteristic function. In the estimate (5.2.15) the definitions of the truncation operators have been applied which result for $\underline{u}_k(x, t) - u(x, t) > 0$ in $T_0^0 u(x, t) = u_0(x, t)$, $T_k^0 u(x, t) = \underline{u}_k(x, t)$, and $T_0^j u(x, t) = u_0(x, t)$ for all $j = 1, ..., m'$. By using (5.2.13), (5.2.14), and (5.2.15) we obtain from (5.2.11)

$$0 \leq \int_Q [(\underline{u}_k - u)^+]^{p-1}\, dxdt \leq 0,$$

which proves that $\underline{u}_k \leq u$ for any $k = 1, ..., m$.

The proof of $\bar{u}_k \geq u$, for any $k = 1, ..., m'$ can be done in a similar way. This completes the proof of Lemma 5.2.1. $\qquad\qquad\square$

An immediate consequence of Lemma 5.2.1 is the existence of solutions of the IBVP (5.2.1) within the interval $[\underline{u}, \bar{u}]$, which follows by setting $\underline{u}_i = \underline{u}$ and $\bar{u}_j = \bar{u}$. Also the following corollary follows readily from Lemma 5.2.1.

Corollary 5.2.1. *Let S denote the solution set of the IBVP (5.2.1) enclosed by the upper and lower solution $\bar{u}$ and $\underline{u}$, respectively, i.e.,*

$$S = \{u \in \mathcal{W}_0 \mid u \in [\underline{u}, \bar{u}] \text{ and } u \text{ is a solution of the IBVP (5.2.1)}\}.$$

Then S is directed which means that whenever $u_1, u_2 \in S$ there exists an element $u_3 \in S$ such that $u_1 \le u_3$ and $u_2 \le u_3$, and an element $u_4 \in S$ such that $u_1 \ge u_4$ and $u_2 \ge u_4$.

The following result will be used in the proof of Theorem 5.2.1.

Lemma 5.2.2. *The solution set S of the IBVP (5.2.1) has the following properties:*

  (i) *S is bounded in $\mathcal{W}_0$.*
 (ii) *If (u_n) is an increasing sequence in S, it converges weakly in $\mathcal{W}_0$ and strongly in $L^p(Q)$ and its limit belongs to S.*

Proof. We prove the boundedness of S first. To this end let $u \in S$ be arbitrarily given. With the special test function $\varphi = u$ we get

$$\langle \frac{\partial u}{\partial t}, u \rangle + a(u, u) = \int_Q (Fu)u \, dxdt + \langle h, u \rangle. \qquad (5.2.16)$$

Since $u \in [\underline{u}, \bar{u}]$ are uniformly $L^p(Q)$-bounded, we get by means of (A1), (A2) and (H2) and using the estimate (5.2.6) and (5.2.7) the following inequalities

$$a(u, u) = \langle Au, u \rangle \ge \mu \|\nabla u\|^p_{L^p(Q)} - \varepsilon_1 \|\nabla u\|^p_{L^p(Q)} - C(\varepsilon_1) \qquad (5.2.17)$$

and

$$|\langle Fu, u \rangle| = \left| \int_Q (Fu)u \, dxdt \right| \le \varepsilon_2 \|\nabla u\|^p_{L^p(Q)} + C(\varepsilon_2) \qquad (5.2.18)$$

for any $\varepsilon_i > 0$ and with constants $C(\varepsilon_i)$, $i = 1, 2$, not depending on u. Because of

$$\langle \frac{\partial u}{\partial t}, u \rangle = \frac{1}{2} \|u(\cdot, \tau)\|^2_{L^2(\Omega)}, \quad \text{and}$$

$$|\langle h, u \rangle| \le \|h\|_{\mathcal{V}_0^*} \|u\|_{\mathcal{V}_0} \le C(\varepsilon_3) \|h\|^q_{\mathcal{V}_0^*} + \varepsilon_3 \|u\|^p_{\mathcal{V}_0}$$

for any $\varepsilon_3 > 0$, we obtain from (5.2.16) by using (5.2.17) and (5.2.18) and selecting ε_i, $i = 1, 2, 3$, sufficiently small a uniform bound in $\mathcal{V}_0$, i.e.,

$$\|u\|_{\mathcal{V}_0} \le C \quad \text{for all } u \in S. \qquad (5.2.19)$$

By means of (5.2.19) and (A1), (H2) one easily gets

$$\|\frac{\partial u}{\partial t}\|_{V_0^*} = \sup_{\|\varphi\|_{V_0}=1} |\langle\frac{\partial u}{\partial t}, \varphi\rangle| \leq C \quad \text{for all } u \in \mathcal{S}, \qquad (5.2.20)$$

such that (5.2.19) and (5.2.20) imply that $\mathcal{S}$ is uniformly bounded, i.e.,

$$\|u\|_{\mathcal{W}_0} \leq C \quad \text{for all } u \in \mathcal{S}. \qquad (5.2.21)$$

To prove (ii) of the lemma, let $(u_n) \subseteq \mathcal{S}$ be an increasing sequence. Then in view of the boundedness of (u_n) according to (5.2.21) and applying Lebesgue's dominated convergence theorem as well as the compact embedding of $\mathcal{W}_0 \subset L^p(Q)$ there is a function $w \in \mathcal{W}_0$ such that $u_n \rightharpoonup w$ in $\mathcal{W}_0$ and $u_n \to w$ in $L^p(Q)$ as $n \to \infty$. To complete the proof we need to show that the limit w belongs to $\mathcal{S}$. Since $u_n \in D(L)$ and $D(L)$ is closed with respect to the norm of $\mathcal{W}_0$ and convex, it follows that the weak limit $w \in D(L)$. Furthermore, we have (note $L = \partial/\partial t$)

$$\begin{aligned}
\langle(A - F)u_n, u_n - w\rangle &= -\langle Lu_n, u_n - w\rangle + \langle h, u_n - w\rangle \\
&= -\langle L(u_n - w), u_n - w\rangle - \langle Lw, u_n - w\rangle + \langle h, u_n - w\rangle \\
&\leq -\langle Lw, u_n - w\rangle + \langle h, u_n - w\rangle, \qquad (5.2.22)
\end{aligned}$$

due to

$$\langle L(u_n - w), u_n - w\rangle = \frac{1}{2}\|(u_n - w)(\cdot, \tau)\|^2_{L^2(\Omega)} \geq 0.$$

Since $-\langle Lw, u_n - w\rangle + \langle h, u_n - w\rangle \to 0$ as $n \to \infty$, we get from (5.2.22)

$$\limsup_{n \to \infty}\langle(A - F)u_n, u_n - w\rangle \leq 0,$$

and thus by the pseudomonotonicity of the operator $A - F : V_0 \to V_0^*$ with respect to the graph norm topology of $D(L)$ it follows that (see E.3), in particular,

$$(A - F)u_n \rightharpoonup (A - F)w \text{ in } V_0^*. \qquad (5.2.23)$$

The convergence properties of the sequence (u_n) and (5.2.23) allow to pass to the limit as $n \to \infty$ in the equation

$$\langle(L + A - F)u_n, \varphi\rangle = \langle h, \varphi\rangle \quad \text{for all } \varphi \in V_0,$$

which proves that the limit w is in $\mathcal{S}$. $\qquad\qquad\qquad\qquad\qquad\qquad$ $\square$

5.2.3 Proof of Theorem 5.2.1 and compactness result. In this subsection we are going to prove our main result, Theorem 5.2.1, and show that the solution set S is compact.

In the proof we focus on the existence of the greatest solution u^* only, since the existence of the least solution u_* can be shown analogously.

Since the Sobolev space W is separable and $S \subset W_0 \subset W$, it follows that S is separable too. Let $U = \{v_n \mid n \in \mathbb{N}\}$ be a countable dense subset of S. By means of Lemma 5.2.1 we can construct a sequence $(u_n) \subseteq S$ in the following way:

Let $u_1 := v_1$. Let u_n be chosen, then select $u_{n+1} \in S$ so that

$$\max(v_n, u_n) \leq u_{n+1} \leq \bar{u}.$$

The existence of such a $u_{n+1} \in S$ follows from Lemma 5.2.1. In this way we obtain an increasing sequence $(u_n) \subseteq S$, which by Lemma 5.2.2 is weakly convergent in W_0 and strongly convergent in $L^p(Q)$ to $w = \sup_n u_n \in S$. Since by construction we have $\max(v_1, ..., v_n) \leq u_{n+1} \leq w$ for all $n = 1, 2, ...,$ it follows that w is an upper bound of $U = \{v_n \mid n \in \mathbb{N}\}$. Thus U is contained in the order interval $[\underline{u}, w]$ which is a closed subset of W. Since U was a dense subset of S we obtain

$$S \subseteq \overline{U} \subseteq \overline{[\underline{u}, w]} = [\underline{u}, w].$$

Hence it follows that the limit $w \in S$ is an upper bound of S which shows that w must be the greatest element of S. This completes the proof of Theorem 5.2.1.   □

Theorem 5.2.2. *Let the hypotheses of Theorem 5.2.1 be satisfied. Then the solution set $S \subset W_0$ is compact.*

Proof. By inspection of the above proof of Theorem 5.2.1 we have the following result. Given any sequence $(u_n) \subseteq S$ then due to the boundedness of S in W_0 there is a subsequence of (u_n) denoted by (u_k) which is weakly convergent to u in W_0, and the weak limit u belongs to S. We are going to show that $u_k \to u$ strongly in W_0. Since any $w \in S$ is contained in the interval $[\underline{u}, \bar{u}]$ it must necessarily be a solution of the equation

$$\frac{\partial w}{\partial t} + A_T w = F_T w + h \quad \text{in } V_0^*, \qquad (5.2.24)$$

where the operator $A_T : V_0 \to V_0^*$ is defined by

$$\langle A_T w, \varphi \rangle = a_T(w, \varphi) = \sum_{i=1}^{N} \int_Q a_i(x, t, Tw, \nabla w) \frac{\partial \varphi}{\partial x_i} \, dx dt$$

and

$$F_T w(x,t) = f(x,t,Tw(x,t),\nabla w(x,t)),$$

with T being the truncation given by

$$Tu = \begin{cases} \bar{u}(x,t) & \text{if} & u > \bar{u}(x,t), \\ u & \text{if} & \underline{u}(x,t) \leq u \leq \bar{u}(x,t), \\ \underline{u}(x,t) & \text{if} & u < \underline{u}(x,t). \end{cases}$$

In a similar way as in subsection 5.1.4 one can show that the coefficients of A_T satisfy the condition

$$\sum_{i=1}^{N} a_i(x,t,Ts,\xi)\xi_i \geq c\,|\xi|^p - k(x,t)$$

for a.e. $(x,t) \in Q$, $s \in \mathbb{R}$ and $\xi \in \mathbb{R}^N$ with some constant $c > 0$ and $k \in L^1(Q)$, which in view of (A1) and (A2) and by Theorem E.3.2 implies that the operator A_T possesses the (S_+)-property with respect to $D(L)$, which means that for any sequence $(u_n) \subset D(L)$ with $u_n \rightharpoonup u$ in $\mathcal{V}_0$, $Lu_n \rightharpoonup Lu$ in $\mathcal{V}_0^*$ and $\limsup_{n\to\infty}\langle A_T u_n, u_n - u \rangle \leq 0$ it follows that $u_n \to u$ strongly in $\mathcal{V}_0$. To prove that the weakly convergent subsequence (u_k) is strongly convergent in $\mathcal{W}_0$ note that $u_k, u \in \mathcal{S}$, and thus for all k we have

$$\begin{aligned}
\langle A_T u_k, u_k - u \rangle &= -\langle \frac{\partial u_k}{\partial t}, u_k - u \rangle + \langle F_T u_k + h, u_k - u \rangle \\
&= -\langle \frac{\partial(u_k - u)}{\partial t}, u_k - u \rangle \\
&\quad -\langle \frac{\partial u}{\partial t}, u_k - u \rangle + \langle F_T u_k + h, u_k - u \rangle \\
&\leq -\langle \frac{\partial u}{\partial t}, u_k - u \rangle + \langle F_T u_k + h, u_k - u \rangle.
\end{aligned}$$

$$(5.2.25)$$

The weak convergence of $u_k \rightharpoonup u$ in $\mathcal{W}_0$ implies its strong convergence in $L^p(Q)$ due to the compact embedding of $\mathcal{W} \subset L^p(Q)$. Since (u_k) is bounded in $\mathcal{W}_0$, it follows that $(F_T u_k)$ is bounded in $L^q(Q)$, and thus the right-hand side of (5.2.25) tends to zero so that we get

$$\limsup_{k\to\infty}\langle A_T u_k, u_k - u \rangle \leq 0,$$

which implies by the (S_+)-property of A_T the strong convergence $u_k \to u$ in $\mathcal{V}_0$. To complete the proof we need to show also the strong convergence

$$\frac{\partial u_k}{\partial t} \to \frac{\partial u}{\partial t} \quad \text{in } \mathcal{V}_0^*,$$

which readily follows by using the equation (5.2.24) in the following way. Subtracting the corresponding equations for u_k and u we get

$$|\langle \frac{\partial u_k - u}{\partial t}, \varphi \rangle| \leq |\langle F_T u_k - F_T u, \varphi \rangle| + |\langle A_T u_k - A_T u, \varphi \rangle|,$$

and thus

$$\|\frac{\partial u_k}{\partial t} - \frac{\partial u}{\partial t}\|_{\mathcal{V}_0^*} = \sup_{\|\varphi\|_{\mathcal{V}_0}=1} |\langle \frac{\partial u_k - u}{\partial t}, \varphi \rangle|$$

$$\leq \|F_T u_k - F_T u\|_{\mathcal{V}_0^*} + \|A_T u_k - A_T u\|_{\mathcal{V}_0^*} \to 0$$

as $k \to \infty$ due to continuity of the operators $A_T, F_T : \mathcal{V}_0 \to \mathcal{V}_0^*$ and the strong convergence of $u_k \to u$ in $\mathcal{V}_0$. $\qquad \square$

Remark 5.2.3. Theorem 5.2.1 and Theorem 5.2.2 remain true if conditions (A2) and (A3) are replaced by (A2)$_1$, (A2)$_2$, and (A3').

5.3 Discontinuous quasilinear problems

The extremality results obtained so far combined with fixed point theorems for increasing operators obtained in chapter 1 will be applied in this section to treat quasilinear elliptic and parabolic problems with discontinuous nonlinearities. In order to emphasize the main idea we are going to concentrate on elliptic problems only, since the corresponding parabolic problems can be handled in an obvious analogous way. By using the notations and definitions of section 5.1 let us consider the following BVP

$$Au = Fu \quad \text{in } \Omega, \quad u = 0 \text{ on } \partial\Omega. \tag{5.3.1}$$

The difference of BVP (5.3.1) compared with problem (5.1.1) is that the Nemytskij operator F may now be discontinuous. More precisely, we assume in this section that F is related with a function $f : \Omega \times \mathbb{R} \times \mathbb{R} \times \mathbb{R}^N \to \mathbb{R}$ by

$$Fu(x) = f(x, u(x), u(x), \nabla u(x)),$$

where $f = f(x, r, s, \xi)$ may be discontinuous in the first and second argument. Let $\bar{u}$ and $\underline{u}$ be upper and lower solutions of (5.3.1), respectively. Instead of hypothesis (H2) of section 5.1 we impose the following hypotheses on the new f:

(H2)$_1$ There exist a function $k_2 \in L^q_+(\Omega)$ and a constant $c_1 \geq 0$ such that

$$|f(x, r, s, \xi)| \leq k_2(x) + c_1 |\xi|^{p-1}$$

for a.e. $x \in \Omega$ and for all $\xi \in \mathbb{R}^N$ and $r, s \in [\underline{u}(x), \bar{u}(x)]$.

(H2)$_2$ The function

(i) $(s, \xi) \to f(x, r, s, \xi)$ is continuous for a.e. $x \in \Omega$ and for all $r \in \mathbb{R}$.

(ii) $r \to f(x, r, s, \xi)$ is increasing for a.e. $x \in \Omega$ and for all $(s, \xi) \in \mathbb{R} \times \mathbb{R}^N$.

(iii) $(x, r) \to f(x, r, s, \xi)$ is sup-measurable for each $(s, \xi) \in \mathbb{R} \times \mathbb{R}^N$.

Remark 5.3.1. Hypothesis (H2)$_1$ corresponds with (H2) of section 5.1. Since f may now depend also discontinuously on u, Carathéodory conditions can no longer be assumed and have to be extended appropriately such that the least regularity requirement of f being sup-measurable is met. This property can easily be deduced from (H2)$_2$, where condition (iii) of (H2)$_2$ means that whenever $v : \Omega \to \mathbb{R}$ is measurable then $x \to f(x, v(x), s, \xi)$ is measurable in Ω for each $(s, \xi) \in \mathbb{R} \times \mathbb{R}^N$. The latter is fulfilled if the function $(x, r) \to f(x, r, s, \xi)$ is a so-called standard function in the sense of Shragin; cf. [12]. In particular, f may be of the form

$$f(x, r, s, \xi) = f_1(x, s, \xi) + f_2(x, r),$$

where $f_1 : \Omega \times \mathbb{R} \times \mathbb{R}^N \to \mathbb{R}$ is a Carathéodory function and $f_2 : \Omega \times \mathbb{R} \to \mathbb{R}$ is sup-measurable and increasing (possibly discontinuous) in r. The special case that $f_2 \equiv 0$ coincides with the situation of section 5.1.

The main result of this section is the following existence and extremality theorem.

Theorem 5.3.1. *Let hypotheses (A1)–(A3) of section 5.1 on the operator A be satisfied, and let there exist upper and lower solutions $\bar{u}$ and $\underline{u}$ of the BVP (5.3.1), respectively, satisfying $\underline{u} \leq \bar{u}$. Then under assumptions (H2)$_1$ and (H2)$_2$ the BVP (5.3.1) possesses extremal solutions within the interval $[\underline{u}, \bar{u}]$.*

Proof. In the proof we focus on the existence of the greatest solution u^* within $[\underline{u}, \bar{u}]$ only and remark that the existence of the least solution within the interval can be shown analogously. The idea of proof is to transform the discontinuous BVP (5.3.1) into a fixed point problem for an increasing (but not necessarily continuous) operator that maps some subset of V into itself, which allows one to apply a fixed point theorem obtained in the first chapter.

(a) Definition of a fixed point operator G.

For any upper solution $v \in [\underline{u}, \bar{u}]$ of the BVP (5.3.1) we consider the following auxiliary BVP

$$Au = F_v u \quad \text{in } \Omega, \quad u = 0 \quad \text{on } \partial\Omega, \qquad (5.3.2)$$

where the Nemytskij operator F_v is defined by

$$F_v u(x) = f(x, v(x), u(x), \nabla u(x)).$$

The BVP (5.3.2) may be considered as the associated "continuous" BVP, since due to hypotheses (H2)$_1$ and (H2)$_2$ for fixed $v \in [\underline{u}, \bar{u}]$ the function $(x, s, \xi) \to f(x, v(x), s, \xi)$ is a Carathéodory function satisfying a certain growth condition, which implies that the operator $F_v : [\underline{u}, \bar{u}] \subset V \to L^q(\Omega) \subset V^*$ is continuous and bounded. One readily observes that v which is supposed to be an upper solution of (5.3.1) is also an upper solution of the BVP (5.3.2), and by the monotonicity of f with respect to r the lower solution $\underline{u}$ of (5.3.1) is also a lower solution of (5.3.2) satisfying $\underline{u} \leq v$. Applying the main result of section 5.1 there exist extremal solutions of the BVP (5.3.2) within the interval $[\underline{u}, v]$. Thus the operator that assigns to each upper solution $v \in [\underline{u}, \bar{u}]$ the greatest solution of the BVP (5.3.2) with respect to the interval $[\underline{u}, v]$ is well defined, and will be denoted by G. Define the partially ordered set $\mathcal{U} \subset V$ by

$$\mathcal{U} = \{v \in V \mid v \in [\underline{u}, \bar{u}] \text{ and } v \text{ is an upper solution of (5.3.1)}\}.$$

Then any fixed point of $G : \mathcal{U} \to V_0$ is obviously a solution of the BVP (5.3.1) within the interval $[\underline{u}, \bar{u}]$. Conversely, let $\tilde{u} \in [\underline{u}, \bar{u}]$ be any solution of (5.3.1) then $\tilde{u}$ is, in particular, also an upper solution of (5.3.1) and thus belongs to $\mathcal{U}$. But $\tilde{u}$ is trivially a solution of (5.3.2) with the right-hand side $F_{\tilde{u}}$ and of course the greatest one with respect to the interval $[\underline{u}, \tilde{u}]$ which shows $\tilde{u} = G\tilde{u}$. Thus $u \in [\underline{u}, \bar{u}]$ is a solution of the BVP (5.3.1) if and only if u is a fixed point of G.

(b) $G : \mathcal{U} \to \mathcal{U}$ is increasing.

By definition of G we always have $\underline{u} \leq Gv \leq v$ for any $v \in \mathcal{U}$ and thus $G: \mathcal{U} \to [\underline{u}, \bar{u}]$. To prove that $Gv \in \mathcal{U}$ for any $v \in \mathcal{U}$ we have to show that Gv is an upper solution of (5.3.1). This, however, readily follows by the monotonicity condition of the function f with respect to the variable r according to hypothesis $(H2)_2$ and in view of $Gv \in [\underline{u}, v]$. To prove the monotonicity of G, let $v_1, v_2 \in \mathcal{U}$, and assume that $v_1 \leq v_2$. Then we have

$$Gv_1 \in [\underline{u}, v_1] \text{ is greatest solution of } \quad Au = F_{v_1}u, \quad u = 0 \text{ on } \partial\Omega$$

and

$$Gv_2 \in [\underline{u}, v_2], \text{ is greatest solution of } \quad Au = F_{v_2}u, \quad u = 0 \text{ on } \partial\Omega. \quad (5.3.3)$$

Since $v_1 \leq v_2$, it follows that $Gv_1 \leq v_2$ and, in addition, Gv_1 is a lower solution and v_2 is an upper solution for the BVP (5.3.3). By means of Theorem 5.1.1 we infer the existence of solutions (even extremal solutions) of (5.3.3) within the interval $[Gv_1, v_2]$. However, since Gv_2 is the greatest solution of (5.3.3) in $[\underline{u}, v_2] \supseteq [Gv_1, v_2]$ it must exceed, in particular, also Gv_1, and thus $Gv_1 \leq Gv_2$.

(c) Applying an abstract fixed point result.

We apply the abstract fixed point result given by Theorem 1.1.1. In order to apply Theorem 1.1.1 to the operator $G : \mathcal{U} \to \mathcal{U}$, note that $\bar{u} \in \mathcal{U}$ is obviously an upper bound of $G[\mathcal{U}]$. Hence for the existence of the greatest fixed point of G in $\mathcal{U}$ it suffices to show that each decreasing sequence (u_n) of $G[\mathcal{U}]$ converges weakly in V to an element of $\mathcal{U}$. To this end note that $G[\mathcal{U}]$ is uniformly bounded in V which has been shown in section 5.1. Thus, given a decreasing sequence $(u_n)_{n=0}^{\infty} \subset G[\mathcal{U}]$ it converges to w weakly in V_0 and strongly in $L^p(\Omega)$. Next we show that the weak limit w belongs to $\mathcal{U}$. By definition, $u_n = Gv_n \in V_0$, where $\underline{u} \leq u_n \leq v_n$ and

$$Au_n = F_{v_n}u_n. \quad (5.3.4)$$

By means of the special test function $u_n - w \in V_0$ and taking the boundedness of (u_n) in V_0 and $u_n \to w$ in $L^p(\Omega)$ into account we get from (5.3.4)

$$\langle Au_n, u_n - w \rangle = \int_\Omega (F_{v_n}u_n)\,(u_n - w)\,dx \to 0, \quad \text{as } n \to \infty. \quad (5.3.5)$$

Since $A : [\underline{u}, \bar{u}] \cap V_0 \to V_0^*$ enjoys the (S_+)-property, we obtain from (5.3.5) along with the weak convergence $u_n \rightharpoonup w$ in V_0 that the sequence (u_n) is even strongly convergent to w in V_0. The monotonicity of f with respect to r, which implies the monotonicity of $F_v u$ with respect to v, yields in view of $\underline{u} \le w \le u_n \le v_n \le \bar{u}$ the inequality

$$\langle Au_n, \varphi \rangle \ge \int_\Omega (F_w u_n)\, \varphi\, dx \quad \text{for all } \varphi \in V_0 \cap L^p_+(\Omega). \tag{5.3.6}$$

Finally, the strong convergence of $u_n \to w$ in V_0 allows one to pass to the limit in inequality (5.3.6) as $n \to \infty$, which shows that $w \in [\underline{u}, \bar{u}]$ satisfies

$$\langle Aw, \varphi \rangle \ge \int_\Omega (F_w w)\, \varphi\, dx \quad \text{for all } \varphi \in V_0 \cap L^p_+(\Omega),$$

and thus w is an upper solution of the BVP (5.3.1) because of $F_w w = Fw$, i.e., $w \in \mathcal{U}$. Hence, Theorem 1.1.1 can be applied, which yields the existence of the greatest fixed point u^* of G in $\mathcal{U}$ which implies that u^* must be greatest solution of (5.3.1) within the order interval $[\underline{u}, \bar{u}]$, since any fixed point of G corresponds with a solution of the BVP (5.3.1) in $[\underline{u}, \bar{u}]$, and vice versa. This completes the proof of Theorem 5.3.1. $\qquad \square$

Remark 5.3.2. In a recent paper by the authors [83] extremal solutions have been studied for reaction-diffusion equations with discontinuous reaction terms under discontinuous and nonlocal flux boundary condition in the form

$$\left.\begin{aligned}
\frac{\partial u(x,t)}{\partial t} + Au(x,t) &= f(x,t,u(x,t),u(x,t),\nabla u(x,t)) \quad \text{in } Q, \\
u(x,0) &= \psi(x) \text{ in } \Omega \\
\frac{\partial u(x,t)}{\partial \nu} &= g(x,t,u(x,t),u(x,t)) + \Phi(u)(x,t) \text{ on } \Gamma,
\end{aligned}\right\} \tag{5.3.7}$$

where A is a second order strongly elliptic differential operator and $\partial/\partial\nu$ denotes the exterior conormal derivative on Γ associated with the operator A. The particularities of the IBVP (5.3.7) are twofold. First, the nonlinearities f and g may depend discontinuously on the unknown function u, and second, the flux on Γ may be determined by an additional nonlocal term $\Phi(u)$ which may also depend discontinuously on u. As an example consider a nonlocal boundary term of the form

$$\Phi(u)(x,t) = \int_\Omega K(x,x',u(x',t))\, dx', \tag{5.3.8}$$

where the nonlinear kernel $K = K(x, x', r)$ with $K : \partial\Omega \times \Omega \times \mathbb{R} \to \mathbb{R}$ may be discontinuous with respect to all its arguments. The following special case of problem (5.3.7) has been treated in [189, 191] within the framework of classical solutions

$$\left.\begin{aligned} \frac{\partial u}{\partial t} + Au &= f(x, u) \quad \text{in } Q, \\ u(x, 0) &= \psi(x) \text{ in } \Omega \\ \frac{\partial u}{\partial \nu} + \alpha u &= \int_{\Omega} K(x, x')\, u(x', t)\, dx' \text{ on } \Gamma, \end{aligned}\right\} \qquad (5.3.9)$$

where α is some nonnegative constant, A is a linear strongly elliptic operator, and all data are assumed to be sufficiently smooth. The nonlocal IBVP (5.3.9) stands, e.g., for a model problem arising from quasi-static thermo-elasticity. It should be noted that due to the discontinuous nonlinearities involved in problem (5.3.7) and due to the nonlinear dependence of f on the gradients the monotone iteration method that is basically used in [189, 191] to deal with the IBVP (5.3.9) cannot be applied to (5.3.7). Thus, the extension of the existence results for the smooth problem (5.3.9) to the discontinuous one is by no means straightforward and requires completely different tools.

5.4 Notes and comments

As we already mentioned in the introduction to this chapter the upper and lower solution method combined with the monotone iterative technique is one way to prove the existence of extremal solutions within a sector of upper and lower solutions. However, this technique is restricted to problems that can be transformed to operator equations whose governing operators have certain monotonicity properties which are compatible with the underlying partial ordering, see, e.g., [46, 47, 48, 63, 165, 189, 205]. Motivated by the well-known Perron method on sub- and superharmonic functions the existence of classical extremal solutions between an ordered pair of upper and lower solutions has been proved for semilinear and quasilinear equations, e.g., in [8, 19, 179]. The upper and lower solution method was later developed by Deuel and Hess in [102, 103] for weak solutions of general quasilinear elliptic and parabolic problems. However, the problem of extremal weak solutions was not considered in [102, 103], and requires a method of proof that is essentially different from that used in [102, 103]. The existence of weak extremal solutions within a sector of weak upper and lower solutions has been proved in [50, 99, 163, 170, 172] for quasilinear problems where A is a monotone operator of divergence type whose coefficients $a_i = a_i(\cdot, \nabla u)$ do not depend on u. Extremality results for general quasilinear elliptic BVP involving nonmonotone operators A with coefficients $a_i = a_i(\cdot, u, \nabla u)$ have been obtained in different ways by Puel in [198] and Carl in [53]. The existence of extremal weak solutions for the corresponding general quasilinear parabolic BVP has been considered under different regularity assumptions on the coefficients a_i of the operator A and the lower order terms by Grenon in [121, 122] and Carl in [56]. Our approach to the extremality problem allows one to treat elliptic and parabolic problems in a unified way. It is based on truncation techniques used in [102] and [163] combined with the special test function technique employed in [45, 94].

Quasilinear discontinuous problems have been treated in section 5.3 by combining the extremality results of sections 5.1 and 5.2 with an abstract fixed point theorem for increasing operators from chapter 1. Discontinuous problems including also periodic boundary value problems as well as discontinuous nonlocal flux conditions have been studied by the authors, e.g., in [66, 68, 69, 70, 79, 83, 88, 123, 141].

Chapter 6

Differential inclusions of hemivariational inequality type

The variational formulation of various boundary value problems in Mechanics and Engineering governed by nonconvex, possibly nonsmooth energy functionals (so-called superpotentials) leads to hemivariational inequalities introduced by Panagiotopoulos; cf., e.g., [101, 181, 187, 188]. An abstract formulation of a hemivariational inequality reads as follows:

Let X be a reflexive Banach space and X^* its dual, let $A : X \to X^*$ be some pseudomonotone and coercive operator (see section D) satisfying certain continuity conditions, and let $h \in X^*$ be some given element. Find $u \in X$ such that

$$\langle Au - h, v - u \rangle + J^o(u; v - u) \geq 0 \quad \text{for all } v \in X, \tag{A}$$

where $J^o(u; v)$ denotes the generalized directional derivative in the sense of Clarke (cf. [96]) of a locally Lipschitz functional $J : X \to \mathbb{R}$. An equivalent multivalued formulation of (A) is given by

$$Au + \partial J(u) \ni h \quad \text{in } X^*, \tag{B}$$

where $\partial J(u)$ denotes Clarke's generalized gradient; cf. [96]. Abstract existence results for (A) (resp. (B)) can be found, e.g., in [181].

In this chapter we consider concrete realizations of (B) and its corresponding dynamic counterpart given in the form

$$\frac{\partial u}{\partial t} + Au \in \partial J(u) + h, \tag{C}$$

where A is assumed to be a quasilinear (in case (C) also time-dependent) elliptic differential operator of Leray-Lions type. We establish the method of upper and lower solutions for problems in the form (B) and (C) to obtain existence and enclosure results. Furthermore, the existence of extremal solutions within a sector of appropriately defined upper and lower solutions as well as compactness of the solution set are investigated.

191

6.1 Quasilinear elliptic inclusions with state-dependent subdifferentials

6.1.1 Notations and hypotheses. Let $\Omega \subset \mathbb{R}^N$ be a bounded domain with Lipschitz boundary $\partial\Omega$. In this section we deal with quasilinear elliptic differential inclusions under Dirichlet boundary conditions in the form

$$Au + \beta(\cdot, u, u) \ni 0 \quad \text{in } \Omega, \qquad u = 0 \quad \text{on } \partial\Omega, \qquad (6.1.1)$$

where

$$Au = -\sum_{i=1}^{N} \frac{\partial}{\partial x_i} a_i(\cdot, u, \nabla u)$$

is an operator of Leray-Lions type from $V_0 = W_0^{1,p}(\Omega)$ into its dual V_0^*, and the multifunction $\beta(x, u, \cdot) : \mathbb{R} \to 2^\mathbb{R} \setminus \emptyset$ is a maximal monotone graph in $\mathbb{R}^2$ (see section F), which is assumed to depend, in addition, on the solution itself.

Various free boundary problems can be reduced to boundary value problems (BVP) with discontinuous nonlinearities, which in turn can successfully be treated in a multivalued setting such as given by the inclusion (6.1.1); cf., e.g., [14, 91, 105].

Our aim is to prove the existence of weak solutions of (6.1.1) lying between appropriately defined upper and lower solutions. To this end the problem under consideration is shown to be equivalent to some quasivariational inequality of the form

$$\langle Au, \varphi - u \rangle + J(u, \varphi) - J(u, u) \geq 0.$$

By applying truncation and comparison techniques this quasi-variational inequality can be solved using fixed point arguments. Finally, the solution set enclosed by upper and lower solutions will be proved to have extremal elements with respect to the natural partial ordering of functions introduced in section 5.1.

Let $V = W^{1,p}(\Omega)$ and $V_0 = W_0^{1,p}$ denote the usual Sobolev spaces as in section 5.1. We assume conditions (A1), (A2), and (A3) of section 5.1 on the coefficients a_i, $i = 1, ...N$, of the operator A to be satisfied, and introduce accordingly the semilinear form a related with A by

$$\langle Au, \varphi \rangle = a(u, \varphi) = \int_\Omega \sum_{i=1}^{N} a_i(x, u, \nabla u) \frac{\partial \varphi}{\partial x_i} \, dx.$$

Let us introduce the notion of a (weak) solution of the BVP (6.1.1).

Definition 6.1.1. A function $u \in V_0$ is called a *solution* of problem (6.1.1) if there exists a function $v \in L^q(\Omega)$ such that

(i) $v(x) \in \beta(x, u(x), u(x))$ for a.e. $x \in \Omega$,

(ii) $a(u, \varphi) + \int_\Omega v \, \varphi \, dx = 0$ for all $\varphi \in V_0$.

We define upper and lower solutions as follows.

Definition 6.1.2. A function $\bar{u} \in V$ is called an *upper solution* to the BVP (6.1.1) if there exists a function $\bar{v} \in L^q(\Omega)$ such that

(i) $\bar{v}(x) \in \beta(x, \bar{u}(x), \bar{u}(x))$ for a.e. $x \in \Omega$,

(ii) $\bar{u} \geq 0$ on $\partial\Omega$,

(iii) $a(\bar{u}, \varphi) + \int_\Omega \bar{v} \, \varphi \, dx \geq 0$ for all $\varphi \in V_0 \cap L^p_+(\Omega)$.

Definition 6.1.3. A function $\underline{u} \in V$ is called a *lower solution* to the BVP (6.1.1) if there exists a function $\underline{v} \in L^q(\Omega)$ such that

(i) $\underline{v}(x) \in \beta(x, \underline{u}(x), \underline{u}(x))$ for a.e. $x \in \Omega$,

(ii) $\underline{u} \leq 0$ on $\partial\Omega$,

(iii) $a(\underline{u}, \varphi) + \int_\Omega \underline{v} \, \varphi \, dx \leq 0$ for all $\varphi \in V_0 \cap L^p_+(\Omega)$.

We shall make the following hypotheses.

(H1) Let there exist an upper and a lower solution $\bar{u}$ and $\underline{u}$ of the BVP (6.1.1), respectively, such that $\underline{u} \leq \bar{u}$.

(H2) The function $f : \Omega \times \mathbb{R} \times \mathbb{R} \to \mathbb{R}$ has the following properties:

(i) $f(x, r, s)$ is measurable in $x \in \Omega$ for all $(r, s) \in \mathbb{R} \times \mathbb{R}$ and continuous in r for a.e. x uniformly with respect to s, i.e., for any $r_0 \in \mathbb{R}$ and $\varepsilon > 0$ there is a $\delta(\varepsilon, r_0, x)$ independent of s such that

$$|f(x, r, s) - f(x, r_0, s)| < \varepsilon \quad \text{whenever} \quad |r - r_0| < \delta(\varepsilon, r_0, x) \, .$$

(ii) $s \to f(x, r, s)$ is increasing (possibly discontinuous) for a.e. x and for each $r \in \mathbb{R}$, and it is related with the multivalued function β by

$$\beta(x, r, s) = [f(x, r, s - 0), f(x, r, s + 0)], \tag{6.1.2}$$

where the closed interval on the right-hand side of (6.1.2) is formed by the one-sided limits

$$f(x, r, s \pm 0) = \lim_{\varepsilon \downarrow 0} f(x, r, s \pm \varepsilon) \, .$$

(iii) $(x, s) \to f(x, r, s)$ is Borel measurable in $\Omega \times \mathbb{R}$ for each $r \in \mathbb{R}$.

(H3) There is a function $k_2 \in L^q_+(\Omega)$ and a constant $\alpha > 0$ such that

$$|f(x, r, s)| \leq k_2(x) \, ,$$

for a.e. $x \in \Omega$ and for all $r \in [\underline{u}(x), \bar{u}(x)]$ and $s \in [\underline{u}(x) - \alpha, \bar{u}(x) + \alpha]$.

Remark 6.1.1. According to (H2)(i) $(x, r) \to f(x, r, s)$ is a Carathéodory function uniformly with respect to $s \in \mathbb{R}$. Conditions (i)–(iii) of hypothesis (H2) imply, in particular, that the function f is sup-measurable, i.e., whenever $u, v : \Omega \to \mathbb{R}$ are measurable functions then the Nemytskij operator F defined by $F(u, v)(x) := f(x, u(x), v(x))$ generates a measurable function. Furthermore, by hypotheses (H2) and (H3) the mapping $u \to F(u, v)$ is continuous and bounded from $[\underline{u}, \bar{u}] \subset L^p(\Omega)$ to $L^q(\Omega)$ uniformly with respect to $v \in [\underline{u} - \alpha, \bar{u} + \alpha]$. By (H3) we only impose a local $L^q(\Omega)$-boundedness of the nonlinearity f which, for example, is trivially satisfied if f fulfills a (global) growth condition of the form $|f(r, s)| \leq c\,(1 + |r|^{p-1} + |s|^{p-1})$. This last growth condition can further be relaxed if the upper and lower solutions happen to be essentially bounded.

Remark 6.1.2. One possibility to determine upper and lower solutions to the inclusion (6.1.1) is to replace the inclusion by a single-valued BVP of the form

$$Au + \tilde{f}(x, u, u) = 0 \quad \text{in } \Omega, \quad u = 0 \quad \text{on } \partial\Omega, \qquad (6.1.3)$$

where $\tilde{f} : \Omega \times \mathbb{R} \times \mathbb{R} \to \mathbb{R}$ may be any single-valued selection of the multifunction β such as for example the generating function f of β itself. Then obviously any upper (lower) solution of (6.1.3) is also an upper (lower) solution of (6.1.1).

6.1.2 Preparatory results. Throughout this section we assume that hypotheses (A1), (A2) of section 5.1 and (H1)–(H3) of subsection 6.1.1 are satisfied.

For any real number $\eta \geq 0$ we introduce truncation operators T_η by

$$T_\eta u(x) = \begin{cases} \bar{u}(x) + \eta & \text{if} \quad u(x) > \bar{u}(x) + \eta, \\ u(x) & \text{if} \quad \underline{u}(x) - \eta \leq u(x) \leq \bar{u}(x) + \eta, \\ \underline{u}(x) - \eta & \text{if} \quad u(x) < \underline{u}(x) - \eta, \end{cases}$$

where $\underline{u}$ and $\bar{u}$ are the given upper and lower solutions of (6.1.1) satisfying $\underline{u} \leq \bar{u}$. By means of T_0 and T_α we define the function $f_{0,\alpha}$ by

$$f_{0,\alpha}(x, w(x), u(x)) := f(x, T_0 w(x), T_\alpha u(x)),$$

where $\alpha > 0$ is as in hypothesis (H3), and consider the following functional

$$J(w, u) = \int_\Omega \int_0^{u(x)} f_{0,\alpha}(x, w(x), s)\, ds\, dx.$$

Hypotheses (H2) and (H3) ensure that the functional $J : L^p(\Omega) \times L^p(\Omega) \to \mathbb{R}$ is well defined, and for any given $w, u_1, u_2 \in L^p(\Omega)$ we get an estimate of the form

$$|J(w, u_1) - J(w, u_2)| \leq \int_\Omega |k_2(x)| \, |u_1(x) - u_2(x)| \, dx$$
$$\leq \|k_2\|_{L^q(\Omega)} \|u_1 - u_2\|_{L^p(\Omega)} , \qquad (6.1.4)$$

which shows that $J(w, \cdot) : L^p(\Omega) \to \mathbb{R}$ is Lipschitz continuous uniformly in w. The function $(x, s) \to f_{0,\alpha}(x, w(x), s)$ is sup-measurable, increasing in s according to (H2), and satisfies in view of (H3)

$$|f_{0,\alpha}(x, w(x), s)| \leq k_2(x) .$$

Hence the functional $J(w, \cdot) : L^p(\Omega) \to \mathbb{R}$ is also convex (cf. [92]), which implies the existence of the subdifferential (see F) of J with respect to its second argument denoted by $\partial_2 J(w, u)$. Applying results due to Chang ([92, Lemma 2.1 and Theorem 2.3]) in a slightly modified form one gets

$$\partial_2 J(w, u)(x) = [f_{0,\alpha}(x, w(x), u(x) - 0), f_{0,\alpha}(x, w(x), u(x) + 0)], \quad (6.1.5)$$

which holds in the space $L^p(\Omega)$ and in V as well. For any $w, u \in [\underline{u}, \bar{u}]$ from (6.1.5) it follows that

$$\partial_2 J(w, u)(x) = [f(x, w(x), u(x) - 0), f(x, w(x), u(x) + 0)]$$
$$= \beta(x, w(x), u(x)). \qquad (6.1.6)$$

Thus by definition of the subdifferential of the functional $J(w, \cdot) : L^p(\Omega) \to \mathbb{R}$ we obtain the following equivalent formulation of the inclusion

$$v(x) \in \beta(x, u(x), u(x)) \quad \text{for a.e. } x \in \Omega$$

for $v \in L^q(\Omega)$, $u \in L^p(\Omega)$ and $u \in [\underline{u}, \bar{u}]$ in form of the variational inequality

$$\int_\Omega v \, (\psi - u) \, dx \leq J(u, \psi) - J(u, u) \quad \text{for all } \psi \in L^p(\Omega). \qquad (6.1.7)$$

Lemma 6.1.1. *The functional* $J : L^p(\Omega) \times L^p(\Omega) \to \mathbb{R}$ *is continuous.*

Proof. We are going to show that

$$J(w_n, u_n) \to J(w, u) \quad \text{as } n \to \infty \tag{6.1.8}$$

whenever $(w_n, u_n) \to (w, u)$ in $L^p(\Omega) \times L^p(\Omega)$ as $n \to \infty$. By the Lipschitz continuity of J with respect to its second argument according to (6.1.4) we have

$$|J(w_n, u_n) - J(w_n, u)| \to 0 \quad \text{as } n \to \infty, \tag{6.1.9}$$

uniformly with respect to w_n. Further, we consider the following difference

$$|J(w_n, u) - J(w, u)| = \left| \int_\Omega \int_0^{u(x)} \Big(f_{0,\alpha}(x, w_n(x), s) - f_{0,\alpha}(x, w(x), s) \Big) \, ds dx \right|$$

which yields via the substitution $s(t)(x) := t\, u(x)$, $t \in [0, 1]$, and by applying Fubini's Theorem an estimate of the form

$$|J(w_n, u) - J(w, u)|$$

$$= \left| \int_\Omega \int_0^1 \Big(f_{0,\alpha}(x, w_n(x), tu(x)) - f_{0,\alpha}(x, w(x), tu(x)) \Big) \, u(x) \, dt dx \right|$$

$$\leq \left(\int_0^1 \|F_{0,\alpha}(w_n, s(t)) - F_{0,\alpha}(w, s(t))\|_{L^q(\Omega)} dt \right) \|u\|_{L^p(\Omega)}, \tag{6.1.10}$$

where $F_{0,\alpha}$ denotes the Nemytskij operator associated with $f_{0,\alpha}$. The right-hand side of (6.1.10) tends to zero as $n \to \infty$ due to the continuity of $F_{0,\alpha}$ in its first argument, uniformly with respect to its second argument. Hence, (6.1.8) follows from

$$J(w_n, u_n) - J(w, u) = J(w_n, u_n) - J(w_n, u) + J(w_n, u) - J(w, u)$$

by taking into account (6.1.9) and (6.1.10). $\square$

Lemma 6.1.2. *Let* $u \in V_0$ *satisfy* $u \in [\underline{u}, \bar{u}]$. *Then* u *is a solution of the BVP (6.1.1) if and only if it is a solution of the following quasi-variational inequality (QVI for short)*

$$u \in V_0 : \quad a(u, \varphi - u) + J(u, \varphi) - J(u, u) \geq 0 \quad \text{for all } \varphi \in V_0. \tag{6.1.11}$$

Proof. Let $u \in [\underline{u}, \bar{u}]$ be a solution of the QVI (6.1.11), which means by definition of $\partial_2 J(u, u)$ that u satisfies the inclusion

$$-Au \in \partial_2 J(u, u) \quad \text{in } V_0^*. \tag{6.1.12}$$

By (6.1.5) we have $\partial_2 J(u, u) \subset L^q(\Omega) \subset V_0^*$, and thus due to (6.1.12) there is function $v \in L^q(\Omega)$ satisfying

$$-Au = v \quad \text{in } V_0^*$$

and $v(x) \in [f_{0,\alpha}(x, u(x), u(x) - 0), f_{0,\alpha}(x, u(x), u(x) + 0)]$. Since $u \in [\underline{u}, \bar{u}]$ we get $T_0 u = u$ and $T_\alpha(u \pm \varepsilon) = u \pm \varepsilon$ for any $\varepsilon \in (0, \alpha)$ so that the latter implies that $v(x) \in \beta(x, u(x), u(x))$, which shows that u is a solution of the BVP (6.1.1).

Conversely, let $u \in [\underline{u}, \bar{u}]$ be a solution of the BVP (6.1.1). Then there is a function $v \in L^q(\Omega)$ such that $v(x) \in \beta(x, u(x), u(x))$ and

$$a(u, \varphi) + \int_\Omega v \varphi \, dx = 0 \quad \text{for all } \varphi \in V_0. \tag{6.1.13}$$

For $u \in [\underline{u}, \bar{u}]$ the identity

$$\beta(x, u(x), u(x)) = [f_{0,\alpha}(x, u(x), u(x) - 0), f_{0,\alpha}(x, u(x), u(x) + 0)]$$

holds. Since $f_{0,\alpha}(x, u(x), s) : \Omega \times \mathbb{R} \to \mathbb{R}$ is sup-measurable, increasing in s and uniformly bounded, the latter identity implies

$$v(x) \left(\varphi(x) - u(x) \right) \leq \int_{u(x)}^{\varphi(x)} f_{0,\alpha}(x, u(x), s) \, ds \quad \text{for all } \varphi \in L^p(\Omega),$$

and thus by integration

$$\int_\Omega v \left(\varphi - u \right) dx \leq J(u, \varphi) - J(u, u) \quad \text{for all } \varphi \in L^p(\Omega). \tag{6.1.14}$$

In particular (6.1.14) also holds for all $\varphi \in V_0$. Replacing $\varphi \in V_0$ in (6.1.13) by $\varphi - u$ we get in view of (6.1.14)

$$a(u, \varphi - u) + J(u, \varphi) - J(u, u) \geq 0 \quad \text{for all } \varphi \in V_0,$$

which is the QVI (6.1.11). □

6.1.3 Existence and enclosure result. For assumptions (A1)–(A3) we refer to those of section 5.1 imposed on the coefficients a_i of the operator A, and hypotheses (H1)–(H3) are given in subsection 6.1.1. We are going to show that there always exist solutions of the BVP (6.1.1) within the interval $[\underline{u}, \bar{u}]$ of upper and lower solutions.

Theorem 6.1.1. *Let assumptions (A1), (A2), and (H1)–(H3) be satisfied. Then the BVP (6.1.1) has a solution u with $\underline{u} \le u \le \bar{u}$.*

Proof. By Lemma 6.1.2 the BVP (6.1.1) is equivalent with the QVI (6.1.11) as long as we consider solutions within the interval $[\underline{u}, \bar{u}]$. Hence the theorem is proved provided we are able to show the existence of solutions of the QVI (6.1.11) lying in $[\underline{u}, \bar{u}]$. This will be done in steps (a), (b), and (c).

(a) Auxiliary variational inequality.

For fixed $w \in L^p(\Omega)$ we associate with the QVI (6.1.11) the following auxiliary variational inequality:

$$\text{Find } u \in V_0: \quad a_w(u, \varphi - u) + J(w, \varphi) - J(w, u) \ge 0 \quad \text{for all } \varphi \in V_0, \tag{6.1.15}$$

where a_w is the semilinear form defined by

$$a_w(u, \varphi) = \int_\Omega \sum_{i=1}^N a_i(\cdot, T_0 w, \nabla u) \frac{\partial \varphi}{\partial x_i} \, dx,$$

and A_w the operator related with a_w by $\langle A_w u, \varphi \rangle = a_w(u, \varphi)$. Due to (A1) and (A2) the operator $A_w : V_0 \to V_0^*$ can easily be seen to be continuous, bounded, and uniformly monotone, and thus, in particular, coercive. The functional $J(w, \cdot) : V_0 \to \mathbb{R}$ is convex and even Lipschitz continuous. Hence the variational inequality (6.1.15) possesses a uniquely defined solution (cf., e.g., [219, Theorem 54.A]). Let us introduce an operator $S : L^p(\Omega) \to V_0$ which assigns to each $w \in L^p(\Omega)$ the unique solution $u \in V_0$ of the variational inequality (6.1.15). Then any fixed point of S which belongs to the interval $[\underline{u}, \bar{u}]$ is in fact a solution of the QVI (6.1.11). Schauder's fixed point theorem will be used to show the existence of fixed points of S.

(b) Existence of fixed points of S.

By using hypothesis (H3) from the variational inequality (6.1.15) with $\varphi = 0$ we obtain

$$a_w(u, u) \le -J(w, u) \le \|k_2\|_{L^q(\Omega)} \|u\|_{L^p(\Omega)},$$

which yields by means of (A2) and by applying Poincaré-Friedrichs inequality (see B)

$$\mu \|u\|_{V_0}^p \le c \left(\|k_2\|_{L^q(\Omega)} + \sum_{i=1}^N \|a_i(\cdot, T_0 w, 0)\|_{L^q(\Omega)} \right) \|u\|_{V_0}. \tag{6.1.16}$$

By (A1) we have

$$\|a_i(\cdot, T_0w, 0)\|_{L^q(\Omega)} \le c\,(\|k_0\|_{L^q(\Omega)} + \|T_0w\|_{L^p(\Omega)}^{p-1})$$
$$\le c\,(\|k_0\|_{L^q(\Omega)} + \|\bar{u}\|_{L^p(\Omega)}^{p-1} + \|\underline{u}\|_{L^p(\Omega)}^{p-1}),$$

which yields in view of (6.1.16) a uniform bound for the range of S, i.e.,

$$\|Sw\|_{V_0} \le C \quad \text{for all } w \in L^p(\Omega). \tag{6.1.17}$$

This together with the (compact) embedding $V_0 \subset L^p(\Omega)$ shows that S provides a mapping of a ball $B_R \subset L^p(\Omega)$ into itself for sufficiently large radius R. In order to show the continuity of $S : B_R \to B_R$, let $u = Sw$ and $u_0 = Sw_0$. Taking in (6.1.15) as a special test function $\varphi = u_0$ for the solution u and $\varphi = u$ for the solution u_0, and adding the resulting inequalities we obtain

$$a_w(u, u - u_0) - a_w(u_0, u - u_0) \le a_{w_0}(u_0, u - u_0) - a_w(u_0, u - u_0)$$
$$+ \int_\Omega \int_{u(x)}^{u_0(x)} (f_{0,\alpha}(x, w(x), s) - f_{0,\alpha}(x, w_0(x), s))\, ds\, dx\,,$$

which results in

$$\mu\|u - u_0\|_{V_0}^p \le c\sum_{i=1}^{N} \|a_i(\cdot, T_0w_0, \nabla u_0) - a_i(\cdot, T_0w, \nabla u_0)\|_{L^q(\Omega)}\|u - u_0\|_{V_0}$$
$$+ \left(\int_0^1 \|F_{0,\alpha}(w, s(t)) - F_{0,\alpha}(w_0, s(t))\|_{L^q(\Omega)}\, dt\right)\|u - u_0\|_{L^p(\Omega)}, \tag{6.1.18}$$

where the following substitution has been used for the second term on the right-hand side of (6.1.18):

$$s(t)(x) = tu_0(x) + (1 - t)u(x)\,, \quad t \in [0, 1]\,.$$

The equicontinuity of $w \to F_{0,\alpha}(w, v)$ (cf. Remark 6.1.1) and the continuity of $w \to a_i(\cdot, T_0w, \nabla u_0)$ due to (A1) imply by (6.1.18) the continuity of the operator $S : B_R \to B_R$ which is compact due to the compact embedding $V_0 \subset L^p(\Omega)$. Thus Schauder's fixed point theorem can be applied which ensures the existence of a fixed point of the operator S, i.e., there exists a $u \in V_0$ such that

$$a_u(u, \varphi - u) + J(u, \varphi) - J(u, u) \ge 0 \quad \text{for all } \varphi \in V_0. \tag{6.1.19}$$

(c) Completion of the proof by comparison.

To complete the proof we only need to show that any solution u of (6.1.19) belongs to the interval $[\underline{u}, \bar{u}]$, since then $T_0 u = u$ and $a_u(u, \varphi) = a(u, \varphi)$, and thus u is a solution of the QVI (6.1.11) within the interval of upper and lower solutions and also a solution of the BVP (6.1.1) in view of Lemma 6.1.2. We are going to prove that any solution of (6.1.19) satisfies $u \leq \bar{u}$. The proof for $u \geq \underline{u}$ can be done in a similar way.

The upper solution $\bar{u}$ of the BVP (6.1.1) satisfies by definition $\bar{u} \geq 0$ on $\partial\Omega$ and

$$a(\bar{u}, \varphi) + \int_\Omega \bar{v}\, \varphi\, dx \geq 0 \quad \text{for all } \varphi \in V_0 \cap L_+^p(\Omega), \qquad (6.1.20)$$

where $\bar{v} \in \beta(\cdot, \bar{u}, \bar{u})$. This latter inclusion implies

$$\int_\Omega \bar{v}\,(\varphi - \bar{u})\, dx \leq J(\bar{u}, \varphi) - J(\bar{u}, \bar{u}) \qquad (6.1.21)$$

for all $\varphi \in L^p(\Omega)$. Using the special test functions $\varphi = (u-\bar{u})^+ \in V_0 \cap L_+^p(\Omega)$ in (6.1.20) and $\varphi = \bar{u} + (u - \bar{u})^+$ in (6.1.21), respectively, we get

$$a(\bar{u}, (u - \bar{u})^+) + J(\bar{u}, \bar{u} + (u - \bar{u})^+) - J(\bar{u}, \bar{u}) \geq 0. \qquad (6.1.22)$$

Taking in (6.1.19) the special test function $\varphi = u - (u - \bar{u})^+ \in V_0$ we obtain in view of (6.1.22)

$$a_u(u, (u - \bar{u})^+) - a(\bar{u}, (u - \bar{u})^+)$$
$$\leq J(\bar{u}, \bar{u} + (u - \bar{u})^+) - J(\bar{u}, \bar{u}) + J(u, u - (u - \bar{u})^+) - J(u, u). \qquad (6.1.23)$$

Applying the definition of the functional J one can show by elementary calculations that the right-hand side of (6.1.23) is equal to zero. Thus we obtain

$$0 \geq a_u(u, (u - \bar{u})^+) - a(\bar{u}, (u - \bar{u})^+)$$
$$= \int_\Omega \sum_{i=1}^N (a_i(\cdot, T_0 u, \nabla u) - a_i(\cdot, \bar{u}, \nabla\bar{u})) \frac{\partial(u - \bar{u})^+}{\partial x_i}$$
$$\geq \mu \|\nabla(u - \bar{u})^+\|_{L^p(\Omega)}^p,$$

which implies by Poincaré-Friedrichs inequality that $(u - \bar{u})^+ = 0$, i.e., $u \leq \bar{u}$. This completes the proof of Theorem 6.1.1. $\qquad\qquad \square$

Remark 6.1.3. It should be noted that Theorem 6.1.1 still remains true if hypothesis (A2) is replaced by (A2)$_1$ and (A2)$_2$. One possible way to prove Theorem 6.1.1 under (A2)$_1$ and (A2)$_2$ instead of (A2) will be demonstrated in section 6.4 where we provide the parabolic version of Theorem 6.1.1. The proof given there, which applies likewise also to the elliptic case, is based on a regularization technique and allows one to take into account also nonlinear lower order terms satisfying certain growth conditions without any additional difficulties.

6.1.4 Extremality and compactness of the solution set. Let S denote the set of all solutions of the BVP (6.1.1) within the interval $[\underline{u}, \bar{u}]$ of upper and lower solutions. In this subsection we show that S possesses extremal elements and that S is compact. The proof of the extremality of S is heavily based on the following lemma which will provide the directedness of S.

Lemma 6.1.3. *Let hypotheses (A1)–(A3) of section 5.1 and (H1)–(H3) of subsection 6.1.1 be satisfied. If u_1, $u_2 \in S$, then $\max(u_1, u_2)$ is a lower solution and $\min(u_1, u_2)$ is an upper solution of the BVP 6.1.1.*

Proof. In the proof we make use of the special test function technique introduced in section 5.1 which, however, has to be adapted appropriately to the present situation.

Let $u = \max(u_1, u_2)$. We define a function $v : \Omega \to \mathbb{R}$ by

$$
v(x) = \begin{cases} v_1(x) & \text{if} \quad x \in \{u_1 \geq u_2\}, \\ v_2(x) & \text{if} \quad x \in \{u_2 > u_1\}, \end{cases}
$$

where according to Definition 6.1.1 the functions $v_j \in L^q(\Omega)$ satisfy $v_j \in \beta(\cdot, u_j, u_j)$ as well as

$$
a(u_j, \varphi) + \int_\Omega v_j \, \varphi \, dx = 0 \quad \text{for all } \varphi \in V_0, \ j = 1, 2 \,. \tag{6.1.24}
$$

Obviously $v \in L^q(\Omega)$ and $v(x) \in \beta(x, u(x), u(x))$ for a.e. $x \in \Omega$ and $u \in V_0$ so that the assertion of the lemma is proved provided we are able to show

$$
a(u, \varphi) + \int_\Omega v \, \varphi \, dx \leq 0 \quad \text{for all } \varphi \in V_0 \cap L^p_+(\Omega) \,. \tag{6.1.25}
$$

Let the function $\theta_\varepsilon : \mathbb{R} \to [0, \infty)$ be as in section 5.1 which is related with the modulus of continuity ω as follows

$$
\theta_\varepsilon(t) = \begin{cases} 0 & \text{if} \quad t < \delta(\varepsilon), \\[2mm] \displaystyle\int_{\delta(\varepsilon)}^{t} \frac{dr}{\omega^q(r)} & \text{if} \quad \delta(\varepsilon) \le t \le \varepsilon, \\[2mm] 1 & \text{if} \quad t > \varepsilon. \end{cases}
$$

Then θ_ε is Lipschitz continuous for any $\varepsilon > 0$, increasing, and satisfies

$$
\theta_\varepsilon(t) \to \chi_{\{t>0\}} \quad \text{as} \quad \varepsilon \to 0,
$$

where $\chi_{\{t>0\}}$ denotes the characteristic function of the set of the positive real line $\{t > 0\}$. Let us introduce the following set of nonnegative smooth functions

$$
\mathcal{D}_+ := \{\psi \in C_0^\infty(\Omega) \mid \psi \ge 0\}.
$$

Let $\psi \in \mathcal{D}_+$. Taking as special test function in (6.1.24) for $j = 1$ the function $\varphi = (1 - \theta_\varepsilon(u_2 - u_1))\psi \in V_0 \cap L_+^p(\Omega)$ and for $j = 2$ the function $\varphi = \theta_\varepsilon(u_2 - u_1)\psi \in V_0 \cap L_+^p(\Omega)$, and adding the resulting equations we obtain

$$
a(u_1, (1 - \theta_\varepsilon(u_2 - u_1))\psi) + a(u_2, \theta_\varepsilon(u_2 - u_1)\psi)
$$
$$
= -\int_\Omega [v_1(1 - \theta_\varepsilon(u_2 - u_1))\psi + v_2\theta_\varepsilon(u_2 - u_1)\psi]\, dx.
$$
$$(6.1.26)$$

By applying (A2) the left-hand side of the last equation can be estimated below as follows

$$
a(u_1, (1 - \theta_\varepsilon(u_2 - u_1))\psi) + a(u_2, \theta_\varepsilon(u_2 - u_1)\psi)
$$
$$
= a(u_1, \psi) + a(u_2, \theta_\varepsilon(u_2 - u_1)\psi) - a(u_1, \theta_\varepsilon(u_2 - u_1)\psi)
$$
$$
\ge a(u_1, \psi) + \mu \int_\Omega |\nabla(u_2 - u_1)|^p\, \psi\theta_\varepsilon'(u_2 - u_1)\, dx
$$
$$
+ \int_\Omega \sum_{i=1}^{N}(a_i(\cdot, u_2, \nabla u_1) - a_i(\cdot, u_1, \nabla u_1))\frac{\partial(u_2 - u_1)}{\partial x_i}\, \psi\theta_\varepsilon'(u_2 - u_1)\, dx
$$
$$
+ \int_\Omega \sum_{i=1}^{N}(a_i(\cdot, u_2, \nabla u_2) - a_i(\cdot, u_1, \nabla u_1))\frac{\partial\psi}{\partial x_i}\theta_\varepsilon(u_2 - u_1)\, dx.
$$
$$(6.1.27)$$

Using hypothesis (A3) and the property of θ_ε we obtain by means of Young's inequality an estimate of the second integral on the right-hand side of (6.1.27) in the form

$$
\left| \int_\Omega \sum_{i=1}^N (a_i(\cdot, u_2, \nabla u_1) - a_i(\cdot, u_1, \nabla u_1)) \frac{\partial(u_2 - u_1)}{\partial x_i} \, \psi \, \theta_\varepsilon'(u_2 - u_1) \, dx \right|
$$

$$
\leq \frac{\mu}{2} \int_\Omega |\nabla(u_2 - u_1)|^p \theta_\varepsilon'(u_2 - u_1) \psi \, dx
$$

$$
+ \int_{\{\delta(\varepsilon) < u_2 - u_1 < \varepsilon\}} c_\mu \left[|k_1| + |u_1|^{p-1} + |u_2|^{p-1} + |\nabla u_1|^{p-1} \right]^q \psi \, dx \, .
\tag{6.1.28}
$$

Finally, the estimates (6.1.27) and (6.1.28) along with equation (6.1.26) yield for any $\varepsilon > 0$ the inequality

$$
- \int_\Omega [v_1(1 - \theta_\varepsilon(u_2 - u_1))\psi + v_2 \theta_\varepsilon(u_2 - u_1)\psi] \, dx
$$

$$
\geq a(u_1, \psi) + \int_\Omega \sum_{i=1}^N (a_i(\cdot, u_2, \nabla u_2) - a_i(\cdot, u_1, \nabla u_1)) \frac{\partial \psi}{\partial x_i} \theta_\varepsilon(u_2 - u_1) \, dx
$$

$$
- \int_{\{\delta(\varepsilon) < u_2 - u_1 < \varepsilon\}} c_\mu \left[|k_1| + |u_1|^{p-1} + |u_2|^{p-1} + |\nabla u_1|^{p-1} \right]^q \psi \, dx,
\tag{6.1.29}
$$

where the last integral on the right-hand side of (6.1.29) tends to zero as $\varepsilon \to 0$. By Lebesgue's dominated convergence theorem we may pass to the limit in (6.1.29), which yields due to

$$
\theta_\varepsilon(u_2 - u_1) \to \chi_{\{u_2 > u_1\}} \quad \text{as } \varepsilon \to 0
$$

the inequality

$$
- \int_\Omega [v_1(1 - \chi_{\{u_2 > u_1\}})\, \psi + v_2 \, \chi_{\{u_2 > u_1\}} \, \psi] \, dx
$$

$$
\geq a(u_1, \psi) + \int_\Omega \sum_{i=1}^N (a_i(\cdot, u_2, \nabla u_2) - a_i(\cdot, u_1, \nabla u_1)) \frac{\partial \psi}{\partial x_i} \chi_{\{u_2 > u_1\}} \, dx
$$

$$
= \int_\Omega \sum_{i=1}^N a_i(\cdot, u_1, \nabla u_1) \frac{\partial \psi}{\partial x_i} (1 - \chi_{\{u_2 > u_1\}}) \, dx
$$

$$
+ \int_\Omega \sum_{i=1}^N a_i(\cdot, u_2, \nabla u_2) \frac{\partial \psi}{\partial x_i} \chi_{\{u_2 > u_1\}} \, dx,
$$

which is nothing else than

$$- \int_{\Omega} v \, \psi \, dx \geq a(u, \psi) \quad \text{for all } \psi \in \mathcal{D}_+ . \qquad (6.1.30)$$

Since the completion of $\mathcal{D}_+$ with respect to the norm of V yields $V_0 \cap L_+^p(\Omega)$, from (6.1.30) we get (6.1.25) and thus the assertion, i.e., $u = \max(u_1, u_2)$ is a lower solution. The proof for $\min(u_1, u_2)$ being an upper solution follows by obvious modifications and is left to the reader. □

Lemma 6.1.4. *Let the assumption of Lemma 6.1.3 be fulfilled. Then any increasing (decreasing) sequence of S converges weakly in V_0 and strongly in $L^p(\Omega)$ to an element of S.*

Proof. The solution set S is obviously $L^p(\Omega)$-bounded because of $S \subseteq [\underline{u}, \bar{u}]$. By Lemma 6.1.2 any $u \in S$ satisfies the QVI (6.1.11) from which one readily obtains the boundedness of S in V_0. Let $(u_n) \subseteq S$ be an increasing (decreasing) sequence. Then by Lebesgue's dominated convergence theorem and due to the compact embedding one easily shows that $u_n \rightharpoonup w$ weakly in V_0 and $u_n \to w$ strongly in $L^p(\Omega)$. To complete the proof we need to show that $w \in S$. To this end note that each $u_n \in S$ belongs to $[\underline{u}, \bar{u}]$ and u_n is a solution of the QVI (6.1.11), i.e.,

$$u_n \in V_0 : \quad a(u_n, \varphi - u_n) + J(u_n, \varphi) - J(u_n, u_n) \geq 0 \quad \text{for all } \varphi \in V_0. \qquad (6.1.31)$$

Taking $\varphi = w$ as a special test function in (6.1.31) and noting that $u_n \rightharpoonup w$ in V_0 and $u_n \to w$ in $L^p(\Omega)$ we obtain by the continuity of the functional $J : L^p(\Omega) \times L^p(\Omega) \to \mathbb{R}$ according to Lemma 6.1.1 the inequality

$$\limsup_{n \to \infty} a(u_n, u_n - w) \leq 0. \qquad (6.1.32)$$

The pseudomonotonicity of the operator $A : V_0 \to V_0^*$ related with the semilinear form a by $\langle Au, \varphi \rangle = a(u, \varphi)$ (see D.2) along with $u_n \rightharpoonup w$ in V_0 and (6.1.32) imply that

$$A u_n \rightharpoonup A w \quad \text{in } V_0^*, \quad \text{and} \quad \langle A u_n, u_n \rangle \to \langle A w, w \rangle.$$

This allows the passage to the limit in (6.1.31) as $n \to \infty$ showing that w is a solution of the QVI (6.1.11) and thus also of the BVP (6.1.1), since obviously $w \in [\underline{u}, \bar{u}]$. Hence $w \in S$ which proves the lemma. □

By means of the preceding lemmas the following extremality result can now be proved.

Theorem 6.1.2. *Under the hypotheses of Lemma 6.1.3 the BVP (6.1.1) possesses extremal solutions within the interval $[\underline{u}, \bar{u}]$ formed by upper and lower solutions, that is, the set S has extremal elements with respect to the underlying natural partial ordering.*

Proof. We prove the existence of the greatest solution u^* only, since the existence of the least solution can be shown analogously.

The solution set $S \subset V_0$ is separable, since V_0 is separable. Let $U = \{v_n \mid n \in \mathbb{N}\}$ be a countable dense subset of S. Then by means of Theorem 6.1.1 and Lemma 6.1.3 we are able to construct an increasing sequence $(u_n) \subseteq S$ in the following way:

Let $u_1 := v_1$. Let $u_n \in S$ be chosen, then select $u_{n+1} \in S$ so that

$$\max(v_n, u_n) \leq u_{n+1} \leq \bar{u}.$$

Since by Lemma 6.1.3 $\max(v_n, u_n) \in [\underline{u}, \bar{u}]$ is a lower solution, the existence of such an element $u_{n+1} \in S$ follows from Theorem 6.1.1. By induction we obtain an increasing sequence $(u_n) \subseteq S$, which by Lemma 6.1.4 is weakly convergent in V_0 and strongly convergent in $L^p(\Omega)$ to $w = \sup_n u_n \in S$. By construction we have $\max(v_1, ..., v_n) \leq u_{n+1} \leq w$ for all $n = 1, 2, ...$, so that $w \in S$ is an upper bound of $U = \{v_n \mid n \in \mathbb{N}\}$. Thus U is contained in the order interval $[\underline{u}, w]$ which is a closed subset of V. Since U was a dense subset of S, we obtain

$$S \subseteq \overline{U} \subseteq \overline{[\underline{u}, w]} = [\underline{u}, w].$$

Hence it follows that the limit $w \in S$ is an upper bound of S, which shows that w must be the greatest element of S. $\qquad\qquad\square$

The extremality result of Theorem 6.1.2 remains true if instead of hypotheses (A2) and (A3) we assume (A2)$_1$, (A2)$_2$, and (A3').

Corollary 6.1.1. *Let hypotheses (A1),(A2)$_1$, (A2)$_2$, (A3') of section 5.1 and hypotheses (H1)–(H3) of this section be satisfied. Then the BVP (6.1.1) possesses extremal solutions within the interval $[\underline{u}, \bar{u}]$ formed by the upper and lower solutions $\bar{u}$ and $\underline{u}$, respectively.*

Remark 6.1.4. In Remark 6.1.3 we already mentioned that Theorem 6.1.1 remains true if (A2) is replaced by (A2)$_1$ and (A2)$_2$. To prove Corollary 6.1.1 one has to show that Lemma 6.1.3 still remains valid when using assumptions (A2)$_1$, (A2)$_2$, (A3') instead of (A2) and (A3). This can be done following the proof of Lemma 6.1.3 step by step. Here again the interrelation between the monotonicity with respect to ξ and the modulus of continuity can be seen.

Theorem 6.1.3. *Let the hypothesis of Lemma 6.1.3 be satisfied. Then the solution set S is compact.*

Proof. Let us be given any sequence $(u_n) \subseteq S$. Since S is bounded in V_0, there exists a subsequence of (u_n) denoted by (u_k) converging weakly to $u \in V_0$, i.e., $u_k \rightharpoonup u$. We are going to prove that (u_k) is strongly convergent in V_0 to u and $u \in S$. The weak convergence of $u_k \rightharpoonup u$ in V_0 implies that $u_k \to u$ strongly in $L^p(\Omega)$ due to the compact embedding of $V \subset L^p(\Omega)$, and thus by the pseudomonotonicity of the operator A and the continuity of the functional J we conclude in just the same way as in the proof of Lemma 6.1.4 that the weak limit u belongs to S. To show the strong convergence of $u_k \to u$ in V_0 we observe that any $v \in S \subseteq [\underline{u}, \bar{u}]$ is necessarily a solution of the following modified quasi-variational inequality

$$v \in V_0: \quad a_{T_0}(v, \varphi - v) + J(v, \varphi) - J(v, v) \geq 0 \quad \text{for all } \varphi \in V_0. \quad (6.1.33)$$

where (cf. subsection 5.1.4) the semilinear form a_{T_0} is given by

$$a_{T_0}(v, \varphi) := \sum_{i=1}^{N} \int_{\Omega} a_i(x, T_0 v, \nabla v) \frac{\partial \varphi}{\partial x_i} \, dx,$$

with T_0 being the truncation operator that truncates between $\underline{u}$ and $\bar{u}$. The semilinear form a_{T_0} generates an operator $A_{T_0} : V_0 \to V_0^*$ by

$$\langle A_{T_0} v, \varphi \rangle = a_{T_0}(v, \varphi) \quad \text{for all } \varphi \in V_0,$$

which, in addition, possesses the (S_+)-property (see D.2). Since the u_k satisfy the QVI (6.1.33), we get by taking as special test function its weak limit u, and by applying the continuity of the functional J

$$\limsup_{k \to \infty} \langle A_{T_0} u_k, u_k - u \rangle = \limsup_{k \to \infty} a_{T_0}(u_k, u_k - u) \leq 0,$$

which implies due to the (S_+)-property of A_{T_0} the strong convergence of $u_k \to u$ in V_0, and thus the assertion. $\qquad\square$

Remark 6.1.5. For the compactness of the solution set S only the conditions of Theorem 6.1.1 that ensure the existence of solutions within the interval of upper and lower solutions are actually needed. Thus either conditions (A1), (A2) and (H1)–(H3) of Theorem 6.1.1 or conditions (A1), (A2)$_1$, (A2)$_2$, and (H1)–(H3) according to Remark 6.1.4 are sufficient for the compactness of S.

In the following section we employ an abstract fixed point result provided in chapter 1 and the extremality result of this section to prove the existence of extremal solutions of quasilinear elliptic inclusions involving an additional single-valued discontinuous nonlinear term.

6.2 State-dependent subdifferentials perturbed by discontinuous nonlinearities

6.2.1 Problem and some notation. Keeping all assumptions and notations of section 6.1 in this section we extend the BVP (6.1.1) by a nonmonotone discontinuous lower order term g, and consider the following BVP

$$Au + \beta(\cdot, u, u) \ni g(\cdot, u, u) \quad \text{in } \Omega, \quad u = 0 \quad \text{on } \partial\Omega, \tag{6.2.1}$$

where the nonlinearity $g : \Omega \times \mathbb{R} \times \mathbb{R} \to \mathbb{R}$ may depend discontinuously on u in a similar way like the nonlinearity f which generates the multifunction β. Various models in applications may be described by boundary value problems of the form (6.2.1).

Example 6.2.1. The model for Joule heating of a body subjected to an electric current whose boundary is kept at a constant temperature (which can be assumed to be zero by choosing an appropriate temperature scale) can be described by a BVP in the form

$$-\sum_{i=1}^{N} \frac{\partial}{\partial x_i} \left(k(x, u(x)) \frac{\partial u(x)}{\partial x_i} \right) = \sigma(x, u(x)), \quad x \in \Omega, \quad u(x) = 0 \quad x \in \partial\Omega,$$

$$\tag{6.2.2}$$

where $u(x)$ is the unknown temperature distribution, $\sigma(x, u)$ and $k(x, u)$ are, respectively, the electrical and thermal conductivity at the point x and temperature u; cf. [160, 161]. The electrical conductivity σ is allowed to change discontinuously at certain temperatures where, for example, a phase change takes place. Realistic assumptions on σ due to [160] are the following:

(i) $\sigma(x, s) \geq \sigma_0(x) > 0$,

(ii) $\sigma(x, s) = \sigma_1(x, s) + \psi_1(x, s) - \psi_2(x, s)$,

where σ_1 is a Carathéodory function, and $\psi_i(x, s)$, $i = 1, 2$, are functions which are continuous in x for fixed $s \in \mathbb{R}$ and increasing (possibly discontinuous) in s for fixed x. Furthermore, the thermal conductivity k is assumed to be a $C^1(\bar{\Omega} \times \mathbb{R})$ function.

A multivalued version of (6.2.2) can be transformed to a BVP of the form (6.2.1) by setting, for example, either

$$f(x, r, s) = -\sigma_1(x, r) + \psi_2(x, s), \quad \text{and} \quad g(x, r, s) = \psi_1(x, s),$$

or

$$f(x, r, s) = \psi_2(x, s), \quad \text{and} \quad g(x, r, s) = \sigma_1(x, r) + \psi_1(x, s),$$

where f is the function generating the multifunction β. The restrictive C^1 regularity on the conductivity k can be relaxed in our treatment by assuming only Carathéodory conditions on k which is also a desirable assumption in applications.

Example 6.2.2. In [11] the following BVP has been studied by means of a dual variational principle

$$-\Delta u = H(u - a)p(u) \quad \text{in } \Omega, \quad u = 0 \quad \text{on } \partial\Omega, \qquad (6.2.3)$$

where H is the Heaviside step function, and $p(s) \geq 0$ for all $s \in \mathbb{R}$. Thus the BVP (6.2.3) is a very special case of (6.2.1) to be considered in this section by setting

$$f(x, r, s) \equiv 0, \quad g(x, r, s) = H(s - a)p(r) \quad \text{and} \quad A = -\Delta.$$

The BVP (6.2.3) describes, for instance, a model in plasma physics.

The advantage of the extended problem (6.2.1) is that it is more flexible to cover a wide range of models in applications.

The notions of solution as well as of upper and lower solutions of the BVP (6.2.1) that will be defined next are natural extensions of those defined in the last section.

Definition 6.2.1. A function $u \in V_0$ is called a *solution* of problem (6.2.1) if there exists a function $v \in L^q(\Omega)$ such that

(i) $v(x) \in \beta(x, u(x), u(x))$ for a.e. $x \in \Omega$,
(ii) $a(u, \varphi) + \int_\Omega v\, \varphi\, dx = \int_\Omega g(\cdot, u, u)\, \varphi\, dx$ for all $\varphi \in V_0$.

Definition 6.2.2. A function $\bar{u} \in V$ is called an *upper solution* to the BVP (6.2.1) if there exists a function $\bar{v} \in L^q(\Omega)$ such that

(i) $\bar{v}(x) \in \beta(x, \bar{u}(x), \bar{u}(x))$ for a.e. $x \in \Omega$,
(ii) $\bar{u} \geq 0$ on $\partial\Omega$,
(iii) $a(\bar{u}, \varphi) + \int_\Omega \bar{v}\, \varphi\, dx \geq \int_\Omega g(\cdot, \bar{u}, \bar{u})\, \varphi\, dx$ for all $\varphi \in V_0 \cap L^p_+(\Omega)$.

A *lower solution* $\underline{u}$ is defined analogously. We make the following assumptions on g:

(G1) The function $g : \Omega \times \mathbb{R} \times \mathbb{R} \to \mathbb{R}$ satisfies:

   (i) $(x, r) \to g(x, r, s)$ is a Carathéodory function for each $s \in \mathbb{R}$.

   (ii) $s \to g(x, r, s)$ is increasing (possibly discontinuous) for a.e. x and for each $r \in \mathbb{R}$.

   (iii) $(x, s) \to g(x, r, s)$ is Borel measurable in $\Omega \times \mathbb{R}$ for each $r \in \mathbb{R}$.

(G2) There is a $k_3 \in L^q_+(\Omega)$ such that

$$|g(x, r, s)| \le k_3(x),$$

for a.e. $x \in \Omega$ and for all $r, s \in [\underline{u}(x), \bar{u}(x)]$.

Remark 6.2.1. Assumption (G1) implies that g is sup-measurable, i.e., whenever $u, v : \Omega \to \mathbb{R}$ are measurable, then also $x \to g(x, u(x), v(x))$ is a measurable function in Ω. This is because $x \to g(x, r, v(x))$ is measurable in x for each $r \in \mathbb{R}$ due (iii) of (G1), and $r \to g(x, r, v(x))$ is continuous for a.e. $x \in \Omega$. Note also that both $u \to g(x, u, u)$ and $u \to f(x, u, u)$ are, in general, nonmonotone and discontinuous function.

Remark 6.2.2. Let G denote the Nemytskij operator associated with g by

$$G(u, v)(x) = g(x, u(x), v(x)).$$

Then due to (G1) and (G2) for fixed $v \in [\underline{u}, \bar{u}]$ the mapping $u \to G(u, v)$ is continuous and bounded from the interval $[\underline{u}, \bar{u}]$ to $L^q(\Omega)$ and for fixed $u \in [\underline{u}, \bar{u}]$ the mapping $v \to G(u, v)$ is increasing and bounded from $[\underline{u}, \bar{u}]$ to $L^q(\Omega)$.

The aim of this section is to prove the existence of extremal solutions of the BVP (6.2.1) within the interval $[\underline{u}, \bar{u}]$ of upper and lower solutions. The proof of this extremality result strongly relies on the result obtained in the previous section.

6.2.2 Existence of extremal solutions. We assume the hypotheses of Theorem 6.1.2 of the last section and (G1) and (G2) to hold throughout this section, and consider first an auxiliary BVP which arises from the original problem (6.2.1) by freezing the variable of the third argument of the discontinuous nonlinearity g. Let $z \in [\underline{u}, \bar{u}]$ be given, we consider the auxiliary BVP

$$Au + \beta(\cdot, u, u) \ni g(\cdot, u, z) \quad \text{in } \Omega, \quad u = 0 \quad \text{on } \partial\Omega. \tag{6.2.4}$$

Lemma 6.2.1. *Let $z \in [\underline{u}, \bar{u}]$ be any upper solution of the BVP (6.2.1). Then the auxiliary BVP (6.2.4) has extremal solutions within the interval $[\underline{u}, z]$.*

Proof. The monotonicity of $s \to g(x, r, s)$ readily implies that the upper solution z and the lower solution $\underline{u}$ of the BVP (6.2.1) are also upper and lower solutions of the auxiliary BVP (6.2.4). Let $f^z : \Omega \times \mathbb{R} \times \mathbb{R} \to \mathbb{R}$ be defined by

$$f^z(x, r, s) := f(x, r, s) - g(x, r, z(x)).$$

Since $g(x, r, z(x))$ is a Carathéodory function, f^z satisfies the same regularity and structure conditions as f, and, moreover, one readily observes that the multifunction β^z generated by f^z is given by

$$\beta^z(x, r, s) = \beta(x, r, s) - g(x, r, z(x)), \tag{6.2.5}$$

which shows that the BVP (6.2.4) is equivalent with

$$Au + \beta^z(\cdot, u, u) \ni 0 \quad \text{in } \Omega, \quad u = 0 \quad \text{on } \partial\Omega. \tag{6.2.6}$$

Since z is an upper solution of (6.2.4), there is a $v \in L^q(\Omega)$ such that

(i) $v(x) \in \beta(x, z(x), z(x))$ for a.e. $x \in \Omega$,
(ii) $a(z, \varphi) + \int_\Omega v \, \varphi \, dx \geq \int_\Omega g(\cdot, z, z) \, \varphi \, dx$ for all $\varphi \in V_0 \cap L^p_+(\Omega)$.

Setting $v^z(x) = v(x) - g(x, z(x), z(x))$ one readily verifies that

(iii) $v^z(x) \in \beta^z(x, z(x), z(x))$ for a.e. $x \in \Omega$, and
(iv) $a(z, \varphi) + \int_\Omega v^z \, \varphi \, dx \geq 0$ for all $\varphi \in V_0 \cap L^p_+(\Omega)$,

and thus z is also an upper solution of (6.2.6). Obviously also the converse is true, i.e., any upper solution of (6.2.6) within $[\underline{u}, \bar{u}]$ is also an upper solution of (6.2.4). Similarly, w is a lower solution of (6.2.4) if and only if it is a lower solution of (6.2.6). Now we can apply Theorem 6.1.2 which ensures the existence of extremal solutions of the BVP (6.2.6), and thus of the BVP (6.2.4) within the interval $[\underline{u}, z]$ which proves the lemma.   □

Remark 6.2.3. In the same way one can show that if $z \in [\underline{u}, \bar{u}]$ is any lower solution of the BVP (6.2.1), then the auxiliary BVP (6.2.4) has extremal solutions within $[z, \bar{u}]$.

We are now able to prove the existence of extremal solutions of the original BVP (6.2.1).

Theorem 6.2.1. *Let the hypotheses of Theorem 6.1.2 and hypotheses (G1), (G2) be satisfied. Then the BVP (6.2.1) has extremal solutions within the interval $[\underline{u}, \bar{u}]$.*

Proof. By using Lemma 6.2.1 we are going to show the existence of the greatest solution only. The existence of the least solution can be proved in a similar way by using Remark 6.2.3.

Let the set $\mathcal{Z}$ be given by

$$\mathcal{Z} = \{z \in [\underline{u}, \bar{u}] \mid z \text{ is an upper solution of the BVP } (6.2.1)\}.$$

By Lemma 6.2.1 we define the following operator P:
P assigns to each upper solution $z \in \mathcal{Z}$ the greatest solution Pz of the BVP (6.2.4) within the interval $[\underline{u}, z]$. Since $\underline{u} \le Pz \le z$, we readily observe in view of the monotonicity of g in its last argument that Pz is again an upper solution of the original BVP (6.2.1), and thus $Pz \in \mathcal{Z}$ which shows that

$$P: \ \mathcal{Z} \to \mathcal{Z}.$$

Obviously any fixed point of P is a solution of the original BVP (6.2.1) within $[\underline{u}, \bar{u}]$, and vice versa. Thus the assertion of the theorem is proved provided P can be shown to have the greatest fixed point. To this end we apply the abstract fixed point result given by Theorem 1.1.1 to the mapping $P: \ \mathcal{Z} \to \mathcal{Z}$, and show first that $P: \ \mathcal{Z} \to \mathcal{Z}$ is increasing. Let $z_1, \ z_2 \in \mathcal{Z}$ satisfy $z_1 \le z_2$. By definition Pz_1 is the greatest solution in $[\underline{u}, z_1]$ of the BVP

$$Au + \beta(\cdot, u, u) \ni g(\cdot, u, z_1) \quad \text{in } \Omega, \quad u = 0 \quad \text{on } \partial\Omega,$$

and Pz_2 is the greatest solution in $[\underline{u}, z_2]$ of the BVP

$$Au + \beta(\cdot, u, u) \ni g(\cdot, u, z_2) \quad \text{in } \Omega, \quad u = 0 \quad \text{on } \partial\Omega. \tag{6.2.7}$$

Since $Pz_1 \le z_1 \le z_2$, it follows that Pz_1 is a lower solution of the problem (6.2.7) and trivially z_2 is an upper solution of (6.2.7). Thus there exist solutions of (6.2.7) within the interval $[Pz_1, z_2]$. But Pz_2 is the greatest solution of (6.2.7) within $[\underline{u}, z_2] \supseteq [Pz_1, z_2]$ which implies $Pz_1 \le Pz_2$, and thus the monotonicity of P. In order to apply Theorem 1.1.1 we still have to verify that any decreasing sequence (u_n) of $P[\mathcal{Z}]$ converges weakly in $\mathcal{Z}$. Note $\bar{u}$ is an upper bound of $P[\mathcal{Z}]$.

The $L^q(\Omega)$-boundedness of $\beta(\cdot, u, v)$ and $G(u, v)$ within $[\underline{u}, \bar{u}]$ and (A1) and (A2) imply that the sequence (u_n) of $P[\mathcal{Z}]$ is bounded in V_0. Thus by the monotonicity of the sequence we have $u_n \rightharpoonup w$ in V_0 and $u_n \to w$ in

$L^p(\Omega)$. We shall show that $w \in \mathcal{Z}$. By definition $u_n = Pz_n$ satisfy the BVP (6.2.4), i.e.,

$$a(u_n, \varphi) + \int_\Omega v_n \, \varphi \, dx = \int_\Omega G(u_n, z_n) \, \varphi \, dx \qquad (6.2.8)$$

for all $\varphi \in V_0$, where $v_n(x) \in \beta(x, u_n(x), u_n(x))$ for a.e. $x \in \Omega$. The latter implies that

$$\int_\Omega v_n \, (\varphi - u_n) \, dx \le J(u_n, \varphi) - J(u_n, u_n), \quad \text{for all } \varphi \in L^p(\Omega).$$

Since (v_n) is bounded in $L^q(\Omega)$, there exists a weakly convergent subsequence (which is again denoted by (v_n)) with weak limit v. The continuity of the functional $J : L^p(\Omega) \times L^p(\Omega) \to \mathbb{R}$ and $v_n \rightharpoonup v$ yields

$$\int_\Omega v \, (\varphi - w) \, dx \le J(w, \varphi) - J(w, w) \quad \text{for all } \varphi \in L^p(\Omega),$$

which implies $v \in \beta(\cdot, w, w)$. Since $w \le u_n \le z_n$ holds for all n, we get from (6.2.8) and by using the monotonicity of $z \to G(u, z)$ the inequality

$$a(u_n, \varphi) + \int_\Omega v_n \, \varphi \, dx \ge \int_\Omega G(u_n, w) \, \varphi \, dx \qquad (6.2.9)$$

for all $\varphi \in V_0 \cap L_+^p(\Omega)$. With the $L^q(\Omega)$-boundedness of the sequences (v_n) and $(G(u_n, z_n))$ we get from (6.2.8) by taking $\varphi = u_n - w$

$$\lim_{n \to \infty} a(u_n, u_n - w) = 0,$$

which due the pseudomonotonicity of the operator A related with the semilinear form a, and due to $v_n \rightharpoonup v$ in $L^q(\Omega)$ as well as the continuity of the Nemytskij operator $G(u, z)$ in its first argument allows the passage to the limit in (6.2.9) as $n \to \infty$, and thus

$$a(w, \varphi) + \int_\Omega v \, \varphi \, dx \ge \int_\Omega G(w, w) \, \varphi \, dx \quad \text{for all } \varphi \in V_0 \cap L_+^p(\Omega), \quad (6.2.10)$$

where $v \in \beta(\cdot, w, w)$. This shows that w is an upper solution of (6.2.1), i.e., $w \in \mathcal{Z}$. Now Theorem 1.1.1 can be applied which yields the existence of the greatest fixed point u^* of P in $\mathcal{Z}$, which is the greatest solution of the BVP (6.2.1) within the interval $[\underline{u}, \bar{u}]$. $\qquad \square$

Remark 6.2.4. The extremality result of Theorem 6.2.1 remains true if condition (A2) is replaced by (A2)$_1$ and (A2)$_2$, and condition (A3) is replaced by (A3').

Remark 6.2.5. If the function $s \to g(x, r, s)$ is, in addition, one-sided continuous, then one can get the extremal solutions by monotone iteration. To be more precise, in case that $s \to g(x, r, s)$ is right-sided continuous then by means of the fixed point operator P introduced above the iteration

$$u_0 := \bar{u} : \quad u_{n+1} = P u_n, \quad n = 0, 1, ...,$$

yields a decreasing sequence (u_n) which converges from above to the greatest solution u^*. In case $s \to g(x, r, s)$ is left-sided continuous, then starting the iteration with the lower solution $\underline{u}$ one gets a sequence of iterates that converges monotonically from below to the least solution u^*. According to the definition of the operator P each iteration step consists in finding the extremal solutions of a quasilinear elliptic BVP, whose existence is guaranteed by Lemma 6.2.1.

Remark 6.2.6. The method developed in this section allows one to treat also BVP with multivalued and discontinuous flux conditions in the form

$$Au + \beta_1(\cdot, u, u) \ni g_1(\cdot, u, u) \text{ in } \Omega, \quad \frac{\partial u}{\partial \nu} + \beta_2(\cdot, u, u) \ni g_2(\cdot, u, u) \text{ on } \partial\Omega,$$

where β_i and g_i $i = 1, 2$, are of the same type as β and g of problem (6.2.1).

6.3 Elliptic inclusions with generalized gradients

In this section we extend the variational approach combined with the method of upper and lower solutions to treat quasilinear elliptic inclusions whose multivalued term is generated by a function $f : \Omega \times \mathbb{R} \to \mathbb{R}$ which is only assumed to be a Baire measurable function; cf. [12]. The main goal is to prove the existence of solutions between appropriately defined upper and lower solutions. The tools used here are truncation techniques and variational methods for some nonsmooth functional whose critical points are related with the solution of the inclusion problem under consideration. The existence of critical points will be proved by minimization of an associated functional which turns out to be locally Lipschitzian. In solving the minimization problem compactness arguments of Palais-Smale type are applied which are suggested by Ekeland's well-known variational principle; cf. [108, 109]. In this respect one of the key steps is to verify some generalized Palais-Smale condition.

6.3.1   Problem, notations, and assumptions. Let $\Omega \subset \mathbb{R}^N$ be a bounded domain with Lipschitz boundary $\partial\Omega$. We consider the following BVP

$$Au(x) \in \mathcal{F}(x, u(x)), \ x \in \Omega, \quad u(x) = 0, \ x \in \partial\Omega, \tag{6.3.1}$$

where A is a quasilinear differential operator in divergence form given by

$$Au(x) = -\sum_{i=1}^{N} \frac{\partial}{\partial x_i} a_i(x, \nabla u(x)).$$

Unlike in previous sections the coefficients a_i do not depend on u itself, i.e., the operator A is of special form, and in fact we will assume that A is a potential operator. The multivalued term $\mathcal{F}$ on the right-hand side is supposed to be generated by a Baire measurable function $f : \Omega \times \mathbb{R} \to \mathbb{R}$ in the following way

$$\mathcal{F}(x, s) = [\underline{f}(x, s), \bar{f}(x, s)], \tag{6.3.2}$$

where

$$\underline{f}(x, s) = \liminf_{t \to s} f(x, t) \text{ and } \bar{f}(x, s) = \limsup_{t \to s} f(x, t).$$

The coefficients a_i are assumed to satisfy conditions (A1) and (A2) of section 5.1, i.e.,

(A1) Each $a_i : \Omega \times \mathbb{R}^N \to \mathbb{R}$ is a Carathéodory function, and there exist a constant $c_0 > 0$ and a function $k_0 \in L^q(\Omega)$, $1/p + 1/q = 1$, such that
$$|a_i(x, \xi)| \le k_0(x) + c_0 \, |\xi|^{p-1}$$
for a.e. $x \in \Omega$ and for all $\xi \in \mathbb{R}^N$.

(A2) $\sum_{i=1}^{N}(a_i(x, \xi) - a_i(x, \xi'))(\xi_i - \xi_i') \ge \mu|\xi - \xi'|^p$
for a.e. $x \in \Omega$, and for all $\xi, \xi' \in \mathbb{R}^N$ with μ being some positive constant.

In addition we assume that the operator A is a *potential operator* which is guaranteed by the following assumption.

(P) The operator $A : V_0 \to V_0^*$ is supposed to satisfy the relation

$$\int_0^1 \langle A(tu), u \rangle \, dt - \int_0^1 \langle A(tv), v \rangle \, dt = \int_0^1 \langle A(v + t(u - v)), u - v \rangle \, dt \tag{6.3.3}$$

for all $u, v \in V_0$, and its *potential* P is given by

$$P(u) = \int_0^1 \langle A(tu), u \rangle dt. \tag{6.3.4}$$

Remark 6.3.1. (i) For example, the p-Laplacian ($p \geq 2$), i.e.,

$$Au = -\sum_{i=1}^{N} \frac{\partial}{\partial x_i} \left(|\nabla u|^{p-2} \frac{\partial u}{\partial x_i} \right)$$

and the uniformly linear elliptic differential operator, i.e.,

$$Au = -\sum_{i,j=1}^{N} \frac{\partial}{\partial x_i} \left(a_{ij} \frac{\partial u}{\partial x_j} \right)$$

with $a_{ij}(x) = a_{ji}(x)$ fulfill hypotheses (A1), (A2), and (P).

(ii) A sufficient condition for A being a potential operator is the following: $a_j(x, \cdot) \in C^1(\mathbb{R}^N)$, $j = 1, \ldots, N$, and for each fixed x the family $\{a_j(x, \xi)\}_{j=1}^{N}$ is supposed to be completely integrable with respect to $\xi \in \mathbb{R}^N$, i.e., the relations $\frac{\partial}{\partial \xi_k}(a_j(x, \xi)) = \frac{\partial}{\partial \xi_j}(a_k(x, \xi))$, $k, j = 1, \ldots, N$, are satisfied for all $\xi \in \mathbb{R}^N$; cf. [185].

Let a be the semilinear form associated with the operator A by

$$\langle Au, \varphi \rangle = a(u, \varphi) = \int_\Omega \sum_{i=1}^{N} a_i(x, \nabla u) \frac{\partial \varphi}{\partial x_i} \, dx.$$

Definition 6.3.1. A function $u \in V_0$ is called a *solution* of the BVP (6.3.1) if there is a function $v \in L^q(\Omega)$ such that

  (i) $v(x) \in \mathcal{F}(x, u(x))$ for a.e. $x \in \Omega$,
  (ii) $a(u, \varphi) = \int_\Omega v \, \varphi \, dx$ for all $\varphi \in V_0$.

Definition 6.3.2. A function $\bar{u} \in V$ is called an *upper solution* of the BVP (6.3.1) if $\bar{f}(\cdot, \bar{u}) \in L^q(\Omega)$ and

  (i) $\bar{u} \geq 0$ on $\partial\Omega$,
  (ii) $a(\bar{u}, \varphi) \geq \int_\Omega \bar{f}(\cdot, \bar{u}) \, \varphi \, dx$ for all $\varphi \in V_0 \cap L^p_+(\Omega)$.

Definition 6.3.3. A function $\underline{u} \in V$ is called a *lower solution* of the BVP (6.3.1) if $\underline{f}(\cdot, \underline{u}) \in L^q(\Omega)$ and

  (i) $\underline{u} \leq 0$ on $\partial\Omega$,
  (ii) $a(\underline{u}, \varphi) \leq \int_\Omega \underline{f}(\cdot, \underline{u}) \, \varphi \, dx$ for all $\varphi \in V_0 \cap L^p_+(\Omega)$.

We assume the following hypotheses on the function f generating the multifunction $\mathcal{F}$:

  (F1) $f : \Omega \times \mathbb{R} \to \mathbb{R}$ as well as the functions $\underline{f}, \bar{f} : \Omega \times \mathbb{R} \to \mathbb{R}$ are Baire measurable and satisfy the inequality
  $\underline{f}(x, s) \leq f(x, s) \leq \bar{f}(x, s)$ for a.e. $x \in \Omega$ and for all $s \in \mathbb{R}$.

(F2) There exists a function $k \in L^q(\Omega)$ $(q = p/(p-1))$ such that

$$|f(x,s)| \leq k(x)$$

for a.e. $x \in \Omega$ and for $s \in [\underline{u}(x), \bar{u}(x)]$.

Remark 6.3.2. Baire measurable functions $f : \Omega \times \mathbb{R} \to \mathbb{R}$ are in particular sup-measurable, i.e., the composed function $x \to f(x, u(x))$ is measurable whenever the function u is measurable; cf. [12]. For example, any function that belongs to some Baire-Carathéodory class B_α (α a countable ordinal number) is sup-measurable; cf. [12]. Let B_0 denote the class of all Carathéodory functions; then $f \in B_\alpha$ if it admits a representation of the form

$$f(x,s) = \lim_{n \to \infty} f_n(x,s) \quad \text{with } x \in \Omega \setminus D_0, \ s \in \mathbb{R},$$

where the functions f_n $(n = 1, 2, \ldots)$ belong to classes B_{α_n} with $\alpha_n < \alpha$, and D_0 is some null set. For example $f(x,s) = \text{sign}(x-s)$ with $\Omega = [0,1]$ is of Baire-Carathéodory class B_1, since it is the limit of Carathéodory functions of the form $f(x,s) = \lim_{n \to \infty} \frac{2}{\pi} \arctan n(x-s)$; cf. [12]. If $f = f(x,s)$ does not depend on x, then $\bar{f} : \mathbb{R} \to \mathbb{R}$ is an upper semicontinuous function and $\underline{f} : \mathbb{R} \to \mathbb{R}$ is a lower semicontinuous function; cf. [92]. Semicontinuous functions are Baire measurable, and thus sup-measurable. Another example for f satisfying hypotheses (F1) is a Baire measurable function $f = f(x,s)$ that is increasing in s, since in this case $\bar{f}$ and $\underline{f}$ can be represented by the limits $\lim_{n \to \infty} f(x, s \pm 1/n)$, and thus $\bar{f}, \underline{f}$ are Baire measurable.

The main result of this section reads as follows.

Theorem 6.3.1. *Let $\bar{u}$ and $\underline{u}$ be upper and lower solutions, respectively, of the BVP (6.3.1) satisfying $\underline{u} \leq \bar{u}$, and let hypotheses (A1),(A2), (P), and (F1),(F2) be satisfied. Then the BVP (6.3.1) has a solution u with $\underline{u} \leq u \leq \bar{u}$.*

Before proving Theorem 6.3.1 some preliminary results are needed which will be proved in the next subsection.

6.3.2 Preliminary results. Throughout this section we shall assume that the hypotheses of Theorem 6.3.1 are satisfied. Let us introduce the "truncated" function $g : \Omega \times \mathbb{R} \to \mathbb{R}$ which is related to the nonlinearity f by

$$g(x, u(x)) = \begin{cases} f(x, \bar{u}(x)) & \text{if} \quad u(x) > \bar{u}(x), \\ f(x, u(x)) & \text{if} \quad \underline{u}(x) \leq u(x) \leq \bar{u}(x), \\ f(x, \underline{u}(x)) & \text{if} \quad u(x) < \underline{u}(x), \end{cases}$$

If $T : V \to [\underline{u}, \bar{u}]$ is the truncation operator that truncates u between the upper and the lower solution, then as in previous sections g may be represented as the composition $g = f \circ T$ of f with T, and one readily verifies that g is sup-measurable due to (F1). By (F2) an estimate of the form

$$|g(x, u(x))| \leq k(x) \qquad (6.3.5)$$

holds for a.e. $x \in \Omega$ and for all $u \in L^p(\Omega)$.

Consider the following auxiliary BVP

$$Au \in \mathcal{G}(\cdot, u) \quad \text{in } \Omega, \quad u = 0 \quad \text{on } \partial\Omega, \qquad (6.3.6)$$

where the multifunction $\mathcal{G}$ is related with g by

$$\mathcal{G}(x, s) = [\underline{g}(x, s), \bar{g}(x, s)],$$

and $\underline{g}(x, s)$, $\bar{g}(x, s)$ are given by

$$\underline{g}(x, s) = \liminf_{t \to s} g(x, t) \text{ and } \bar{g}(x, s) = \limsup_{t \to s} g(x, t).$$

We introduce the functional $\Phi : V_0 \to \mathbb{R}$ by

$$\Phi(u) = P(u) - \hat{\Psi}(u), \qquad (6.3.7)$$

where P is defined by (6.3.4) and $\hat{\Psi} : V_0 \to \mathbb{R}$ is the restriction of the functional $\Psi : L^p(\Omega) \to \mathbb{R}$ defined by

$$\Psi(u) = \int_\Omega \int_0^{u(x)} g(x, s) \, ds \, dx. \qquad (6.3.8)$$

The aim of this section is to show that the functional Φ is locally Lipschitzian ($\Phi \in \text{Lip}_{loc}(V_0, \mathbb{R})$), and that there exist critical points of Φ in the following sense.

Definition 6.3.4. Let $\Phi \in \text{Lip}_{loc}(V_0, \mathbb{R})$. A point $u_0 \in V_0$ is said to be a *critical point* of the functional Φ if $0 \in \partial\Phi(u_0)$, where $\partial\Phi(\cdot)$ denotes the generalized gradient due to Clarke. The value $c \in \mathbb{R}$ is a *critical value* of Φ if there exists a critical point $u_0 \in V_0$ such that $\Phi(u_0) = c$.

For the definition of the generalized gradient in the sense of Clarke we refer to [96]; see also section F.

Hint. One readily verifies that a local minimum of the functional Φ is a critical point; see [96, Proposition 2.3.2].

For the reader's convenience we recall an abstract result proved in [98] which will be used in solving the auxiliary BVP (6.3.6).

Theorem 6.3.2. *Let X be a reflexive Banach space and let $\Phi : X \to \mathbb{R}$ be a locally Lipschitz functional which is bounded from below, and $c = \inf_X \Phi$. If Φ satisfies the Palais-Smale condition $(PS)^*_{c,+}$ then c is attained, that is, there exists $u_0 \in X$ such that $\Phi(u_0) = c$. In particular, u_0 is a critical point of Φ, i.e., $0 \in \partial\Phi(u_0)$.*

The Palais-Smale condition $(PS)^*_{c,+}$ that appears in the Theorem reads as follows (see [98]):

$(PS)^*_{c,+}$. *Whenever $(u_n) \subset X$ and (ε_n), $(\delta_n) \subset \mathbb{R}_+$ are sequences with $\varepsilon_n \to 0$, $\delta_n \to 0$ and such that*

$$\Phi(u_n) \to c, \tag{6.3.9}$$

$$\Phi(u_n) \leq \Phi(u) + \varepsilon_n \|u_n - u\| \quad if \quad \|u_n - u\| \leq \delta_n, \tag{6.3.10}$$

then (u_n) possesses a convergent subsequence $u_{n'} \to \hat{u}$.

Similarly the Palais-Smale condition $(PS)^*_{c,-}$ is defined by interchanging u and u_n in inequality (6.3.10), and Φ is said to satisfy the condition $(PS)^*_c$ provided both $(PS)^*_{c,+}$ and $(PS)^*_{c,-}$ are fulfilled. We also make use of the the *weak* Palais-Smale condition denoted by $(PS)^*_{c,w,+}$ (respectively, $(PS)^*_{c,w,-}$) which differs from $(PS)^*_{c,+}$ (respectively, $(PS)^*_{c,-}$) in that only the existence of a weakly convergent subsequence is implied by (6.3.9) and (6.3.10). The functional Φ satisfies the weak Palais-Smale condition $(PS)^*_{c,w}$ if both $(PS)^*_{c,w,+}$ and $(PS)^*_{c,w,-}$ hold.

As will be seen a critical point of the functional Φ given by (6.3.7) turns out to be a solution of the auxiliary BVP (6.3.6). In order to apply Theorem 6.3.2 to prove the existence of a critical point of Φ we have to verify that $\Phi : V_0 \to \mathbb{R}$ is locally Lipschitzian and satisfies $(PS)^*_{c,+}$.

Lemma 6.3.1. *Both functionals $\Psi : L^p(\Omega) \to \mathbb{R}$ and $\hat{\Psi} : V_0 \to \mathbb{R}$ are (globally) Lipschitzian.*

Proof. Applying (6.3.5) we obtain

$$|\Psi(u) - \Psi(v)| = \left| \int_\Omega \int_{v(x)}^{u(x)} g(x,s) \, ds dx \right|$$

$$\leq \int_\Omega |k(x)||u(x) - v(x)| \, dx \leq \|k\|_{L^q(\Omega)} \|u - v\|_{L^p(\Omega)},$$

that is, $\Psi \in \mathrm{Lip}(L^p(\Omega), \mathbb{R})$. The assertion for $\hat{\Psi}$ follows by the continuous embedding $V_0 \subset L^p(\Omega)$. $\qquad\square$

Lemma 6.3.2. *The potential $P : V_0 \to \mathbb{R}$ is locally Lipschitzian.*

Proof. By using hypothesis (P) on the operator A, i.e.,

$$\int_0^1 \left(\langle A(tu), u \rangle - \langle A(tv), v \rangle \right) dt = \int_0^1 \langle A(v + t(u - v)), u - v \rangle \, dt,$$

its potential P given by

$$P(u) = \int_0^1 \langle A(tu), u \rangle \, dt,$$

can be estimated as follows

$$
\begin{aligned}
|P(u) - P(v)| &= \left| \int_0^1 \left(\langle A(tu), u \rangle - \langle A(tv), v \rangle \right) dt \right| \\
&= \left| \int_0^1 \langle A(tu + (1-t)v), u - v \rangle \, dt \right| \\
&\leq \left(\int_0^1 \|A(tu + (1-t)v)\|_{V_0^*} \, dt \right) \|u - v\|_{V_0} .
\end{aligned}
$$
(6.3.11)

Since $A : V_0 \to V_0^*$ is continuous and bounded, the assertion of the lemma follows from (6.3.11). $\qquad \Box$

As a consequence of the Lemmas 6.3.1 and 6.3.2 we obtain the following corollary.

Corollary 6.3.1. *The functional $\Phi : V_0 \to \mathbb{R}$ given by (6.3.7) is locally Lipschitzian.*

In order to show that Φ fulfills the Palais-Smale condition $(PS)^*_{c,+}$, we are going to prove first that Φ satisfies the weak Palais-Smale condition $(PS)^*_{c,w}$, and provide after that a general result which implies that the conditions $(PS)^*_{c,w}$ and $(PS)^*_c$ are in fact equivalent.

Lemma 6.3.3. *The functional $\Phi : V_0 \to \mathbb{R}$ satisfies $(PS)^*_{c,w,+}$ and $(PS)^*_{c,w,-}$, that is, $(PS)^*_{c,w}$.*

Proof. Let $(u_n) \subset V_0$, (ε_n), $(\delta_n) \subset \mathbb{R}$ be sequences with $\varepsilon_n \to 0$, $\delta_n \to 0$ and such that
$\Phi(u_n) \to c$, and

$\Phi(u_n) \le \Phi(u) + \varepsilon_n \|u_n - u\|_{V_0}$ if $\|u - u_n\|_{V_0} \le \delta_n$,

which yields by substituting Φ according to (6.3.7)

$$\int_0^1 \langle A(su_n), u_n \rangle \, ds - \int_0^1 \langle A(su), u \rangle \, ds \le \hat{\Psi}(u_n) - \hat{\Psi}(u) + \varepsilon_n \|u - u_n\|_{V_0}.$$

Since A is a potential operator and $\hat{\Psi}$ is even globally Lipschitzian we get the following estimate

$$\int_0^1 \langle A(u + s(u_n - u)), u_n - u \rangle \, ds \le (L + \varepsilon_n)\|u - u_n\|_{V_0}, \qquad (6.3.12)$$

where L denotes the Lipschitz constant of $\hat{\Psi}$. Assume that $(\|u_n\|_{V_0})$ is unbounded. Setting $u = u_n - t_n u_n$ with $t_n > 0$ such that $\|u - u_n\|_{V_0} = t_n\|u_n\|_{V_0} \le \delta_n$ we get in view of (6.3.12)

$$\int_0^1 \langle A((1 + (s-1)t_n)u_n) - A(0), (1 + (s-1)t_n)u_n - 0 \rangle \frac{ds}{1 + (s-1)t_n}$$
$$+ \langle A(0), u_n \rangle \le (L + \varepsilon_n)\|u_n\|_{V_0}. \qquad (6.3.13)$$

The assumption on $(\|u_n\|_{V_0})$ implies that the sequence (t_n) must be bounded, and t_n can always be chosen in such a way that $0 < t_n < 1$ holds. By the uniform monotonicity of the operator A due to (A2) from (6.3.13) it follows

$$\mu \int_0^1 (1 + (s-1)t_n)^{p-1}\|\nabla u_n\|_{L^p(\Omega)}^p \, ds - \|A(0)\|_{V_0^*}\|u_n\|_{V_0}$$
$$\le (L + \varepsilon_n)\|u_n\|_{V_0}. \qquad (6.3.14)$$

Obviously for any $t \in (0,1)$ the inequality $1 \ge (1-t)^{p-1}$ holds (note $p > 1$) which implies immediately the inequality

$$\frac{1 - (1-t)^p}{t} \ge 1, \quad \text{for all } t \in (0,1).$$

Using this last inequality we get by elementary calculations

$$\int_0^1 (1 + (s-1)t_n)^{p-1} \, ds = \frac{1}{pt_n}(1 - (1-t_n)^p) \ge \frac{1}{p}, \quad \text{for all } t_n \in (0,1),$$

and thus by applying Poincaré-Friedrichs inequality and the boundedness of (ε_n) we obtain from (6.3.14)

$$\frac{\mu}{p}\|u_n\|_{V_0}^{p-1} \leq C \left(\|A(0)\|_{V_0^*} + L + 1\right),$$

for some positive constant C which implies the boundedness of $\|u_n\|_{V_0}$. This is a contradiction to our assumption, and hence $\|u_n\|_{V_0}$ must be bounded. By the reflexivity of V_0 the latter implies the existence of a weakly convergent subsequence $(u_{n'})$ which completes the proof of the lemma, since the proof for $(PS)_{c,w,-}^*$ is done analogously.   □

Theorem 6.3.3. *Let X, Y be reflexive real Banach spaces with $X \subset Y$ compactly embedded and such that X is dense in Y. Let $\Phi : X \to \mathbb{R}$ be given in the form*

$$\Phi(u) = \int_0^1 \langle A(tu), u \rangle\, dt - \hat{\Psi}(u), \quad u \in X, \qquad (6.3.15)$$

where $A : X \to X^*$ is an uniformly monotone and hemicontinuous potential operator satisfying

$$\langle Au - Av, u - v \rangle \geq a(\|u - v\|_X)\|u - v\|_X$$

for all $u, v \in X$ with a function $a : [0, \infty) \to [0, \infty)$ being continuous and strongly increasing, and satisfying $a(0) = 0$, $\lim_{s \to \infty} a(s) = \infty$. The functional $\hat{\Psi} = \Psi|_X$ is the restriction of a locally Lipschitzian functional $\Psi : Y \to \mathbb{R}$. Then Φ satisfies $(PS)_c^*$ if and only if Φ satisfies $(PS)_{c,w}^*$.

Proof. Obviously condition $(PS)_c^*$ implies $(PS)_{c,w}^*$, so that it remains to show that $(PS)_{c,w}^*$ implies $(PS)_c^*$. Taking into account that monotone and hemicontinuous operators are locally bounded (cf. [218]), the local Lipschitz continuity of the potential P (see (6.3.4)) can be proved in just the same way as in Lemma 6.3.2. Thus the functional Φ given by (6.3.15) is locally Lipschitzian. From [98, Corollary 3] we know that $\Phi \in \mathrm{Lip}_{loc}(X, \mathbb{R})$ satisfies $(PS)_{c,w}^*$ if and only if Φ satisfies the condition $(PS)_{c,w}$ which due to Chang [92] reads as follows:

$(PS)_{c,w}$. *Whenever $(u_n) \subset X$ is such that $\Phi(u_n) \to c$, and $m(u_n) = \min\{\|u_n^*\|_{X^*} \mid u_n^* \in \partial\Phi(u_n)\} \to 0$, then (u_n) possesses a weakly convergent subsequence $(u_{n'})$.*

Thus Theorem 6.3.3 is proved provided we are able to show that $(PS)_{c,w}$ implies $(PS)_c^*$. Let Φ satisfy $(PS)_{c,w}$, and denote the weakly convergent subsequence again by (u_n). Then we have (see [96, Proposition 2.3.3])

$$\partial\Phi(u_n) \subseteq Au_n - \partial\hat{\Psi}(u_n). \qquad (6.3.16)$$

Let $w_n \in \partial \Phi(u_n)$ be such that

$$m(u_n) = \|w_n\|_{X^*}, \tag{6.3.17}$$

and let $I : X \subset Y$ be the embedding operator. Then $\hat{\Psi}(u) = \Psi \circ I(u)$ and by applying the chain rule for the generalized gradient (see [96, Theorem 2.3.10]) we obtain

$$\partial \hat{\Psi}(u) \subseteq I^* \circ \partial \Psi(Iu) \subset X^*,$$

where $I^* : Y^* \subset X^*$ is the adjoint operator to I. Therefore from (6.3.16) and (6.3.17) we get

$$w_n = Au_n - I^* \varrho_n \quad \text{with} \quad \varrho_n \in \partial \Psi(Iu_n) \subset Y^*. \tag{6.3.18}$$

By the compact embedding $X \subset Y$ the weak convergence of (u_n) in X implies the strong convergence in Y, and hence the boundedness of (Iu_n) in Y. This yields the boundedness of the set $\bigcup_n \partial \Psi(Iu_n) \subset Y^*$. Consequently the sequence $(\varrho_n) \subset Y^*$ possesses a weakly convergent subsequence (still denoted by (ϱ_n)). By the compactness of the adjoint operator I^* the sequence $(I^* \varrho_n)$ is strongly convergent in X^*. The assumptions imposed on A imply the existence and continuity of the inverse operator A^{-1}, such that from (6.3.18) we get

$$u_n = A^{-1}(w_n + I^* \varrho_n). \tag{6.3.19}$$

Since $\|w_n\|_{X^*} \to 0$, i.e., $w_n \to 0$ in X^*, and $I^* \varrho_n \to I^* \hat{\varrho}$ in X^*, by (6.3.19) and the continuity of A^{-1} the sequence (u_n) (already being some subsequence) is strongly convergent in X, which completes the proof. $\square$

The results obtained so far allow one to apply Theorem 6.3.2 which leads to the following corollary.

Corollary 6.3.2. *Let $\Phi : V_0 \to \mathbb{R}$ be the functional given by (6.3.7). Then $c = \inf_{V_0} \Phi$ is attained, that is, there exists an element $u_0 \in V_0$ such that $c = \Phi(u_0)$ and u_0 is a critical point of Φ, that is, $0 \in \partial \Phi(u_0)$.*

Proof. Let us verify the suppositions of Theorem 6.3.2. By definition the functional Φ is given in the form $\Phi(u) = P(u) - \hat{\Psi}(u)$, with

$$P(u) = \int_0^1 \langle A(tu), u \rangle \, dt, \quad \text{and} \quad \hat{\Psi}(u) = \int_\Omega \int_0^{u(x)} g(x, s) \, ds \, dx.$$

By means of hypothesis (A2) and Young's inequality we get an estimate below for P as follows:

$$P(u) = \int_0^1 \frac{1}{t}\langle A(tu) - A(0), tu \rangle \, dt + \int_0^1 \langle A(0), u \rangle \, dt$$

$$\geq \int_0^1 \mu t^{p-1} \|\nabla u\|_{L^p(\Omega)}^p \, dt - \|A(0)\|_{V_0^*} \|u\|_{V_0}$$

$$\geq \frac{\mu}{p}\|\nabla u\|_{L^p(\Omega)}^p - \varepsilon \|u\|_{V_0}^p - C(\varepsilon)\|A(0)\|_{V_0^*}^q$$

$$\text{(6.3.20)}$$

for all $\varepsilon > 0$. The functional $\hat{\Psi}$ can be estimated by means of (6.3.5) in the form

$$\left| \int_\Omega \int_0^{u(x)} g(x,s) \, ds \, dx \right| \leq \int_\Omega |k(x)||u(x)| \, dx \leq \|k\|_{L^q(\Omega)}\|u\|_{L^p(\Omega)}$$

$$\leq \eta\|u\|_{V_0}^p + C(\eta)\|k\|_{L^q(\Omega)}^q \qquad \text{(6.3.21)}$$

for all $\eta > 0$. Taking Poincaré-Friedrichs inequality into account and selecting ε and η sufficiently small, estimates (6.3.20) and (6.3.21) show that the functional Φ is bounded from below, i.e., there exists $c := \inf_{V_0} \Phi$, and according to Corollary 6.3.1 Φ is locally Lipschitz continuous. By Lemma 6.3.3 and Theorem 6.3.3 it follows that the Palais-Smale condition $(PS)_c^*$ is satisfied, which allows to apply Theorem 6.3.2 to Φ, and thus proves the assertion of the corollary. $\qquad \square$

6.3.3 Proof of the main result.

In this subsection we are going to prove Theorem 6.3.1.

Proof of Theorem 6.3.1. By Corollary 6.3.2 there is a critical point $u_0 \in V_0$ such that

$$0 \in \partial\Phi(u_0) \subseteq Au_0 + \partial(-\hat{\Psi})(u_0) = Au_0 - \partial\hat{\Psi}(u_0),$$

and hence

$$Au_0 \in \partial\hat{\Psi}(u_0). \qquad \text{(6.3.22)}$$

Let $I : V_0 \subset L^p(\Omega)$ denote the embedding operator. Then the generalized gradient of $\hat{\Psi}$ at $u \in V_0$ can be calculated by applying the chain rule (cf. [96, Theorem 2.3.10]), which yields

$$\partial\hat{\Psi}(u) = \partial(\Psi \circ I)(u) \subseteq I^* \circ \partial\Psi(Iu) \subset V_0^*, \qquad \text{(6.3.23)}$$

where $I^* : L^q(\Omega) \subset V_0^*$ $(q = p/(p-1))$ denotes the adjoint operator to I. Since $\partial\Psi(u_0) \subset L^q(\Omega)$, there is an element $v_0 \in \partial\Psi(u_0)$ such that in view of (6.3.22) and (6.3.23) we have

$$Au_0 = v_0 \quad \text{in} \quad V_0^* \,,$$

or equivalently

$$a(u_0, \varphi) = \int_\Omega v_0 \, \varphi \, dx \quad \text{for all } \varphi \in V_0 \,. \tag{6.3.24}$$

Next let us calculate the generalized gradient $\partial\Psi(u)$ of the functional $\Psi : L^p(\Omega) \to \mathbb{R}$ given by

$$\Psi(u) = \int_\Omega \int_0^{u(x)} g(x, t) \, dt dx \,.$$

By definition we have

$$\partial\Psi(u) = \{ w \in L^q(\Omega) \mid \Psi^\circ(u; \varphi) \geq \int_\Omega w \, \varphi \, dx \quad \text{for all } \varphi \in L^p(\Omega) \} \,, \tag{6.3.25}$$

where the generalized directional derivative Ψ° is defined by

$$\Psi^\circ(u; \varphi) = \limsup_{\substack{h \to 0 \\ \lambda \to +0}} \frac{1}{\lambda} [\Psi(u + h + \lambda\varphi) - \Psi(u + h)] \tag{6.3.26}$$

with $u, \varphi, h \in L^p(\Omega)$. From (6.3.26) we obtain by substituting Ψ the following relation

$$\Psi^\circ(u; \varphi) = \limsup_{\substack{h \to 0 \\ \lambda \to +0}} \frac{1}{\lambda} \int_\Omega \int_{(u+h)(x)}^{(u+h+\lambda\varphi)(x)} g(x, t) \, dt dx \,. \tag{6.3.27}$$

Changing the variables according to $t = t(s)$ with $t(s) = u(x) + h(x) + s\lambda\varphi(x)$ and applying Fatou's lemma to (6.3.27) yields

$$\Psi^\circ(u; \varphi) \leq \int_\Omega \limsup_{\substack{h \to 0 \\ \lambda \to +0}} \int_0^1 g(x, u + h + s\lambda\varphi) \, \varphi \, ds dx$$

$$\leq \int_{\{\varphi > 0\}} \bar{g}(x, u(x)) \, \varphi(x) \, dx + \int_{\{\varphi < 0\}} \underline{g}(x, u(x)) \, \varphi(x) \, dx \,. \tag{6.3.28}$$

By definition (6.3.25) and in view of (6.3.28) we find that $w \in \partial\Psi(u)$ if and only if

$$\int_\Omega w\,\varphi\,dx \leq \Psi^o(u;\varphi) \leq \int_{\{\varphi>0\}} \bar{g}(\cdot,u)\,\varphi\,dx + \int_{\{\varphi<0\}} \underline{g}(\cdot,u)\,\varphi\,dx$$

for all $\varphi \in L^p(\Omega)$, which implies

$$w(x) \in [\underline{g}(x,u(x)), \bar{g}(x,u(x))] = \mathcal{G}(x,u(x)), \quad \text{for a.e. } x \in \Omega. \qquad (6.3.29)$$

The critical point u_0 satisfies $Au_0 = v_0$ in V_0^* with $v_0 \in \partial\Psi(u_0)$, which yields according to (6.3.29) that u_0 satisfies

$$u_0 \in V_0: \quad Au_0 \in \mathcal{G}(\cdot,u_0) \quad \text{in } \Omega, \qquad (6.3.30)$$

and thus the critical point u_0 of the functional Φ is a solution of the auxiliary BVP (6.3.6).

To complete the proof of Theorem 6.3.1 we shall show that any solution of the auxiliary BVP (6.3.6) is in fact a solution of the original BVP (6.3.1). To this end we only need to show that any solution u_0 of the BVP (6.3.6) belongs to the interval $[\underline{u}, \bar{u}]$, since then we have by the definition of g the identity $g(x,u_0(x)) \equiv f(x,u_0(x))$, which implies due to (F1) that $v_0 \in \mathcal{F}(\cdot,u_0)$, and thus

$$u_0 \in V_0: \quad Au_0 \in \mathcal{F}(\cdot,u_0) \quad \text{in } \Omega,$$

that is, u_0 solves the original BVP (6.3.1) which proves the theorem.
Let $u_0 \in V_0$ be a solution of the auxiliary BVP (6.3.6). We are going to prove: $u_0 \leq \bar{u}$.
By definition u_0 satisfies (6.3.30), i.e., $Au_0 = v_0$, where

$$v_0 \in \partial\Psi(u_0) \subseteq \mathcal{G}(\cdot,u_0) = [\underline{g}(\cdot,u_0), \bar{g}(\cdot,u_0)]. \qquad (6.3.31)$$

Further, $\bar{u}$ is assumed to be an upper solution of the BVP (6.3.1), that is, $\bar{u}$ satisfies

$$a(\bar{u},\varphi) \geq \int_\Omega \bar{f}(\cdot,\bar{u})\,\varphi\,dx \quad \text{for all } \varphi \in V_0 \cap L^p_+(\Omega), \quad \bar{u} \geq 0 \text{ on } \partial\Omega. \qquad (6.3.32)$$

From (6.3.30) and (6.3.32) we get

$$a(u_0,\varphi) - a(\bar{u},\varphi) \leq \int_\Omega (v_0 - \bar{f}(\cdot,\bar{u}))\,\varphi\,dx \qquad (6.3.33)$$

for all $\varphi \in V_0 \cap L^p_+(\Omega)$. Taking as special test function $\varphi = (u_0 - \bar{u})^+$ in (6.3.33) yields

$$\int_\Omega \sum_{i=1}^N (a_i(\cdot, \nabla u_0) - a_i(\cdot, \nabla \bar{u})) \frac{\partial (u_0 - \bar{u})^+}{\partial x_i} dx \le \int_\Omega (v_0 - \bar{f}(\cdot, \bar{u}))(u_0 - \bar{u})^+ dx$$

$$= \int_{\{u_0 > \bar{u}\}} (v_0 - \bar{f}(\cdot, \bar{u}))(u_0 - \bar{u}) dx. \qquad (6.3.34)$$

According to the definition of the function g and in view of hypothesis (F1) we obtain in the case that $u_0 > \bar{u}$ the following inclusion:

$$v_0(x) \in [\underline{f}(x, \bar{u}(x)), \bar{f}(x, \bar{u}(x))] \quad \text{for a.e. } x \in \Omega,$$

which implies that the right-hand side of the inequality (6.3.34) is nonpositive. Thus from (6.3.34) and by means of hypothesis (A2) we obtain

$$0 \le \mu \int_\Omega |\nabla (u_0 - \bar{u})^+|^p \, dx \le 0,$$

which implies $(u_0 - \bar{u})^+ = 0$, i.e., $u_0 \le \bar{u}$. The proof of $\underline{u} \le u_0$ where $\underline{u}$ is the lower solution of the BVP (6.3.1) can be done analogously. This completes the proof of our main result. $\qquad \square$

6.3.4 Compactness result. Let $\mathcal{S}$ denote the set of all solutions of the BVP (6.3.1) that belong to the interval $[\underline{u}, \bar{u}]$ formed by the upper and lower solutions. In this subsection we are going to show that $\mathcal{S}$ is compact in V_0 provided that the multifunction $\mathcal{G}$ which is the restriction of $\mathcal{F}$ to the interval $[\underline{u}, \bar{u}]$ can be expressed by Clarke's generalized gradient of some regular (in the sense of Clarke, see section F) and locally Lipschitz function.

According to (6.3.29) the generalized gradient $\partial \Psi(u)$ of the functional $\Psi : L^p(\Omega) \to \mathbb{R}$, given by

$$\Psi(u) = \int_\Omega \int_0^{u(x)} g(x, s) \, ds dx,$$

satisfies

$$\partial \Psi(u)(x) \subseteq \mathcal{G}(x, u(x)) \quad \text{for a.e. } x \in \Omega.$$

We impose the following additional hypothesis

(F3) The multifunction $\mathcal{G} : \Omega \times \mathbb{R} \to 2^{\mathbb{R}} \setminus \emptyset$ satisfies

$$\partial \Psi(u)(x) = \mathcal{G}(x, u(x)),$$

which means that $v \in \partial \Psi(u)$ if and only if $v \in L^q(\Omega)$ and $v(x) \in \mathcal{G}(x, u(x))$ for a.e. $x \in \Omega$.

Theorem 6.3.4. *Let the hypotheses of Theorem 6.3.1 and (F3) be satisfied. Then the solution set S is compact in V_0.*

Proof. Given any sequence $(u_n) \subseteq S$, i.e., there is a sequence $(v_n) \subset L^q(\Omega) \subset V_0^*$ with $v_n(x) \in \mathcal{G}(x, u_n(x))$ such that for all n the equation

$$Au_n = v_n \quad \text{in } V_0^* \qquad (6.3.35)$$

holds. In view of (F2) the sequence (v_n) is bounded in $L^q(\Omega)$, and thus by means of (A2) from (6.3.35) we immediately obtain the boundedness of (u_n) in V_0. Hence, there exist subsequences of (u_n) and (v_n) denoted by (u_m) and (v_m), respectively, so that $u_m \rightharpoonup u$ in V_0 and $v_m \rightharpoonup v$ in $L^q(\Omega)$, and due to the compact embedding $V_0 \subset L^p(\Omega)$ we have $u_m \to u$ in $L^p(\Omega)$ as $m \to \infty$. By (F3) it follows that $v_m \in \partial\Psi(u_m)$, which by definition of the generalized gradient yields

$$\Psi^o(u_m; \varphi) \geq \int_\Omega v_m \varphi \, dx \quad \text{for all } \varphi \in L^p(\Omega). \qquad (6.3.36)$$

The upper semicontinuity of the generalized directional derivative Ψ^o allows one to pass to the limit in (6.3.36) as $m \to \infty$, which shows that the weak limits u and v satisfy $v \in \partial\Psi(u)$, and thus by (F3)

$$v(x) \in \mathcal{G}(x, u(x)). \qquad (6.3.37)$$

Using the identity

$$Au_m - Au = v_m - Au \quad \text{in } V_0^*,$$

and taking the special test function $\varphi = u_m - u$, we get in view of (A2)

$$\mu \|\nabla(u_m - u)\|_{L^p(\Omega}^p \leq \langle Au_m - Au, u_m - u \rangle$$

$$= -\langle Au, u_m - u \rangle + \int_\Omega v_m (u_m - u) \, dx \to 0,$$

which implies the strong convergence $u_m \to u$ in V_0. It remains to show that $u \in S$. Obviously $u \in [\underline{u}, \bar{u}]$. Since $u_m \to u$ in V_0 implies $Au_m \to Au$ in V_0^* by the continuity of $A : V_0 \to V_0^*$, and $v_m \rightharpoonup v$ in $L^q(\Omega)$ implies $v_m \to v$ in V_0^* by the compact embedding of $L^q(\Omega) \subset V_0^*$, we may pass to the limit as $m \to \infty$ in the equation

$$Au_m = v_m \quad \text{in } V_0^*,$$

which yields

$$Au = v \quad \text{in } V_0^*,$$

where v satisfies (6.3.37), and thus $v(x) \in \mathcal{F}(x, u(x))$ since $u \in [\underline{u}, \bar{u}]$. This completes the proof. □

Remark 6.3.3. In order to show that the weak limit u belongs to S one can alternatively use also the pseudomonotonicity of the operator $A : V_0 \to V_0^*$, which implies due to

$$\limsup_{m \to \infty} \langle Au_m, u_m - u \rangle = \lim_{m \to \infty} \int_\Omega v_m \left(u_m - u \right) dx \to 0, \qquad (6.3.38)$$

the weak convergence $Au_m \rightharpoonup Au$ in V_0^*, which is enough to pass to the limit in the equation $Au_m = v_m$, as $m \to \infty$, and the strong convergence of the subsequence (u_m) in V_0 is an easy consequence of the (S_+)-property of A; see D.2.

Remark 6.3.4. Hypothesis (F3) can be shown to be fulfilled if in addition to (F1) and (F2) the following two conditions are imposed:

(i) For fixed $x \in \Omega$ the function $s \to f(x,s)$ has the one-sided limits $f(x, s \pm 0)$ for each $s \in \mathbb{R}$.

(ii) If T denotes the truncation between upper and lower solution $\bar{u}$ and $\underline{u}$, respectively, then the primitive j of $f \circ T = g$ given by

$$j(x, s) = \int_0^s f(x, T\zeta) \, d\zeta$$

is regular with respect to s in the sense of Clarke (cf. [96, Chapter 2.3], see also section F).

This is because for fixed $x \in \Omega$ the function j is Lipschitz in s due to (F2), and the generalized gradient of j with respect to s, denoted by $\partial j(x,s)$, yields by using assumption (i) above $\partial j(x,s) = \mathcal{G}(x,s)$. By means of (ii) and applying [96, Theorem 2.7.5] one gets

$$\partial \Psi(u) = \partial \int_\Omega \int_0^{u(x)} g(x, s) \, ds dx = \partial \int_\Omega j(x, u(x)) \, dx = \int_\Omega \partial j(x, u(x)) \, dx,$$

which means (F3).

Remark 6.3.5. In order to get extremality results for the solution set enclosed by upper and lower solutions, additional structure conditions on the nonlinearities that generate the multifunction are needed such as those of sections 6.1 and 6.2 .

In the following section we deal with quasilinear parabolic inclusions which may be considered as the time-dependent version of the corresponding stationary problem of section 6.1.

6.4 Quasilinear parabolic inclusions with state-dependent subdifferentials

In the preceding sections truncation techniques combined with variational methods such as quasi-variational inequalities and critical point theory for nonsmooth, locally Lipschitz functionals were the main tools to treat elliptic inclusions. However, these methods, in general, can no longer be applied to parabolic inclusions considered here. This section deals with the parabolic counterpart of Theorem 6.1.1. We present a regularization and truncation method which allows one to obtain existence and comparison results in a unified way for both quasilinear elliptic and parabolic differential inclusions.

6.4.1 Problem, notations, and hypotheses. Throughout this section we use the notations of section 5.2. Thus let $\Omega \subset \mathbb{R}^N$ be a bounded domain with Lipschitz boundary $\partial\Omega$, $Q = \Omega \times (0, \tau)$ and $\Gamma = \partial\Omega \times (0, \tau)$. This section deals with the existence and enclosure of weak solutions of the following quasilinear initial-Dirichlet boundary value problem (IBVP for short)

$$\left.\begin{array}{l} \dfrac{\partial u}{\partial t} + Au + \beta(\cdot, \cdot, u, u) \ni h \quad \text{in } Q, \\[2mm] u = 0 \text{ in } \Omega \times \{0\} \quad \text{and} \quad u = 0 \text{ on } \Gamma, \end{array}\right\} \tag{6.4.1}$$

where A is a second order quasilinear differential operator in divergence form of Leray-Lions type given by

$$Au(x, t) = -\sum_{i=1}^{N} \frac{\partial}{\partial x_i} a_i(x, t, u(x, t), \nabla u(x, t)),$$

and the multifunction β of the IBVP (6.4.1) is assumed to be a state-dependent subdifferential, i.e., $\beta(x, t, u, \cdot) : \mathbb{R} \to 2^{\mathbb{R}} \setminus \emptyset$ is a maximal monotone graph in $\mathbb{R}^2$ that, in addition, depends on the space-time variable (x, t) and the solution u itself so that the multifunction $u \to \beta(x, t, u, u)$ need not necessarily be monotone. As in section 5.2 we introduce the spaces $V = L^p(0, \tau; V)$ and $V_0 = L^p(0, \tau; V_0)$, with $V = W^{1,p}(\Omega)$ and $V_0 = W_0^{1,p}(\Omega)$ being the usual Sobolev spaces. We assume $p \geq 2$ and denote by q the dual real satisfying $1/p + 1/q = 1$. Denote by V^* and V_0^* the corresponding dual spaces of V and V_0, respectively, and let the spaces W and W_0 be given by

$$W = \{w \in V \mid \frac{\partial w}{\partial t} \in V^*\}, \quad \text{and} \quad W_0 = \{w \in V_0 \mid \frac{\partial w}{\partial t} \in V_0^*\},$$

respectively. For properties of these function spaces we refer to sections E.1 and E.2. Also we note that only for the sake of simplicity and without loss of generality homogeneous initial and boundary conditions have been assumed in the IBVP (6.4.1). Inhomogeneous initial and boundary conditions can be reduced to homogeneous ones by translation in just the same way as in Remark 5.2.1.

We assume that $h \in \mathcal{V}_0^*$, and impose the following conditions of Leray-Lions type on the coefficient functions a_i, $i = 1, ..., N$.

(A1) Each $a_i : Q \times \mathbb{R} \times \mathbb{R}^N \to \mathbb{R}$ satisfies the Carathéodory conditions, i.e., $a_i(x, t, s, \xi)$ is measurable in $(x, t) \in Q$ for all $(s, \xi) \in \mathbb{R} \times \mathbb{R}^N$ and continuous in (s, ξ) for a.e. $(x, t) \in Q$. There exist a constant $c_0 > 0$ and a function $k_0 \in L^q(Q)$, $1/p + 1/q = 1$, such that

$$|a_i(x, t, s, \xi)| \le k_0(x, t) + c_0(|s|^{p-1} + |\xi|^{p-1})$$

for a.e. $(x, t) \in Q$ and for all $(s, \xi) \in \mathbb{R} \times \mathbb{R}^N$.

(A2)$_1$ $\sum_{i=1}^{N}(a_i(x, t, s, \xi) - a_i(x, t, s, \xi'))(\xi_i - \xi_i') > 0$
for a.e. $(x, t) \in Q$, for all $s \in \mathbb{R}$, and for all $\xi, \xi' \in \mathbb{R}^N$ with $\xi \ne \xi'$.

(A2)$_2$ $\sum_{i=1}^{N} a_i(x, t, s, \xi)\xi_i \ge \nu|\xi|^p - k(x, t)$
for a.e. $(x, t) \in Q$, for all $s \in \mathbb{R}$, and for all $\xi \in \mathbb{R}^N$ with some constant $\nu > 0$ and some function $k \in L^1(Q)$.

As usual $\langle \cdot, \cdot \rangle$ denotes the duality pairing between $\mathcal{V}_0^*$ and $\mathcal{V}_0$. Then as a consequence of (A1) the semilinear form a associated with the operator A by

$$\langle Au, \varphi \rangle = a(u, \varphi) = \sum_{i=1}^{N} \int_Q a_i(x, t, u, \nabla u) \frac{\partial \varphi}{\partial x_i} \, dx dt, \quad \varphi \in \mathcal{V}_0$$

is well defined for any $u \in \mathcal{V}$, and the operator $A : \mathcal{V} \to \mathcal{V}^* \subset \mathcal{V}_0^*$ is continuous and bounded. As in section 5.2 we introduce the natural partial ordering by means of the positive cone $L_+^p(Q)$. The notion of weak solution (respectively, upper and lower solutions) of the IBVP (6.4.1) is defined as follows:

Definition 6.4.1. A function $u \in \mathcal{W}_0$ is called a *solution* of the IBVP (6.4.1) if there is a function $v \in L^q(Q)$ such that

(i) $u = 0$ in $\Omega \times \{0\}$,
(ii) $v(x, t) \in \beta(x, t, u(x, t), u(x, t))$ for a.e. $(x, t) \in Q$,
(iii) $\langle \frac{\partial u}{\partial t}, \varphi \rangle + a(u, \varphi) + \int_Q v \varphi \, dx dt = \langle h, \varphi \rangle$ for all $\varphi \in \mathcal{V}_0$.

Definition 6.4.2. A function $\bar{u} \in \mathcal{W}$ is called an *upper solution* of the IBVP (6.4.1) if there is a function $\bar{v} \in L^q(Q)$ such that

    (i) $\bar{u} \geq 0$ in $\Omega \times \{0\}$ and $\bar{u} \geq 0$ on Γ,

    (ii) $\bar{v}(x,t) \in \beta(x,t,\bar{u}(x,t),\bar{u}(x,t))$ for a.e. $(x,t) \in Q$,

    (iii) $\langle \frac{\partial \bar{u}}{\partial t}, \varphi \rangle + a(\bar{u}, \varphi) + \int_Q \bar{v}\varphi \, dx dt \geq \langle h, \varphi \rangle$ for all $\varphi \in V_0 \cap L^p_+(Q)$.

Similarly a function $\underline{u} \in \mathcal{W}$ is a *lower solution* of (6.4.1) if the reversed inequalities hold in Definition 6.4.2 with $\bar{u}$ and $\bar{v}$ replaced by $\underline{u}$ and $\underline{v}$, respectively. The multifunction β is assumed to be generated by a function $f : Q \times \mathbb{R} \times \mathbb{R} \to \mathbb{R}$, and similar to the elliptic case we impose the following hypotheses on f.

  (H1) Let there exist an upper and a lower solution $\bar{u}$ and $\underline{u}$ of the IBVP (6.4.1), respectively, such that $\underline{u} \leq \bar{u}$.

  (H2) The function $f : Q \times \mathbb{R} \times \mathbb{R} \to \mathbb{R}$ satisfies:

    (i) f is measurable in $(x,t) \in Q$ for all $(r,s) \in \mathbb{R} \times \mathbb{R}$ and continuous in r for a.e. $(x,t) \in Q$, uniformly with respect to s.

    (ii) $s \to f(x,t,r,s)$ is increasing (possibly discontinuous) for a.e. $(x,t) \in Q$ and for each $r \in \mathbb{R}$, and it is related with the multifunction β by

$$\beta(x,t,r,s) = [f(x,t,r,s-0), f(x,t,r,s+0)],$$

    where

$$f(x,t,r,s \pm 0) = \lim_{\varepsilon \downarrow 0} f(x,t,r,s \pm \varepsilon),$$

    are the one-sided limits of f with respect to its last argument.

    (iii) $(x,t,s) \to f(x,t,r,s)$ is Borel measurable in $Q \times \mathbb{R}$ for each $r \in \mathbb{R}$.

  (H3) There is a function $k_1 \in L^q_+(Q)$ and a constant $\alpha > 0$ such that

$$|f(x,t,r,s)| \leq k_1(x,t)$$

for a.e. $(x,t) \in Q$, for all $r \in [\underline{u}(x,t), \bar{u}(x,t)]$ and $s \in [\underline{u}(x,t) - 2\alpha, \bar{u}(x,t) + 2\alpha]$.

Remark 6.4.1. By (H2)(i) $(x,t,r) \to f(x,t,r,s)$ is a Carathéodory function uniformly with respect to $s \in \mathbb{R}$. Hypothesis (H2) implies that the function f is sup-measurable, which means that the Nemytskij operator F defined by

$$F(u,v)(x,t) := f(x,t,u(x,t),v(x,t))$$

generates a measurable function whenever $u, v : Q \to \mathbb{R}$ are measurable. By hypotheses (H2) and (H3) the mapping $u \to F(u, v)$ is continuous and bounded from $[\underline{u}, \bar{u}] \subset L^p(Q)$ to $L^q(Q)$ uniformly with respect to $v \in [\underline{u} - 2\alpha, \bar{u} + 2\alpha]$. The local $L^q(Q)$-boundedness condition (H3) is less restrictive than, for instance, a growth condition of the form $|f(x, t, r, s)| \le c(k_2(x, t) + |r|^{p-1} + |s|^{p-1})$ with some $k_2 \in L^q(Q)$.

Remark 6.4.2. One possibility to determine upper and lower solutions of the inclusion (6.4.1) is to replace the problem by an IBVP of the form

$$
\left.
\begin{aligned}
\frac{\partial u}{\partial t} + Au + \tilde{f}(\cdot, \cdot, u, u) = h \quad &\text{in } Q, \\
u = 0 \text{ in } \Omega \times \{0\} \quad \text{and} \quad u = 0 \text{ on } \Gamma,
\end{aligned}
\right\}
\tag{6.4.2}
$$

where $\tilde{f} : Q \times \mathbb{R} \times \mathbb{R} \to \mathbb{R}$ may be any single-valued selection of the multifunction β such as for example the generating function f of β itself. Then obviously any upper (lower) solution of (6.4.2) is also an upper (lower) solution of (6.4.1).

The main result of this section is the following existence, comparison, and compactness result.

Theorem 6.4.1. *Let hypotheses (A1), (A2)$_1$, (A2)$_2$, and (H1)-(H3) be satisfied. Then the IBVP (6.4.1) admits at least one solution u within the order interval $[\underline{u}, \bar{u}]$ formed by the given upper and lower solutions $\bar{u}$ and $\underline{u}$, respectively. The set S of all solutions of the IBVP (6.4.1) within $[\underline{u}, \bar{u}]$ is weakly compact in $\mathcal{W}_0$, and compact in $\mathcal{V}_0$.*

Note that $\mathcal{W}_0 \subset \mathcal{V}_0$ is not compactly embedded so that the last assertion of Theorem 6.4.1 is not trivial. Truncation and regularization techniques as well as existence results for evolutionary equations governed by pseudomonotone operators are the main tools in the proof of Theorem 6.4.1.

6.4.2 Truncation and regularization. Throughout this subsection we shall assume that all the hypotheses of Theorem 6.4.1 are satisfied. The proof of this Theorem, which will be given in subsection 6.4.3, is based on some truncated and regularized auxiliary IBVP which is associated with the original one and which will be studied first. As in previous sections let $\eta \ge 0$ be any nonnegative constant and denote by T_η the truncation operator defined by

$$
T_\eta u(x, t) =
\begin{cases}
\bar{u}(x, t) + \eta & \text{if} \quad u(x, t) > \bar{u}(x, t) + \eta, \\
u(x, t) & \text{if} \quad \underline{u}(x, t) - \eta \le u(x, t) \le \bar{u}(x, t) + \eta, \\
\underline{u}(x, t) - \eta & \text{if} \quad u(x, t) < \underline{u}(x, t) - \eta,
\end{cases}
$$

where $\underline{u}$ and $\bar{u}$ are the given lower and upper solutions, respectively. For later purposes we will use the truncations T_0 and T_α only.

Let $b : Q \times \mathbb{R} \to \mathbb{R}$ be the cut-off function given by

$$
b(x, t, s) = \begin{cases} (s - \bar{u}(x, t))^{p-1} & \text{if} \quad s > \bar{u}(x, t), \\ 0 & \text{if} \quad \underline{u}(x, t) \le s \le \bar{u}(x, t), \\ -(\underline{u}(x, t) - s)^{p-1} & \text{if} \quad s < \underline{u}(x, t). \end{cases}
$$

Then we already know (cf. section 5.2) that b is a Carathéodory function and satisfies a growth condition of the form

$$
|b(x, t, s)| \le k_3(x, t) + c_1|s|^{p-1} \tag{6.4.3}
$$

for some positive constant c_1 and some function $k_3 \in L^q(Q)$, and admits an estimate of the form

$$
\int_Q b(x, t, u(x, t)) u(x, t)\, dx dt \ge c_2 \|u\|^p_{L^p(Q)} - c_3 \tag{6.4.4}
$$

for some positive constants c_2, c_3. According to (6.4.3) the Nemytskij operator B associated with the function b is bounded and continuous from $L^p(Q)$ into $L^q(Q)$, and from (6.4.4) we get

$$
\int_Q (Bu)\, u\, dx dt \ge c_2 \|u\|^p_{L^p(Q)} - c_3 .
$$

Let $\rho : \mathbb{R} \to \mathbb{R}$ be a mollifier, that is, $\rho \in C_0^\infty((-1, +1))$, $\rho \ge 0$ and

$$
\int_{-\infty}^\infty \rho(s)\, ds = 1 .
$$

For any $\varepsilon > 0$ the *regularization* of f with respect to its last argument, denoted by f^ε, is then defined by the convolution

$$
f^\varepsilon(x, t, r, s) = \frac{1}{\varepsilon} \int_{-\infty}^\infty f(x, t, r, s - \zeta) \rho\left(\frac{\zeta}{\varepsilon}\right) d\zeta . \tag{6.4.5}
$$

From hypothesis (H3) it readily follows that for any $u \in [\underline{u}, \bar{u}]$ and for any ε satisfying $0 < \varepsilon < 2\alpha$ we get the estimate

$$
|f^\varepsilon(x, t, u(x, t), u(x, t))| \le k_1(x, t) . \tag{6.4.6}
$$

Let F^ε denote the Nemytskij operator associated with f^ε; then we define

$$F_{0,\alpha}^\varepsilon(u,v)(x,t) := F^\varepsilon(T_0 u, T_\alpha v)(x,t) = f^\varepsilon(x,t,T_0 u(x,t), T_\alpha v(x,t)),$$
(6.4.7)

where α is the constant of hypothesis (H3). Further we introduce a "truncated" operator A_0 and an associated semilinear form a_0 by

$$\langle A_0 u, \varphi \rangle = a_0(u,\varphi) := \sum_{i=1}^N \int_Q a_i(x,t,T_0 u, \nabla u) \frac{\partial \varphi}{\partial x_i}\, dx dt.$$
(6.4.8)

Due to the continuity properties of the truncation mappings T_η the operator $A_0 : V \to V_0^*$ behaves obviously similar like A.

The following auxiliary regularized truncated IBVP plays a crucial role in the proof of the Theorem.

Problem: Find $u \in \mathcal{W}_0$ such that

$$\left.\begin{aligned} \frac{\partial u}{\partial t} + A_0 u + F_{0,\alpha}^\varepsilon(u,u) + Bu = h \quad \text{in } Q, \\ u = 0 \quad \text{in } \Omega \times \{0\}. \end{aligned}\right\} \quad (\text{P}_\varepsilon)$$

Lemma 6.4.1. *The IBVP (P_ε) has a solution $u_\varepsilon \in \mathcal{W}_0$ for any $\varepsilon \in (0,\alpha)$.*

Proof. Let $L = \partial/\partial t$ and its domain $D(L) \subset \mathcal{V}_0$ be given by

$$D(L) = \{u \in \mathcal{W}_0 \mid u(\cdot,0) = 0 \quad \text{in } \Omega\},$$

where $L : D(L) \subset \mathcal{V}_0 \to \mathcal{V}_0^*$ is defined by

$$\langle Lu, \varphi \rangle = \int_0^\tau < \frac{\partial u}{\partial t}(t), \varphi(t) > dt \quad \text{for all } \varphi \in \mathcal{V}_0,$$

and where $< \cdot, \cdot >$ denotes the duality pairing between the Sobolev space V_0 and its dual V_0^*. The linear operator $L : D(L) \subset \mathcal{V}_0 \to \mathcal{V}_0^*$ can be shown to be closed, densely defined, and maximal monotone; cf. [218, Chapter 32] and section E. Now define $\widehat{F} : \mathcal{V}_0 \to \mathcal{V}_0^*$ by $\widehat{F}u := F_{0,\alpha}^\varepsilon(u,u)$, then the IBVP (P_ε) may be reformulated as an operator equation in V_0^* given by:

$$\text{Find } u \in D(L) \subset \mathcal{W}_0 : \quad (L + A_0 + \widehat{F} + B)u = h.$$
(6.4.9)

In view of hypotheses (H2), (H3) and according to Remark 6.4.1 the regularized and truncated Nemytskij operators $F_{0,\alpha}^\varepsilon : L^p(Q) \times L^p(Q) \to L^q(Q)$

are continuous and uniformly bounded. This follows from the fact that $u \to F_{0,\alpha}^{\varepsilon}(u,v)$ is continuous uniformly with respect to $v \in L^p(Q)$ and $v \to F_{0,\alpha}^{\varepsilon}(u,v)$ is continuous for each $u \in L^p(Q)$. The compact embedding $\mathcal{W}_0 \subset L^p(Q)$ implies that $F_{0,\alpha}^{\varepsilon} : \mathcal{W}_0 \times \mathcal{W}_0 \to L^q(Q)$ is compact, and hence $\widehat{F} : D(L) \subset V_0 \to L^q(Q) \subset V_0^*$ is completely continuous with respect to the graph norm topology of $D(L)$. By (6.4.3) the operator $B : L^p(Q) \to L^q(Q)$ is continuous and bounded, and thus due to the compact embedding $\mathcal{W} \subset L^p(Q)$ the operator $B : D(L) \subset V_0 \to L^q(Q) \subset V_0^*$ is also completely continuous with respect to $D(L)$ equipped with its graph norm topology. The hypotheses (A1), (A2)$_1$, (A2)$_2$ along with the properties of the operators $\widehat{F}$ and B derived above imply that the operator $\mathcal{A} := A_0 + \widehat{F} + B : V_0 \to V_0^*$ is continuous, bounded, and pseudomonotone with respect to the graph norm topology of $D(L)$ (see Definition E.3.1 and Theorem E.3.2). Applying Theorem E.3.1 (respectively, Corollary E.3.1) the mapping $L + \mathcal{A} : D(L) \to V_0^*$ is surjective provided that $\mathcal{A} : V_0 \to V_0^*$ is coercive, that is

$$\frac{\langle \mathcal{A}u, u \rangle}{\|u\|_{V_0}} \to \infty \quad \text{as } \|u\|_{V_0} \to \infty. \tag{6.4.10}$$

The coercivity of $\mathcal{A}$ readily follows from (A2)$_2$ and (6.4.4) as well as from the uniform boundedness of $\widehat{F}$. Hence, $L + \mathcal{A} : D(L) \to V_0^*$ is surjective by Theorem E.3.1, which implies the existence of a solution of the auxiliary IBVP (P$_\varepsilon$). $\qquad \square$

Let (ε_n) be any sequence of positive reals satisfying $\varepsilon_n \in (0, \alpha)$, and $\varepsilon_n \searrow 0$ as $n \to \infty$. Assign to each ε_n a solution of the IBVP (P$_{\varepsilon_n}$) denoted by u_n whose existence is guaranteed by Lemma 6.4.1. Then for any sequence (u_n) that can be formed in this way the following result holds.

Lemma 6.4.2. *Let $\varepsilon_n \in (0, \alpha)$ satisfy $\varepsilon_n \searrow 0$ as $n \to \infty$, and let (u_n) be any sequence of solutions of the IBVP (P$_{\varepsilon_n}$). Then there exists a subsequence of (u_n) (again denoted by (u_n)) satisfying*

(i) $u_n \rightharpoonup u$ in $\mathcal{W}_0$, i.e., $u_n \rightharpoonup u$ in V_0, $\frac{\partial u_n}{\partial t} \rightharpoonup \frac{\partial u}{\partial t}$ in V_0^*,

(ii) $u_n \to u$ in $L^p(Q)$,

(iii) $F_{0,\alpha}^{\varepsilon_n}(u_n, u_n) \rightharpoonup v$ in $L^q(Q)$,
    where $v(x,t) \in \beta(x, t, T_0 u(x,t), T_\alpha u(x,t))$.

Proof. Let u_n be any solution of the IBVP (P$_{\varepsilon_n}$). We shall show first that (u_n) is uniformly bounded in $\mathcal{W}_0$. To this end consider the weak formulation of the equation satisfied by u_n and take as special test function $\varphi = u_n$,

which yields

$$\langle \frac{\partial u_n}{\partial t}, u_n \rangle + a_0(u_n, u_n) + \int_Q (F_{0,\alpha}^{\varepsilon_n}(u_n, u_n) + Bu_n) u_n \, dx dt = \langle h, u_n \rangle.$$

By using $(A2)_2$ and (6.4.4) we get an estimate of the form

$$\frac{1}{2} \|u_n(\cdot, \tau)\|_{L^2(\Omega)}^2 + \nu \|\nabla u_n\|_{L^p(Q)}^p + c_2 \|u_n\|_{L^p(Q)}^p$$
$$\leq \|h\|_{\mathcal{V}_0^*} \|u_n\|_{\mathcal{V}_0} + \|F_{0,\alpha}^{\varepsilon_n}(u_n, u_n)\|_{L^q(Q)} \|u_n\|_{L^p(Q)} + c_3 + \|k\|_{L^1(Q)}.$$

By applying Young's inequality to the first and second term on the right-hand side of the last inequality we get due to the uniform boundedness of $\|F_{0,\alpha}^{\varepsilon_n}(u_n, u_n)\|_{L^q(Q)}$ the estimate

$$\frac{1}{2} \|u_n(\cdot, \tau)\|_{L^2(\Omega)}^2 + \nu \|\nabla u_n\|_{L^p(Q)}^p + c_2 \|u_n\|_{L^p(Q)}^p \leq c(\tilde{\varepsilon}) + \tilde{\varepsilon} \|u_n\|_{\mathcal{V}_0}^p \quad (6.4.11)$$

for any $\tilde{\varepsilon} > 0$ and some constant $c(\tilde{\varepsilon})$ not depending on u_n. Since $\|\nabla u_n\|_{L^p(Q)}$ and $\|u_n\|_{\mathcal{V}}$ are equivalent norms in $\mathcal{V}_0$, we obtain from (6.4.11) by selecting $\tilde{\varepsilon}$ sufficiently small a uniform bound of (u_n) in $\mathcal{V}_0$,

$$\|u_n\|_{\mathcal{V}_0} \leq c \quad \text{for all } n. \qquad (6.4.12)$$

By means of (6.4.12) and the boundedness of the operators $A_0, \hat{F}, B : \mathcal{V}_0 \to \mathcal{V}_0^*$ one easily gets

$$\|\frac{\partial u_n}{\partial t}\|_{\mathcal{V}_0^*} = \sup_{\|\varphi\|_{\mathcal{V}_0}=1} |\langle \frac{\partial u_n}{\partial t}, \varphi \rangle| \leq c \quad \text{for all } n, \qquad (6.4.13)$$

and thus (6.4.12) and (6.4.13) imply a uniform bound of (u_n) in $\mathcal{W}_0$, i.e.,

$$\|u_n\|_{\mathcal{W}_0} \leq c \quad \text{for all } n. \qquad (6.4.14)$$

Since $\mathcal{W}_0$ is a reflexive Banach space, (6.4.14) implies the existence of a weakly convergent subsequence of (u_n) (again denoted by (u_n)) to some $u \in \mathcal{W}_0$, which is strongly convergent in $L^p(Q)$ to u due to the compact embedding $\mathcal{W}_0 \subset L^p(Q)$. The uniform boundedness of the sequence $(F_{0,\alpha}^{\varepsilon_n}(u_n, u_n))$ in $L^q(Q)$ implies the existence of a weakly convergent subsequence (again denoted by $(F_{0,\alpha}^{\varepsilon_n}(u_n, u_n))$) in $L^q(Q)$ with weak limit v. Thus it remains to show that $v \in \beta(T_0 u, T_\alpha u)$ which will be done next.

The strong convergence $u_n \to u$ in $L^p(Q)$ and the continuity of the truncation operators T_η imply $T_\eta u_n \to T_\eta u$ in $L^p(Q)$. Due to the monotonicity of $f(x, t, r, s)$ in its last argument we have for any $\varepsilon_n \in (0, \alpha)$ the inequality

$$f(\cdot, \cdot, T_0 u_n, T_\alpha u_n - \varepsilon_n) \leq f^{\varepsilon_n}(\cdot, \cdot, T_0 u_n, T_\alpha u_n) \leq f(\cdot, \cdot, T_0 u_n, T_\alpha u_n + \varepsilon_n).$$
$$(6.4.15)$$

In particular, the strong convergence of $(T_\alpha u_n)$ in $L^p(Q)$ implies the almost everywhere convergence of some subsequence (again denoted by $(T_\alpha u_n)$). By applying Egoroff's theorem, for any $\delta > 0$ there is a measurable set $\omega \subset Q$ with meas(ω) $< \delta$ such that

$$T_\alpha u_n(x, t) \rightrightarrows T_\alpha u(x, t) \quad \text{uniformly in } Q \setminus \omega.$$

Let $\varrho \in (0, \alpha)$ arbitrarily be given. Then due to the uniform convergence of $T_\alpha u_n$ in $Q \setminus \omega$ and because of $\varepsilon_n \to 0$ there is a $n_0(\varrho)$ such that for all $n > n_0(\varrho)$ we have

$$0 < \varepsilon_n < \varrho/2 \quad \text{and} \quad |T_\alpha u_n(x, t) - T_\alpha u(x, t)| < \varrho/2 \quad \text{in } Q \setminus \omega,$$

which yields by (6.4.15) for $(x, t) \in Q \setminus \omega$

$$f^{\varepsilon_n}(x, t, T_0 u_n(x, t), T_\alpha u_n(x, t)) \leq f(x, t, T_0 u_n(x, t), T_\alpha u(x, t) + \varepsilon_n + \varrho/2)$$
$$\leq f(x, t, T_0 u_n(x, t), T_\alpha u(x, t) + \varrho),$$
$$(6.4.16)$$

and analogously

$$f(x, t, T_0 u_n(x, t), T_\alpha u(x, t) - \varrho) \leq f^{\varepsilon_n}(x, t, T_0 u_n(x, t), T_\alpha u_n(x, t)). \quad (6.4.17)$$

Thus (6.4.16) and (6.4.17) result in

$$F(T_0 u_n, T_\alpha u - \varrho) \leq F_{0,\alpha}^{\varepsilon_n}(u_n, u_n) \leq F(T_0 u_n, T_\alpha u + \varrho). \quad (6.4.18)$$

Multiplying (6.4.18) by any $\varphi \in L_+^p(Q \setminus \omega)$, integrating over $Q \setminus \omega$, and taking into account that $u \to F(T_0 u, T_\alpha w)$ is continuous (and bounded) uniformly with respect to w (cf. Remark 6.4.1), and $F_{0,\alpha}^{\varepsilon_n}(u_n, u_n) \rightharpoonup v$ in $L^q(Q)$, we obtain the following inequality by passing to the limit as $n \to \infty$:

$$\int_{Q \setminus \omega} F(T_0 u, T_\alpha u - \varrho) \varphi \, dx \, dt \leq \int_{Q \setminus \omega} v \varphi \, dx \, dt$$

$$\leq \int_{Q \setminus \omega} F(T_0 u, T_\alpha u + \varrho) \, dx \, dt.$$
$$(6.4.19)$$

Since (6.4.19) holds for any $\varrho \in (0, \alpha)$, we get by applying Lebesgue's dominated convergence theorem as $\varrho \to 0$

$$\int_{Q\setminus\omega} F(T_0u, T_\alpha u - 0)\,\varphi\,dx\,dt \le \int_{Q\setminus\omega} v\,\varphi\,dx\,dt \le \int_{Q\setminus\omega} F(T_0u, T_\alpha u + 0)\,dx\,dt$$

for any $\varphi \in L_+^p(Q \setminus \omega)$ which yields

$$v(x,t) \in [\, f(x, t, T_0u(x,t), T_\alpha u(x,t) - 0), f(x, t, T_0u(x,t), T_\alpha u(x,t) + 0)\,]$$

for a.e. $(x, t) \in Q \setminus \omega$ where $\text{meas}(\omega) < \delta$. Since $\delta > 0$ has been chosen arbitrarily, the last inclusion must be true for a.e. $(x, t) \in Q$ such that finally we obtain the assertion, i.e. $v(x,t) \in \beta(x, t, T_0u(x,t), T_\alpha u(x,t))$ for a.e. $(x, t) \in Q$.   □

6.4.3 Proof of the main result–Theorem 6.4.1.

Proof. The proof will be given in two steps.

(a) Existence of solutions of the IBVP (6.4.1).

By Lemma 6.4.2 there is a sequence $(u_n) \subset \mathcal{W}_0$ satisfying the convergence properties (i), (ii), and (iii) of this lemma. Since $u_n \in D(L)$ and $D(L)$ is closed with respect to the norm of $\mathcal{W}_0$ and convex, it follows that it is weakly closed, and thus the limit $u \in D(L)$. By definition, u_n is a solution of the IBVP $(\mathrm{P}_{\varepsilon_n})$, that is,

$$u_n \in D(L): \quad Lu_n + A_0u_n + F_{0,\alpha}^{\varepsilon_n}(u_n, u_n) + Bu_n = h \quad \text{in } \mathcal{V}_0^*,$$

which yields with the test function $\varphi = u_n - u$

$$\langle \frac{\partial u_n}{\partial t}, u_n - u \rangle + a_0(u_n, u_n - u) + \int_Q \Big(F_{0,\alpha}^{\varepsilon_n}(u_n, u_n) + Bu_n \Big)(u_n - u)\,dx\,dt$$
$$= \langle h, u_n - u \rangle.$$

The right-hand side of the last equation tends to zero as $n \to \infty$, and since the sequences (Bu_n), $(F_{0,\alpha}^{\varepsilon_n}(u_n, u_n))$ are bounded in $L^q(Q)$, also the integral on the left-hand side tends to zero as $n \to \infty$. Because

$$-\langle \frac{\partial u_n}{\partial t}, u_n - u \rangle = -\langle \frac{\partial (u_n - u)}{\partial t}, u_n - u \rangle - \langle \frac{\partial u}{\partial t}, u_n - u \rangle$$
$$\le -\langle \frac{\partial u}{\partial t}, u_n - u \rangle \to 0 \quad \text{as } n \to \infty,$$

we obtain

$$\limsup_{n\to\infty} a_0(u_n, u_n - u) = \limsup_{n\to\infty} \langle A_0 u_n, u_n - u \rangle \leq 0.$$

Thus the pseudomonotonicity of the operator A_0 with respect to the graph norm topology of $D(L)$ and the convergence properties due to Lemma 6.4.2 allow the passage to the limit in the corresponding weak form of the IBVP (P_{ε_n}) so that the limits $u \in \mathcal{W}_0$ and $v \in L^q(Q)$ given by Lemma 6.4.2 satisfy the following IBVP

$$\frac{\partial u}{\partial t} + A_0 u + v + Bu = h \quad \text{in } \mathcal{V}_0^*, \quad u = 0 \quad \text{in } \Omega \times \{0\}, \qquad (P_0)$$

where $v(x, t) \in \beta(x, t, T_0 u(x, t), T_\alpha u(x, t))$ for a.e. $(x, t) \in Q$. We shall show that the limit u is in fact a solution of the original IBVP (6.4.1) which is enclosed by the given upper and lower solutions. To this end we only need to show that the solution u of (P_0) satisfies $\underline{u} \leq u \leq \bar{u}$, since then we have $A_0 = A$, $\beta(\cdot, \cdot, T_0 u, T_\alpha u) = \beta(\cdot, \cdot, u, u)$, and $Bu = 0$, so that for this u problem (P_0) coincides with the IBVP (6.4.1).

Let us prove: $u \leq \bar{u}$.

By definition the upper solution $\bar{u}$ satisfies $\bar{u}(x, 0) \geq 0$ in Ω and $\bar{u} \geq 0$ on Γ, as well as

$$\langle \frac{\partial \bar{u}}{\partial t}, \varphi \rangle + a(\bar{u}, \varphi) + \int_Q \bar{v} \varphi \, dx dt \geq \langle h, \varphi \rangle \qquad (6.4.20)$$

for all $\varphi \in \mathcal{V}_0 \cap L_+^p(Q)$, where $\bar{v}(x, t) \in \beta(x, t, \bar{u}(x, t), \bar{u}(x, t))$ for a.e. $(x, t) \in Q$. Since the limit u solves (P_0), it satisfies zero initial and boundary conditions and

$$\langle \frac{\partial u}{\partial t}, \varphi \rangle + a_0(u, \varphi) + \int_Q v \varphi \, dx dt + \int_Q (Bu) \varphi \, dx dt = \langle h, \varphi \rangle \qquad (6.4.21)$$

for all $\varphi \in \mathcal{V}_0$, where $v(x, t) \in \beta(x, t, T_0 u(x, t), T_\alpha u(x, t))$ for a.e. $(x, t) \in Q$. Subtracting (6.4.20) from (6.4.21) and using the special test function $\varphi = (u - \bar{u})^+ \in \mathcal{V}_0 \cap L_+^p(Q)$ we get the inequality

$$\langle \frac{\partial(u - \bar{u})}{\partial t}, (u - \bar{u})^+ \rangle + a_0(u, (u - \bar{u})^+) - a(\bar{u}, (u - \bar{u})^+)$$
$$+ \int_Q (v - \bar{v})(u - \bar{u})^+ \, dx dt + \int_Q Bu \, (u - \bar{u})^+ \, dx dt \leq 0.$$
$$(6.4.22)$$

Since $(u - \bar{u})^+(x, 0) = 0$, we have

$$\langle \frac{\partial(u - \bar{u})}{\partial t}, (u - \bar{u})^+ \rangle \geq 0. \qquad (6.4.23)$$

The second term on the left-hand side of (6.4.22) can be estimated in the following way by using $(A2)_1$

$$a_0(u, (u - \bar{u})^+) - a(\bar{u}, (u - \bar{u})^+)$$

$$= \int_Q \sum_{i=1}^N (a_i(x, t, T_0 u, \nabla u) - a_i(x, t, \bar{u}, \nabla \bar{u})) \frac{\partial(u - \bar{u})^+}{\partial x_i} \, dx dt$$

$$= \int_{\{u > \bar{u}\}} \sum_{i=1}^N (a_i(x, t, \bar{u}, \nabla u) - a_i(x, t, \bar{u}, \nabla \bar{u})) \frac{\partial(u - \bar{u})}{\partial x_i} \, dx dt \geq 0. \qquad (6.4.24)$$

Consider the third term on the left-hand side of (6.4.22). For $u > \bar{u}$ we get $T_0 u = \bar{u}$ which yields $v \in \beta(\cdot, \cdot, \bar{u}, T_\alpha u)$. Since $T_\alpha u > \bar{u}$ for $u > \bar{u}$ and $s \rightarrow \beta(x, t, r, s)$ is maximal monotone, we get $v \geq \bar{v}$ and thus

$$\int_Q (v - \bar{v})(u - \bar{u})^+ \, dx dt = \int_{\{u > \bar{u}\}} (v - \bar{v})(u - \bar{u}) \, dx dt \geq 0. \qquad (6.2.25)$$

From (6.4.23), (6.4.24), and (6.4.25) we finally obtain

$$\int_Q Bu(u - \bar{u})^+ \, dx dt \leq 0. \qquad (6.4.26)$$

By definition of the function b generating the Nemytskij operator B we get

$$\int_Q Bu \, (u - \bar{u})^+ \, dx dt = \int_{\{u > \bar{u}\}} b(x, t, u)(u - \bar{u}) \, dx dt$$

$$= \int_{\{u > \bar{u}\}} (u - \bar{u})^{p-1}(u - \bar{u}) \, dx dt = \int_Q ((u - \bar{u})^+)^p \, dx dt \geq 0,$$

which in view of (6.4.26) results in

$$0 \leq \|(u - \bar{u})^+\|_{L^p(Q)}^p \leq 0, \text{ i.e., } (u - \bar{u})^+ = 0,$$

and hence $u \leq \bar{u}$. This completes the existence part of the proof of Theorem 6.4.1, since $u \geq \underline{u}$ can be proved in a similar way.

(b) Compactness of the solution set $\mathcal{S}$.

Let $(u_n) \subseteq \mathcal{S}$ be any sequence of solutions of the IBVP (6.4.1) within the interval $[\underline{u}, \bar{u}]$, i.e., the u_n satisfy

$$u_n \in \mathcal{W}_0 : \quad \frac{\partial u_n}{\partial t} + A u_n + v_n = h \quad \text{in } \mathcal{V}_0^*, \quad u_n = 0 \quad \text{in } \Omega \times \{0\}, \quad (6.4.27)$$

where $v_n \in L^q(Q) \subset \mathcal{V}_0^*$ and $v_n(x,t) \in \beta(x,t,u_n(x,t),u_n(x,t))$ for a.e. $(x,t) \in Q$. Since $u_n \in [\underline{u}, \bar{u}]$, we get by (H3)

$$|v_n(x,t)| \leq k_1(x,t) \quad \text{for all } n, \quad (6.4.28)$$

where $k_1 \in L_+^q(Q)$. Taking as special test function $\varphi = u_n$ in (6.4.27) we obtain by means of (6.4.28), (A2)$_2$ and

$$\langle \frac{\partial u_n}{\partial t}, u_n \rangle = \frac{1}{2}\|u_n(\cdot,\tau)\|_{L^2(\Omega)}^2 \geq 0,$$

the following estimate

$$\nu \|\nabla u_n\|_{L^p(Q)}^p \leq \|k\|_{L^1(Q)} + \|v_n\|_{L^q(Q)}\|u_n\|_{L^p(Q)} + \|h\|_{\mathcal{V}_0^*}\|u_n\|_{\mathcal{V}_0}. \quad (6.4.29)$$

Young's inequality applied to the last term on the right-hand side of (6.4.29) yields for any $\varepsilon > 0$

$$\|h\|_{\mathcal{V}_0^*}\|u_n\|_{\mathcal{V}_0} \leq \varepsilon\|u_n\|_{\mathcal{V}_0}^p + C(\varepsilon)\|h\|_{\mathcal{V}_0^*}^q,$$

and thus for ε sufficiently small we obtain from (6.4.29) due to the $L^p(Q)$-boundedness of the (u_n) a uniform bound of the u_n in $\mathcal{V}_0$, i.e.,

$$\|u_n\|_{\mathcal{V}_0} \leq c \quad \text{for all } n.$$

By means of (6.4.27) this last estimate and the boundedness of the operator $A : \mathcal{V}_0 \to \mathcal{V}_0^*$ immediately implies the following uniform bound

$$\|\frac{\partial u_n}{\partial t}\|_{\mathcal{V}_0^*} \leq c \quad \text{for all } n,$$

and thus

$$\|u_n\|_{\mathcal{W}_0} \leq c \quad \text{for all } n.$$

Due to the reflexivity of $\mathcal{W}_0$ and the compact embedding $\mathcal{W}_0 \subset L^p(Q)$ there exists a subsequence of (u_n) denoted by (u_k) which is weakly convergent in $\mathcal{W}_0$ and strongly convergent in $L^p(Q)$ to some $u \in \mathcal{W}_0$. We are going to prove that the weak limit u belongs to $\mathcal{S}$ which shows the weak compactness of $\mathcal{S}$ in $\mathcal{W}_0$.

Due to (6.4.28) there exists a subsequence of (v_k) which is weakly convergent in $L^q(Q)$. Let us denote this subsequence again by (v_k), i.e., $v_k \rightharpoonup v$ in $L^q(Q)$ as $k \to \infty$. Note first that the limits u and v of (u_k) and (v_k), respectively, satisfy

$$v(x,t) \in \beta(x,t,u(x,t),u(x,t)) \quad \text{for a.e. } (x,t) \in Q. \tag{6.4.30}$$

The latter can be seen in just the same way as in the elliptic case (see subsection 6.1.2). Equation (6.4.27) for solutions u_k and the special test function $\varphi = u_k - u$ yields

$$\langle \frac{\partial(u_k - u)}{\partial t}, u_k - u \rangle + \langle Au_k, u_k - u \rangle$$
$$= \langle h - \frac{\partial u}{\partial t}, u_k - u \rangle - \int_Q v_k(u_k - u)\, dx dt. \tag{6.4.31}$$

Due to $u_k \rightharpoonup u$ in $\mathcal{W}_0$, $u_k \to u$ in $L^p(Q)$ and $v_k \rightharpoonup v$ in $L^q(Q)$ the right-hand side of (6.4.31) tends to zero, and because of

$$\langle \frac{\partial(u_k - u)}{\partial t}, u_k - u \rangle = \frac{1}{2}\|(u_k - u)(\cdot, \tau)\|^2_{L^2(\Omega)} \geq 0,$$

we get from (6.3.31)

$$\limsup_{k \to \infty} \langle Au_k, u_k - u \rangle \leq 0. \tag{6.4.32}$$

Since the operator $A : \mathcal{V}_0 \to \mathcal{V}_0^*$ is bounded, continuous, and pseudomonotone with respect to the graph norm topology of $D(L)$ $(L = \partial/\partial t)$, it follows from (6.4.32) that $Au_k \rightharpoonup Au$ in $\mathcal{V}_0^*$. This last convergence result along with the convergence properties of the sequences (u_k) and (v_k) allow one to pass to the limit as $k \to \infty$ in the defining equation (6.4.27) of the (u_k), which proves that the limit u belongs to $\mathcal{S}$, and thus the weak compactness of $\mathcal{S}$ in $\mathcal{W}_0$.

To complete the proof we still have to show that $\mathcal{S}$ is compact in $\mathcal{V}_0$ which means that each sequence $(u_n) \subset \mathcal{S}$ possesses a subsequence which is convergent in the norm of $\mathcal{V}_0$ to some $u \in \mathcal{S}$. By the weak compactness of $\mathcal{S}$ in $\mathcal{W}_0$ there is a subsequence (u_k) such that $u_k \rightharpoonup u \in \mathcal{S}$ (weakly) in $\mathcal{W}_0$. Hypotheses (A1), (A2)$_1$, and (A2)$_2$ imply that the operator A enjoys the (S_+)-property with respect to $D(L)$ (see Theorem E.3.2) which implies in view of (6.4.32) that the sequence (u_k) is strongly convergent to u in $\mathcal{V}_0$, which completes the proof.   □

6.4.4 Generalization.

In this subsection we briefly discuss parabolic inclusions which may be considered as the parabolic counterpart of the corresponding elliptic inclusion of section 6.3., i.e.,

$$\left.\begin{array}{l} \dfrac{\partial u}{\partial t} + Au \in \mathcal{F}(u) + h \quad \text{in } Q, \\[2mm] u = 0 \text{ in } \Omega \times \{0\} \quad \text{and} \quad u = 0 \text{ on } \Gamma. \end{array}\right\} \tag{6.4.33}$$

However, unlike in section 6.3 here A may be a general operator of divergence type given by

$$Au(x,t) = -\sum_{i=1}^{N} \frac{\partial}{\partial x_i} a_i(x,t,u(x,t),\nabla u(x,t)).$$

Let $h \in \mathcal{V}_0^*$. The multifunction $\mathcal{F}$ is assumed to be generated by a locally bounded and Baire measurable function $f : \mathbb{R} \to \mathbb{R}$ in the following way:

$$\mathcal{F}(s) = [\underline{f}(s), \bar{f}(s)], \quad s \in \mathbb{R},$$

with $\underline{f}(s)$ and $\bar{f}(s)$ given by

$$\underline{f}(s) = \lim_{\varepsilon \downarrow 0} \underline{f}_\varepsilon(s), \quad \bar{f}(s) = \lim_{\varepsilon \downarrow 0} \bar{f}_\varepsilon(s),$$

where

$$\underline{f}_\varepsilon(s) = \operatorname*{ess\,inf}_{|\xi - s| < \varepsilon} f(\xi), \quad \bar{f}_\varepsilon(s) = \operatorname*{ess\,sup}_{|\xi - s| < \varepsilon} f(\xi).$$

We note that if $j : \mathbb{R} \to \mathbb{R}$ is defined by

$$j(s) = \int_0^s f(\zeta)\, d\zeta$$

with f as given above, i.e. $f \in L_{loc}^\infty(\mathbb{R})$, then j is locally Lipschitz continuous and its generalized gradient ∂j at s (in the sense of Clarke) satisfies

$$\partial j(s) \subseteq \mathcal{F}(s).$$

Moreover, if the one-sided limits $f(s \pm 0)$ at $s \in \mathbb{R}$ exist, then equality holds, i.e., $\partial j(s) = \mathcal{F}(s)$, cf. [92], which means that the multifunction $\mathcal{F}$ is characterized by the generalized gradient of a nonsmooth potential j. The multifunction $\mathcal{F}$ may, in addition, depend also on the space-time variable (x, t) which, however, has been omitted only for simplicity.

The variational approach used in the study of elliptic inclusions which was based on critical point theory for some nonsmooth functional can no longer be applied to the parabolic inclusion problem (6.4.33) involving a nonpotential quasilinear elliptic operator. Instead we apply the regularization technique developed in this section. This method, when applied to elliptic inclusions, allows one to deal with more general nonlinear elliptic operators than those of section 6.3, especially, the elliptic operator need not to be monotone or of potential type. Thus the regularization technique provides an alternative approach to study the problem of section 6.3.

Let the operator A satisfy all the assumptions made in subsection 6.4.1, i.e., (A1), (A2)$_1$, (A2)$_2$.

Definition 6.4.3. A function $u \in \mathcal{W}_0$ is called a *solution* of problem (6.4.33) if there is a function $v \in L^q(Q)$ such that

(i) $u = 0$ in $\Omega \times \{0\}$,
(ii) $v(x, t) \in \mathcal{F}(u(x, t))$ for a.e. $(x, t) \in Q$,
(iii) $\langle \frac{\partial u}{\partial t}, \varphi \rangle + a(u, \varphi) = \int_Q v\,\varphi\,dxdt + \langle h, \varphi \rangle$ for all $\varphi \in \mathcal{V}_0$.

In comparison with the state-dependent subdifferential the more general multifunction $\mathcal{F}$ requires one to modify the notion of upper and lower solutions, which similar to the elliptic case of section 6.3 reads as follows.

Definition 6.4.4. A function $\bar{u} \in \mathcal{W}$ is called an *upper solution* to the IBVP (6.4.33) if

(i) $\bar{u} \geq 0$ on Γ, $\bar{u} \geq 0$ in $\Omega \times \{0\}$,
(ii) $\langle \frac{\partial \bar{u}}{\partial t}, \varphi \rangle + a(\bar{u}, \varphi) \geq \int_Q \bar{f}(\bar{u})\,\varphi\,dxdt + \langle h, \varphi \rangle$ for all $\varphi \in \mathcal{V}_0 \cap L_+^p(Q)$.

Similarly a function $\underline{u} \in \mathcal{W}$ is a *lower solution* of (6.4.33) if the reversed inequalities hold in (i) and (ii) of Definition 6.4.4 where $\bar{f}$ is replaced by $\underline{f}$.

We assume the following hypotheses on the function f generating the multifunction $\mathcal{F}$:

(F) $f : \mathbb{R} \to \mathbb{R}$ is locally bounded and Baire measurable, and satisfies for some constant $\alpha > 0$ and some function $k_1 \in L_+^q(Q)$

$$|f(s)| \leq k_1(x, t)$$

for a.e. $(x, t) \in Q$ and for all $s \in [\underline{u}(x, t) - \alpha, \bar{u}(x, t) + \alpha]$.

The following existence and enclosure result for the IBVP (6.4.33) can be proved by using the truncation and regularization method developed in the preceding subsections.

Theorem 6.4.2. *Let hypotheses (A1), (A2)$_1$, (A2)$_2$, and (F) be satisfied, and assume the existence of upper and lower solutions $\bar{u}$ and $\underline{u}$ of (6.4.33) with $\underline{u} \leq \bar{u}$. Then the IBVP (6.4.33) has a solution within the order interval $[\underline{u}, \bar{u}]$.*

Proof. We sketch the proof only, since most of the details follow by similar arguments as in the proof of Theorem 6.4.1.

(a) Associated auxiliary truncated IBVP.

Let f^ε be the regularization of f which is given according to formula (6.4.5) by

$$f^\varepsilon(s) = \frac{1}{\varepsilon} \int_{-\infty}^{\infty} f(s - \zeta)\rho\left(\frac{\zeta}{\varepsilon}\right) d\zeta,$$

and denote by F^ε its corresponding Nemytskij operator. Using the notations of the preceding subsections we associate with the IBVP (6.4.33) the following auxiliary regularized truncated IBVP

$$\left.\begin{aligned} \frac{\partial u}{\partial t} + A_0 u + Bu &= F^\varepsilon \circ T_0 u + h \quad \text{in } Q, \\ u = 0 \quad \text{in } \Omega \times \{0\}, \quad u &= 0 \quad \text{on } \Gamma. \end{aligned}\right\} \qquad (6.4.34)$$

Following the proof of Lemma 6.4.1 step by step and taking into account that $F^\varepsilon \circ T_0 : \mathcal{V}_0 \to L^q(Q) \subset \mathcal{V}_0^*$ is completely continuous with respect to the graph norm topology of $D(L)$, we obtain that for any $\varepsilon \in (0, \alpha)$ the IBVP (6.4.34) has a solution $u_\varepsilon \in \mathcal{W}_0$. Let u_n be any solution of (6.4.34) with ε replaced by ε_n and assume the sequence $(\varepsilon_n) \subset (0, \alpha)$ to satisfy $\varepsilon_n \searrow 0$ as $n \to \infty$. Then the corresponding results of Lemma 6.4.2 remain true; that is, there is a subsequence of (u_n) (again denoted by (u_n)) such that

(i) $u_n \rightharpoonup u$ in $\mathcal{W}_0$, i.e., $u_n \rightharpoonup u$ in $\mathcal{V}_0$, $\frac{\partial u_n}{\partial t} \rightharpoonup \frac{\partial u}{\partial t}$ in $\mathcal{V}_0^*$,
(ii) $u_n \to u$ in $L^p(Q)$,
(iii) $F^{\varepsilon_n} \circ T_0 u_n \rightharpoonup v$ in $L^q(Q)$, where $v(x, t) \in \mathcal{F}(T_0 u(x, t))$ a.e. in Q.

Furthermore, the limits $u \in \mathcal{W}_0$ and $v \in L^q(Q)$ satisfy

$$\frac{\partial u}{\partial t} + A_0 u + Bu = v + h \quad \text{in } \mathcal{V}_0^*, \quad u = 0 \quad \text{in } \Omega \times \{0\}, \qquad (6.4.35)$$

(b) Existence via comparison.

We are going to show that the limit u, which solves problem (6.4.35), is in fact a solution of the original inclusion problem (6.4.33) which is enclosed by the upper and lower solution $\bar{u}$ and $\underline{u}$, respectively. To this end we only have to show that the limit u belongs to the interval $[\underline{u}, \bar{u}]$, since then we have $T_0 u = u$, $A_0 = A$, $Bu = 0$, and thus u must be a solution of (6.4.33).

Let us prove $u \leq \bar{u}$ in Q. By definition the upper solution $\bar{u}$ satisfies: $\bar{u} \in \mathcal{W}$ and $\bar{u} \geq 0$ on Γ, $\bar{u} \geq 0$ in $\Omega \times \{0\}$, and

$$\langle \frac{\partial \bar{u}}{\partial t}, \varphi \rangle + a(\bar{u}, \varphi) \geq \int_Q \bar{f}(\bar{u})\, \varphi \, dx dt + \langle h, \varphi \rangle \qquad (6.4.36)$$

for all $\varphi \in V_0 \cap L_+^p(Q)$. Subtracting (6.4.36) from (6.4.35) we get for all $\varphi \in V_0 \cap L_+^p(Q)$ the inequality

$$\langle \frac{\partial (u - \bar{u})}{\partial t}, \varphi \rangle + a_0(u, \varphi) - a(\bar{u}, \varphi) + \int_Q Bu\, \varphi \, dx dt$$

$$\leq \int_Q (v - \bar{f}(\bar{u}))\, \varphi \, dx dt. \qquad (6.4.37)$$

By using the special nonnegative test function $\varphi = (u - \bar{u})^+ \in V_0 \cap L_+^p(Q)$ we have

$$\langle \frac{\partial (u - \bar{u})}{\partial t}, (u - \bar{u})^+ \rangle \geq 0,$$

and

$$a_0(u, (u - \bar{u})^+) - a(\bar{u}, (u - \bar{u})^+) \geq 0,$$

so that from (6.4.37) we get the estimate

$$\int_Q Bu\, (u - \bar{u})^+ \, dx dt \leq \int_Q (v - \bar{f}(\bar{u}))\, (u - \bar{u})^+ \, dx dt. \qquad (6.4.38)$$

Since $v \in \mathcal{F}(T_0 u) = [\underline{f}(T_0 u), \bar{f}(T_0 u)]$ and for $u > \bar{u}$ we have $T_0 u = \bar{u}$, it follows that $v \in [\underline{f}(\bar{u}), \bar{f}(\bar{u})]$. Hence the right-hand side of (6.4.38) results in

$$\int_Q (v - \bar{f}(\bar{u}))(u - \bar{u})^+ \, dx dt = \int_{\{u > \bar{u}\}} (v - \bar{f}(\bar{u}))(u - \bar{u})\, dx dt \leq 0,$$

and it follows from (6.4.38) that

$$0 \geq \int_Q Bu\, (u - \bar{u})^+ \, dx dt = \int_{\{u > \bar{u}\}} (u - \bar{u})^{p-1}(u - \bar{u})\, dx dt$$

$$= \int_Q ((u - \bar{u})^+)^p \, dx dt \geq 0,$$

which yields

$$0 \le \|(u - \bar{u})^+\|^p_{L^p(Q)} \le 0, \text{ i.e., } (u - \bar{u})^+ = 0,$$

and hence $u \le \bar{u}$. This completes the proof of Theorem 6.4.2, since $u \ge \underline{u}$ can be proved in a similar way.   □

Remark 6.4.3. In order to obtain compactness results of the solution set S of all solutions of (6.4.33) within the interval $[\underline{u}, \bar{u}]$ analogous to those of Theorem 6.4.1 we have to impose additional assumptions on the function f. These assumptions are needed to ensure that the multifunction $\mathcal{F}$ generated by f restricted to the interval $[\underline{u}, \bar{u}]$ turns out to coincide with the generalized gradient of some locally Lipschitz integral functional $J : V_0 \cap [\underline{u}, \bar{u}] \to \mathbb{R}$. More precisely, the following relation is needed:

$$\partial J(u) = \int_Q \mathcal{F}(u(x,t)) \, dx dt. \qquad (6.4.39)$$

This last relation holds if the one-sided limits of f exist and the function $j : \mathbb{R} \to \mathbb{R}$ given by

$$j(s) = \int_0^s f(T_0\zeta) \, d\zeta$$

is regular in the sense of Clarke (cf. [96]), where the integral functional J is defined by

$$J(u) = \int_Q j(u(x,t)) \, dx dt.$$

This is because then the generalized gradient ∂j of the locally Lipschitz function j satisfies $\partial j(s) = \mathcal{F}(s)$ and due to [96, Theorem 2.7.5] one has the following relation

$$\partial J(u) = \int_Q \partial j(u(x,t)) \, dx dt = \int_Q \mathcal{F}(u(x,t)) \, dx dt,$$

which is equality (6.4.39).

Special case. Let A be a linear operator of the form

$$Au = -\sum_{i,j=1}^N \frac{\partial}{\partial x_i}\left(a_{ij}\frac{\partial u}{\partial x_j}\right),$$

with $a_{ij} \in L^\infty(Q)$, $a_{ij}(x,t)\xi_i\xi_j \geq \mu|\xi|^2$ for a.e. $(x,t) \in Q$ and for all $\xi \in \mathbb{R}^N$, and μ being some positive constant. Then hypotheses (A1), (A2)$_1$, (A2)$_2$ are trivially satisfied for $p = q = 2$. It should be noted that an extended linear operator A of the form

$$Au = -\sum_{i,j=1}^{N} \frac{\partial}{\partial x_i}\left(a_{ij}\frac{\partial u}{\partial x_j}\right) + \sum_{i=1}^{N} b_i \frac{\partial u}{\partial x_i} + cu$$

with a_{ij}, b_i, $c \in L^\infty(Q)$ could be taken into account as well without any difficulties (which has been omitted only for simplicity).

Now the nonlinearity $f : \mathbb{R} \to \mathbb{R}$, $f \in L^\infty_{loc}(\mathbb{R})$ is supposed to satisfy the following growth condition

$$|f(s)| \leq c(1+|s|) \quad \text{for all } s \in \mathbb{R}. \tag{6.4.40}$$

Then it can easily be seen that the upper (resp., lower) semicontinuous function $\bar{f}$ (resp., $\underline{f}$) : $\mathbb{R} \to \mathbb{R}$ satisfies the same growth condition (6.4.40) with a possibly different constant c; i.e., we have

$$-c(1+|s|) \leq \underline{f}(s) \leq \bar{f}(s) \leq c(1+|s|) \tag{6.4.41}$$

for all $s \in \mathbb{R}$.

Consider the following IBVP for functions $\bar{u}$ and $\underline{u}$, respectively, of the form

$$\begin{cases} \dfrac{\partial \bar{u}}{\partial t} + A\bar{u} = c(1+|\bar{u}|) + h & \text{in } Q, \\ \bar{u} = 0 \quad \text{in } \Omega \times \{0\}, \quad \bar{u} = 0 \quad \text{on } \Gamma, \end{cases} \tag{6.4.42}$$

$$\begin{cases} \dfrac{\partial \underline{u}}{\partial t} + A\underline{u} = -c(1+|\underline{u}|) + h & \text{in } Q, \\ \underline{u} = 0 \quad \text{in } \Omega \times \{0\}, \quad \underline{u} = 0 \quad \text{on } \Gamma. \end{cases} \tag{6.4.43}$$

Since the nonlinearities on the right-hand sides of (6.4.42) and (6.4.43), respectively, are uniformly Lipschitz continuous, the existence of a uniquely defined solution for the semilinear IBVP (6.4.42) and (6.4.43), respectively, follows by applying standard theory for evolutionary equations involving strongly monotone, continuous, and coercive operators.

By (6.4.41) the solution $\bar{u}$ of the IBVP (6.4.43) is an upper solution of the IBVP (6.4.33), and the solution $\underline{u}$ of (6.4.43) is a lower solution. Further, the maximum principle implies $\bar{u} \geq \underline{u}$, since the difference $w = \bar{u} - \underline{u}$ satisfies the inequality

$$\frac{\partial w}{\partial t} + Aw = c(1+|\bar{u}|) + c(1+|\underline{u}|) \geq 0$$

and $w = 0$ in $\Omega \times \{0\}$, $w = 0$ on Γ, which results in $w \geq 0$ in Q. In order to show that hypothesis (F) is satisfied let $\alpha > 0$ be given. Then the growth condition (6.4.40) yields

$$|f(s)| \leq c(1 + |s|) \leq c(1 + |\underline{u}(x,t)| + |\bar{u}(x,t)| + \alpha)$$

for all $s \in [\underline{u}(x,t) - \alpha, \bar{u}(x,t) + \alpha]$, and a.e. in Q, where the function on the right-hand side of the last inequality belongs to $L^2(Q)$ (notice $p = q = 2$). Hence Theorem 6.4.2 can be applied which yields the existence and enclosure of solutions within the interval $[\underline{u}, \bar{u}]$. Moreover, by the maximum principle it follows that in this case any solution of the IBVP (6.4.33), i.e., $\frac{\partial w}{\partial t} + Au \in \mathcal{F}(u) + h$ must belong to this interval, so that $\bar{u}$ and $\underline{u}$ are global upper and lower bounds, respectively, of all solutions of (6.4.33).

Remark 6.4.4. The combined use of the upper and lower solution method and regularization techniques as developed in this section applies likewise also to the corresponding elliptic inclusion. Also IBVP involving multivalued flux conditions in the form $\partial u / \partial \nu \in \partial j(u)$ on Γ can be treated.

6.5 Notes and comments

The results on quasilinear elliptic and parabolic inclusions presented in this chapter are mainly based on and extend the authors' recent works on this subject; see [51, 52, 54, 58, 62, 67, 90]. Results in this direction for parabolic inclusions were also obtained in [15, 156]. For semilinear elliptic and parabolic inclusions one can prove existence results in a constructive way by monotone iteration on the basis of which a numerical analysis and related algorithms were established in [49, 64, 65]. For elliptic variational inequalities involving monotone elliptic operators and Carathéodory type lower order terms a different approach to the extremality of solutions with respect to an interval of suitably defined upper and lower solutions has been given recently in [171].

Investigations on the existence of extremal solutions have been extended in recent papers (see [57, 59, 61]) to classes of hemivariational inequalities whose superpotentials are of d.c.-function type which means that these functions may be represented as the difference of two convex (possibly nonsmooth) functions. In particular, in [61] the existence of extremal solutions has been proved for the following mixed boundary hemivariational inequality:

$$Au + Fu = 0 \quad \text{in } \Omega, \tag{6.5.1}$$

$$u = 0 \quad \text{on } \partial\Omega \setminus \Gamma, \quad -\frac{\partial u}{\partial \nu} \in \partial j(u) \quad \text{on } \Gamma, \tag{6.5.2}$$

where $\Omega \subset \mathbb{R}^N$ is a bounded domain with C^1-boundary $\partial\Omega$ and $\Gamma \subset \partial\Omega$ is a relatively open C^1-portion of $\partial\Omega$ having positive surface measure, and $\partial/\partial\nu$ denotes the outer conormal derivative on Γ related with A, which is a quasilinear elliptic operator. F denotes the Nemytskij operator of the lower order terms. The multivalued boundary condition on Γ of (6.5.2) is described by Clarke's generalized gradient $\partial j : \mathbb{R} \to 2^\mathbb{R} \setminus \{\emptyset\}$ of a locally Lipschitz function $j : \mathbb{R} \to \mathbb{R}$. Defining $V_{\Gamma,0}$ by

$$V_{\Gamma,0} = \{u \in W^{1,p}(\Omega) \mid \gamma u = 0 \quad \text{on } \partial\Omega \setminus \Gamma\},$$

where $\gamma : W^{1,p}(\Omega) \to L^p(\partial\Omega)$ denotes the trace operator (see C.3), the corresponding weak formulation of (6.5.1) and (6.5.2) leads to the following hemivariational inequality:

Find $u \in V_{\Gamma,0}$ such that

$$\langle Au + Fu, v - u \rangle + \int_\Gamma j^o(\gamma u; \gamma v - \gamma u) \, d\Gamma \geq 0 \quad \text{for all } v \in V_{\Gamma,0}, \tag{6.5.3}$$

where $j^o(r; s)$ is the generalized directional derivative in the sense of Clarke of the function j at r in the direction s. Extremality results for (6.5.3) have been obtained in [61] under the assumption that j is regular (in the sense of Clarke) and of d.c.-type, which means that j is of the form

$$j(s) = j_1(s) - j_2(s) \quad \text{for all } s \in \mathbb{R}$$

where $j_k : \mathbb{R} \to \mathbb{R}$, $k = 1, 2$, are convex functions.

Unbounded domain problems have been considered in [60] where the upper and lower solution method has been established for quasilinear elliptic differential inclusions of hemivariational type in all of $\mathbb{R}^N$. The following inclusion of the form (6.1.1) but now defined in all of $\mathbb{R}^N$ is treated in [60]

$$Au + \beta(\cdot, u, u) \ni 0 \quad \text{in } \mathcal{D}', \tag{6.5.4}$$

where

$$Au = -\sum_{i=1}^{N} \frac{\partial}{\partial x_i} a_i(\cdot, u, \nabla u)$$

is a general operator of Leray-Lions type, and $\mathcal{D}'$ denotes the dual of the space $\mathcal{D} := C_0^\infty(\mathbb{R}^N)$. Usually such kinds of problems are treated by variational methods. However, these methods, in general, require differential operators of potential type and additional growth conditions imposed on the solution such as $u(x) \to 0$ as $|x| \to \infty$. In our approach neither the elliptic operator nor the multivalued lower order terms need to be of monotone or of potential type. Furthermore, no condition at infinity on the solution we are searching for is imposed. Therefore, standard variational methods applied to associated energy functionals defined on the space of finite energy functions such as, e.g., in [2, 119, 120, 193] cannot be employed here. Moreover, due to the unboundedness of the domain as well as the nonmonotone behavior of the operators involved there are no comparison results available which even more complicates the treatment of problem (6.5.4). In [60] not only the existence of solutions but the existence of extremal ones with respect to an ordered interval of appropriately defined upper and lower solutions has been proved where the underlying solution space is the local Sobolev space $W^{1,p}_{loc}(\mathbb{R}^N)$, i.e., the space of all functions $u \colon \mathbb{R}^N \to \mathbb{R}$ which belong to $W^{1,p}(\Omega)$ for every domain Ω with compact closure in $\mathbb{R}^N$. Applying recent results on both linear and nonlinear eigenvalue problems in $\mathbb{R}^N$ obtained, for example, in [3, 9, 30, 31, 106], there is much potential for the construction of the needed upper and lower solutions by using the so-called eigenfunction approach; cf. [60]. It should be noted that inclusions in the form (6.5.4) may be also treated in unbounded domains of different kind such as, e.g., exterior domains.

Quasilinear elliptic (single-valued) problems with discontinuous nonlinearities in unbounded domains were studied by the authors recently in [55, 72, 75]. It should be noted that the treatment of differential inclusions or discontinuous problems in unbounded domains requires new tools. For example, Theorem 1.1.1 which is a fixed point theorem in ordered normed spaces can no longer be applied, and a generalization of it in the form of Theorem A.2.1 which is fixed point theorem for increasing mappings in posets is needed to treat unbounded domain problems.

Finally we remark that by means of the upper and lower solution method semilinear elliptic problems in unbounded domains have been treated by Pao in [189, 190] within the framework of classical solutions which, however, requires sufficiently smooth data of the problem under consideration. An extension of his approach to quasilinear and discontinuous problems is by no means trivial and requires completely different tools.

Chapter 7
Discontinuous implicit elliptic and parabolic problems

In this chapter we provide existence results for implicitly given parabolic differential equations of the form $H(x, t, u, \Lambda u) = 0$ subject to initial-Dirichlet boundary conditions, where Λ stands for some semilinear parabolic differential operator. The peculiarity of these implicit equations is that their governing outer function H may be discontinuous in all its arguments and that the parabolic operator Λ is not necessarily inverse monotone. The corresponding implicit elliptic problems are also discussed briefly.

The main tools used to treat such problems are on the one hand existence results in partially ordered sets for abstract operator equations of the form $Lu = Nu$ which have been proved in chapter 1, and on the other hand extremality results for nonmonotone nonlinear parabolic and elliptic equations proved in chapter 5.

7.1 Statement of the problem and notations

We are going to study the following implicit parabolic initial-boundary value problem (IBVP)

$$H(x, t, u, \Lambda u) = 0 \ \text{ in } \ Q, \quad u = 0 \ \text{ on } \ \Gamma, \quad u = \psi \ \text{ in } \ \Omega \times \{0\}, \quad (7.1.1)$$

where $\Omega \subset \mathbb{R}^N$ is a bounded domain with Lipschitz boundary $\partial\Omega$, $Q = \Omega \times (0, \tau)$, and $\Gamma = \partial\Omega \times (0, \tau)$. The operator Λ is assumed to be a semilinear parabolic differential operator in the form

$$\Lambda u(x, t) = \frac{\partial u(x, t)}{\partial t} + Au(x, t) + g(x, t, u(x, t)), \quad (7.1.2)$$

where A is a second order strongly elliptic differential operator given by

$$Au(x, t) = -\sum_{i,j=1}^{N} \frac{\partial}{\partial x_i}\left(a_{ij}(x, t)\frac{\partial u}{\partial x_j}\right),$$

with coefficients $a_{ij} \in L^\infty(Q)$ satisfying for all $\xi = (\xi_1, ..., \xi_N) \in \mathbb{R}^N$

$$\sum_{i,j=1}^{N} a_{ij}(x,t)\xi_i\xi_j \geq \mu|\xi|^2 \quad \text{for a.e. } (x,t) \in Q \quad \text{with some constant } \mu > 0.$$

Implicit ordinary differential equations have been studied in chapters 2, 3, and 4. Here we extend these investigations also to implicit parabolic IBVP. Implicit partial differential equations have been treated in the literature mainly by reducing the implicit equation to a differential inclusion of the form

$$\Lambda u \in \vartheta(x,t,u),$$

where ϑ is required to be an appropriate semicontinuous multiselection of the multifunction Θ which is related with H by

$$\Theta(x,t,r) = \{\zeta \in \mathbb{R} \mid H(x,t,r,\zeta) = 0\},$$

cf., e.g., [27, 177]. However, since the governing nonlinearity $H: Q \times \mathbb{R} \times \mathbb{R} \rightarrow \mathbb{R}$ of the implicit IBVP (7.1.1) may be discontinuous in all its arguments, in general, this method cannot be applied to the problem considered here, such that a new method has to be developed. We shall provide an example which shows that the multifunction Θ has neither a lower nor an upper semicontinuous multiselection, but which, nevertheless, can be treated by our method. Thus the method to be developed here may be considered as an alternative approach which complements existing methods (see references above).

Our main goal is to prove the existence of *extremal* solutions of the discontinuous implicit IBVP (7.1.1) by using the abstract existence result for the operator equation

$$Lu = Nu, \tag{7.1.3}$$

given by Corollary 1.1.2.

Concerning the underlying function spaces for problem (7.1.1) we keep the notations of section 5.2, where now $p = q = 2$ is assumed. Thus in what follows the spaces of vector-valued functions $\mathcal{V}$, $\mathcal{W}$ and $\mathcal{V}_0$, $\mathcal{W}_0$ as introduced in section 5.2 are related with the Sobolev spaces $W^{1,2}(\Omega)$ and $W_0^{1,2}(\Omega)$, respectively. Let $\langle \cdot, \cdot \rangle$ denote the duality pairing between $\mathcal{V}_0^*$ and $\mathcal{V}_0$; then the bilinear form a associated with the operator A by

$$\langle Au, \varphi \rangle = a(u,\varphi) = \sum_{i,j=1}^{N} \int_Q a_{ij}(x,t)\frac{\partial u}{\partial x_i}\frac{\partial \varphi}{\partial x_j}\,dxdt, \quad \varphi \in \mathcal{V}_0,$$

is well defined for any $u \in V$, and the operator $A : V \to V_0^*$ is linear and bounded. The underlying partial ordering in $L^2(Q)$ is defined by the cone $L_+^2(Q)$ of all nonnegative elements of $L^2(Q)$, which induces a corresponding partial ordering also in the subspace W of $L^2(Q)$.

Further, we introduce the following subspaces W and W_0 defined by

$$W := \{u \in \mathcal{W} \mid \frac{\partial u}{\partial t} + Au \in L^2(Q)\}$$

and

$$W_0 := \{u \in \mathcal{W}_0 \mid u(\cdot, 0) = 0 \text{ and } \frac{\partial u}{\partial t} + Au \in L^2(Q)\}.$$

Consider the implicit parabolic IBVP (7.1.1) with initial data $\psi \in L^2(\Omega)$. Without loss of generality we may assume that $\psi = 0$. Otherwise problem (7.1.1) can always be reduced to homogeneous initial values by a simple translation $u \to u - \hat{u}$, where $\hat{u} \in \mathcal{W}_0$ is, for example, the unique solution of the linear IBVP

$$\frac{\partial \hat{u}}{\partial t} + A\hat{u} = 0 \quad \text{in } Q, \quad \hat{u} = 0 \text{ on } \Gamma, \quad \hat{u} = \psi \text{ in } \Omega \times \{0\}.$$

Moreover, let $\nu : Q \times \mathbb{R} \times \mathbb{R} \to (0, \infty)$ be any positive function, and let f be defined by $f(x, t, r, \zeta) := \zeta - (\nu \cdot H)(x, t, r, \zeta)$; then $H(x, t, r, \zeta) = 0$ is equivalent with $\zeta = f(x, t, r, \zeta)$. Thus in what follows we consider instead of the IBVP (7.1.1) the following equivalent one

$$\Lambda u = f(x, t, u, \Lambda u) \text{ in } Q, \quad u = 0 \text{ on } \Gamma, \quad u = 0 \text{ in } \Omega \times \{0\}. \quad (7.1.4)$$

Definition 7.1.1. A function $u \in W_0$ is called a *solution* of the IBVP (7.1.4) if

$$\Lambda u(x, t) = f(x, t, u(x, t), \Lambda u(x, t)) \text{ for a.e. } (x, t) \in Q.$$

7.2 Preliminaries

In this section we provide some existence and comparison results for semi-linear parabolic equations which will be used in later sections. We consider first the parabolic operator Λ under the following Lipschitz condition on its nonlinearity g:

(G1) $g : Q \times \mathbb{R} \to \mathbb{R}$ is a Carathéodory function satisfying $g(\cdot, \cdot, 0) \in L^2(Q)$ and a Lipschitz condition in the form

$$|g(x, t, s_1) - g(x, t, s_2)| \le l(x, t)|s_1 - s_2|,$$

for all $(x, t, s_1), (x, t, s_2) \in Q \times \mathbb{R}$ with some function $l \in L_+^\infty(Q)$.

The next lemma provides a comparison result for the semilinear parabolic operator Λ.

Lemma 7.2.1. *Let u, $v \in W$, and assume $\Lambda u \leq \Lambda v$ in Q, $u \leq v$ on Γ and $u(x,0) \leq v(x,0)$ for $x \in \Omega$. Then under hypothesis (G1) it follows $u \leq v$ in Q.*

Proof. Define $Q_t := \Omega \times (0,t) \subseteq Q$ for any $t \in (0,\tau]$. Then the inequality $\Lambda u \leq \Lambda v$ restricted to the subdomain Q_t yields

$$
\langle \frac{\partial(u-v)}{\partial s}, \varphi \rangle + a((u-v), \varphi)
$$
$$
+ \int_{Q_t} \left(g(x,s,u(x,s)) - g(x,s,v(x,s)) \right) \varphi(x,s)\, dx ds \leq 0
$$

for all $\varphi \in \mathcal{V}_0 \cap L_+^2(Q)$. In particular, this inequality holds for the special test function $\varphi = (u-v)^+$ which yields because of $(u-v)^+(x,0) = 0$ and the hypothesis (G1) the inequality

$$
\frac{1}{2}\, \|(u-v)^+(\cdot,t)\|_{L^2(\Omega)}^2 + \mu\, \|\nabla(u-v)^+\|_{L^2(Q_t)}^2
$$
$$
\leq \|l\|_{L^\infty(Q)} \int_{Q_t} ((u-v)^+)^2(x,s)\, dx ds,
$$

which is true for any $t \in [0,\tau]$. Hence, by setting

$$
y(t) := \int_\Omega ((u-v)^+)^2(x,t)\, dx = \|(u-v)^+(\cdot,t)\|_{L^2(\Omega)}^2
$$

we get

$$
y(t) \leq 2\|l\|_{L^\infty(Q)} \int_0^t y(s)\, ds, \quad t \in [0,\tau].
$$

Applying Gronwall's lemma (see section B) and taking into account the continuous embedding $W \subset C([0,\tau]; L^2(\Omega))$, it follows that $y(t) = 0$ for any $t \in [0,\tau]$, and thus $(u-v)^+(x,t) = 0$ for a.e. $(x,t) \in Q$, that is, $u \leq v$, which proves the lemma.   □

Corollary 7.2.1. *Let hypothesis (G1) be satisfied. Then for any $h \in L^2(Q)$ the IBVP*

$$
\textit{Find } u \in W_0 : \quad \Lambda u = h
$$

has a unique solution.

Proof. The existence follows readily by applying the results on first order evolution equation (cf., e.g., [218, Chapter 30], see also E.3), and the uniqueness is a consequence of Lemma 7.2.1.   □

Corollary 7.2.2. *Denote by $\Lambda^{-1}\colon L^2(Q) \to W_0$ the operator that assigns to each $h \in L^2(Q)$ the unique solution $u \in W_0$ according to Corollary 7.2.1. Then $\Lambda^{-1}\colon L^2(Q) \to W_0$ is increasing and continuous.*

Proof. **(a)** $\Lambda^{-1}\colon L^2(Q) \to W_0$ is increasing.

Let $h_1, h_2 \in L^2(Q)$ satisfy $h_1 \leq h_2$, and let $u_1, u_2 \in W_0$ be the corresponding unique solutions of

$$\Lambda u_i = h_i, \quad i = 1, 2.$$

Then by Lemma 7.2.1 it readily follows that $u_1 \leq u_2$, i.e., Λ^{-1} is increasing.

(b) $\Lambda^{-1}\colon L^2(Q) \to W_0$ is continuous.

To this end let h_i and u_i, $i = 1, 2$, be as in part (a). Similar as in the proof of Lemma 7.2.1 we obtain for the difference of the resulting equations in its weak formulation, by using as special test function the difference of the solutions u_i, i.e., $\varphi = u_1 - u_2$, the following estimate

$$\frac{1}{2} \, \|(u_1 - u_2)(\cdot, t)\|^2_{L^2(\Omega)} + \mu \, \|\nabla(u_1 - u_2)\|^2_{L^2(Q_t)}$$

$$\leq \|l\|_{L^\infty(Q)} \int_{Q_t} (u_1 - u_2)^2(x, t) \, dx ds + \frac{1}{2} \int_{Q_t} (u_1 - u_2)^2(x, t) \, dx ds$$

$$+ \frac{1}{2} \int_{Q_t} (h_1 - h_2)^2(x, t) \, dx ds \tag{7.2.1}$$

for any $t \in [0, \tau]$. By means of Gronwall's lemma we get from (7.2.1)

$$\|u_1 - u_2\|^2_{L^2(Q)} \leq c(\tau, l)\|h_1 - h_2\|^2_{L^2(Q)},$$

which implies by using again inequality (7.2.1) the gradient estimate

$$\|\nabla(u_1 - u_2)\|^2_{L^2(Q)} \leq c(\tau, l)\|h_1 - h_2\|^2_{L^2(Q)},$$

and thus

$$\|u_1 - u_2\|_{V_0} \leq c(\tau, l)\|h_1 - h_2\|_{L^2(Q)}, \tag{7.2.2}$$

where $c(\tau, l)$ is some positive constant depending on τ and $\|l\|_{L^\infty(Q)}$. Finally, we get from the difference of the equations in u_1 and u_2, respectively, the following estimate

$$|\langle \frac{\partial(u_1 - u_2)}{\partial t}, \varphi \rangle| \leq |a((u_1 - u_2), \varphi)| + \int_Q |(g(\cdot, \cdot, u_1) - g(\cdot, \cdot, u_2)) \, \varphi| \, dx dt$$

$$+ \int_Q |(h_1 - h_2) \, \varphi| \, dx dt,$$

which yields due to the boundedness of the coefficients a_{ij} and in view of (7.2.2) and (G1) an estimate in the form

$$|\langle \frac{\partial(u_1 - u_2)}{\partial t}, \varphi \rangle| \le c \, \|h_1 - h_2\|_{L^2(Q)}, \quad \text{for all } \varphi \in \mathcal{V}_0 : \|\varphi\|_{\mathcal{V}_0} \le 1.$$

This shows that

$$\|\frac{\partial(u_1 - u_2)}{\partial t}\|_{\mathcal{V}_0^*} \le c \, \|h_1 - h_2\|_{L^2(Q)}$$

holds, and hence it follows

$$\|u_1 - u_2\|_{W_0} \le c \, \|h_1 - h_2\|_{L^2(Q)},$$

where c is some positive constant depending only on the data. This completes the proof.   □

We now consider the case that the nonlinearity g of the operator Λ satisfies the following weaker growth condition:

(G2) The function $g: Q \times \mathbb{R} \to \mathbb{R}$ is a Carathéodory function which satisfies for some $m \in L^2_+(Q)$ and $l \in L^\infty_+(Q)$ the growth condition:
$|g(x, t, r)| \le m(x, t) + l(x, t)|r|$ for a.e $(x, t) \in Q$, and all $r \in \mathbb{R}$.

Let us consider the semilinear IBVP of Corollary 7.2.1, i.e.,

$$\text{Find } u \in W_0 : \quad \Lambda u = h \tag{7.2.3}$$

but now under the hypothesis (G2). Then, in general, uniqueness does not hold for (7.2.3). Instead we are going to show the following result.

Lemma 7.2.2. *Let the nonlinearity g of the operator Λ satisfy hypothesis (G2). Then for any $h \in L^2(Q)$ the IBVP (7.2.3) possesses extremal solutions in W_0; i.e., there exist the least and greatest solution u_* and u^*, respectively, with respect to the natural partial ordering. Furthermore, these extremal solutions are increasing with respect to h.*

Proof. **(a)** Existence of extremal solutions.

Let w be the unique solution of the following linear IBVP

$$\text{Find } w \in W_0 : \quad \frac{\partial w(x, t)}{\partial t} + Aw(x, t) - m(x, t) - l(x, t)w = |h(x, t)|. \tag{7.2.4}$$

Then for the linear operator $\Lambda_l u = u_t + Au - lu$ we have

$$\Lambda_l w = m + |h| \geq 0 = \Lambda_l 0$$

and thus it follows from Lemma 7.2.1 that $w \geq 0$. One also readily observes that w is an upper solution and $-w$ is a lower of the IBVP (7.2.3). Hence, by applying Theorem 5.2.1 we ensure the existence of the least and greatest solution u_* and u^*, respectively, within the order interval $[-w, w]$. Next we shall show that any solution of the IBVP (7.2.3) belongs to $[-w, w]$, and thus the extremal solutions u_* and u^* within $[-w, w]$ must be extremal ones among all solutions of (7.2.3).

To this end let u be any solution of (7.2.3). Hypothesis (G2) implies that $g(x, t, r) \geq -m(x, t) - l(x, t)|r|$, which yields by subtracting equation (7.2.4) from (7.2.3) the following differential inequality

$$\frac{\partial(u - w)}{\partial t} + A(u - w) + l(w - |u|) \leq h - |h|. \tag{7.2.5}$$

Since $u, w \in W_0$, then $(u - w)^+ \in V_0 \cap L_+^2(Q)$ and $(u - w)^+ = 0$ in $\Omega \times \{0\}$, so that from the weak formulation of (7.2.5) with the special test function $\varphi = (u - w)^+$ we get for any $t \in (0, \tau]$

$$\frac{1}{2}\|(u-w)^+(\cdot, t)\|_{L^2(\Omega)}^2 + \mu\|\nabla(u-w)^+\|_{L^2(Q_t)}^2 \leq \int_{Q_t} l\,(|u|-w)(u-w)^+\,dxdt. \tag{7.2.6}$$

Since $w \geq 0$, the right-hand side of (7.2.6) can be written as

$$\int_{Q_t} l\,(|u| - w)(u - w)^+\,dxdt = \int_{\{u>w\}} l\,(u - w)^2\,dxdt$$

$$\leq \|l\|_{L^\infty(Q)} \int_{Q_t} ((u - w)^+)^2\,dxdt,$$

which yields by applying Gronwall's lemma in a similar way as in the proof of Lemma 7.2.1 that for any $t \in [0, \tau]$ we have $\|(u - w)^+\|_{L^2(Q_t)}^2 = 0$, and thus $u \leq w$ in Q. Analogously one can show that any solution u of (7.2.3) satisfies also $u \geq -w$, which shows that the extremal solutions of (7.2.3) within the interval $[-w, w]$ must be the extremal ones among all solutions of the IBVP (7.2.3).

We show next the monotone behavior of these extremal solutions with respect to the right-hand side h.

(b) Monotonicity of the extremal solutions with respect to h.

To show that the extremal solutions of (7.2.3) are increasing with respect to h, let h_1, $h_2 \in L^2(Q)$ satisfy $h_1 \leq h_2$, and consider the corresponding greatest solutions u_1^* and u_2^*, respectively. Then $u_i^* \in W_0$, $i = 1, 2$ satisfy

$$\frac{\partial u_i^*}{\partial t} + A u_i^* + g(\cdot, \cdot, u_i^*) = h_i, \qquad (\mathrm{P}_i)$$

and the greatest solution u_1^* of (P_1) is a lower solution for problem (P_2). If w is the unique solution of the linear IBVP

$$w \in W_0: \quad \frac{\partial w}{\partial t} + A w - l \, w - m = |h_2|,$$

then $w \geq 0$ and w is an upper solution for problem (P_2). Subtracting the last equation from (P_1) one gets

$$\frac{\partial(u_1^* - w)}{\partial t} + A(u_1^* - w) = h_1 - |h_2| - g(\cdot, \cdot, u_1^*) - lw - m,$$

which by using the growth condition of (G2) results in the inequality

$$\frac{\partial(u_1^* - w)}{\partial t} + A(u_1^* - w) \leq l(|u_1^*| - w).$$

Since both u_1^* and w belong to W_0, from the last inequality we obtain in just the same way as in step (a) that $u_1^* \leq w$. Applying again Theorem 5.2.1 there exist solutions (even extremal ones) of (P_2) within the interval $[u_1^*, w]$. Since u_2^* is by definition the greatest solution of (P_2) among all its solution, it follows that $u_1^* \leq u_2^*$, and thus proves the assertion of (b). The corresponding monotone behavior of the least solutions can be shown similarly, which proves the assertion of the lemma.   □

7.3 Main results

In this section we prove existence results for the implicit parabolic problem (7.1.4) which is

$$\Lambda u = f(x, t, u, \Lambda u) \text{ in } Q, \quad u = 0 \text{ on } \Gamma, \quad u = 0 \text{ in } \Omega \times \{0\}, \quad (7.3.1)$$

under the following assumption on the outer function $f \colon Q \times \mathbb{R} \times \mathbb{R} \to \mathbb{R}$:

(F1) The function $(x, t) \to f(x, t, u(x, t), \Lambda u(x, t))$ is measurable in Q for each $u \in W$, and the function $(x, t, r, \zeta) \to f(x, t, r, \zeta)$ is increasing with respect to r and ζ for a.e. $(x, t) \in Q$.

It should be noted that according to hypothesis (F1) no continuity assumptions have been imposed on the nonlinearity f describing the implicit equation (7.3.1). The main tools used in the study of the IBVP (7.3.1) will be the existence results for abstract operator equations in partially ordered sets of the form $Lu = Nu$ obtained in chapter 1. In dependence on what kind of growth condition is imposed on the nonlinearities g and f we consider different cases which will be treated separately.

7.3.1 Existence of extremal solutions in order intervals.
In this subsection we study the existence of extremal solutions of the implicit IBVP (7.3.1) by a variant of the upper and lower solution method.

Definition 7.3.1. A function $\bar{u} \in W$ is called an *upper solution* of the IBVP (7.3.1) if

   (i) $\bar{u} \geq 0$ on Γ and $\bar{u} \geq 0$ on $\Omega \times \{0\}$;
   (ii) $\Lambda\bar{u}(x,t) \geq f(x,t,\bar{u}(x,t),\Lambda\bar{u}(x,t))$ for a.e. $(x,t) \in Q$.

Similarly, a *lower solution* $\underline{u}$ of the IBVP (7.3.1) is defined by reversing the inequality sign in (i) and in (ii), and replacing $\bar{u}$ by $\underline{u}$.

In addition to hypothesis (F1) we shall assume throughout this subsection the following hypotheses:

   (H1) The IBVP (7.3.1) has a lower solution $\underline{u}$ and an upper solution $\bar{u}$ according to Definition 7.3.1 satisfying $\underline{u} \leq \bar{u}$ and $\Lambda\underline{u} \leq \Lambda\bar{u}$.
   (H2) The function g of the operator Λ is a Carathéodory function which is L^2-bounded with respect to the order interval $[\underline{u}, \bar{u}]$; i.e., there exists $m \in L^2_+(Q)$ such that
   $|g(x,t,r)| \leq m(x,t)$ for a.e. $(x,t) \in Q$ and for all $r \in [\underline{u}(x,t), \bar{u}(x,t)]$.

Let $\mathcal{B}$ denote the subset of W_0 defined by

$$\mathcal{B} = \{u \in W_0 \mid \underline{u} \leq u \leq \bar{u} \text{ and } \Lambda\underline{u} \leq \Lambda u \leq \Lambda\bar{u}\}.$$

Then we are going to prove the following existence result.

Theorem 7.3.1. *Let hypotheses (F1), (H1), and (H2) be satisfied. Then the discontinuous implicit parabolic IBVP (7.3.1) has extremal solutions within the set $\mathcal{B}$.*

Proof. Let T be the truncation operator related with $\underline{u}, \bar{u}$ and given by

$$Tu(x,t) = \begin{cases} \bar{u}(x,t) & \text{if} \quad u(x,t) > \bar{u}(x,t), \\ u(x,t) & \text{if} \quad \underline{u}(x,t) \leq u(x,t) \leq \bar{u}(x,t), \\ \underline{u}(x,t) & \text{if} \quad u(x,t) < \underline{u}(x,t), \end{cases}$$

and let the operator $\Lambda_T \colon W \to L^2(Q)$ be defined by

$$\Lambda_T u(x,t) = \frac{\partial u(x,t)}{\partial t} + Au(x,t) + g(x,t,Tu(x,t)). \qquad (7.3.2)$$

We set $h_- := \Lambda \underline{u}$ and $h_+ := \Lambda \bar{u}$, and prove first that for each $h \in [h_-, h_+]$ the following auxiliary "truncated" IBVP:

$$\text{Find } u \in W_0 \colon \quad \Lambda_T u = h \qquad (7.3.3)$$

has extremal solutions in $\mathcal{B}$, which are increasing with respect to h. By (H2) it follows that $|g(x,t,Tv(x,t))| \leq m(x,t)$ for any $v \in L^2(Q)$, and thus Lemma 7.2.2 implies the existence of extremal solutions, which are increasing with respect to h. We still have to verify that these extremal solutions are extremal ones of the set $\mathcal{B}$. To this end we show that any solution u of (7.3.3) with the right-hand side $h \in [h_-, h_+]$ belongs to $\mathcal{B}$. Let $u \in W_0$ be a solution of the IBVP (7.3.3); then $h_+ = \Lambda \bar{u} \geq h = \Lambda_T u$ which yields

$$\frac{\partial(u - \bar{u})}{\partial t} + A(u - \bar{u}) + g(\cdot,\cdot,Tu) - g(\cdot,\cdot,\bar{u}) \leq 0. \qquad (7.3.4)$$

Since $(u - \bar{u})^+ \in V_0 \cap L_+^2(Q)$ and $(u - \bar{u})^+ = 0$ in $\Omega \times \{0\}$, we get from the weak formulation of (7.3.4) with the special test function $\varphi = (u - \bar{u})^+$ the following inequality

$$\frac{1}{2}\|(u - \bar{u})^+(\cdot,\tau)\|_{L^2(\Omega)}^2 + \mu\|\nabla(u - \bar{u})^+\|_{L^2(Q)}^2$$
$$+ \int_Q (g(x,t,Tu(x,t)) - g(x,t,\bar{u}(x,t)))(u(x,t) - \bar{u}(x,t))^+ \, dxdt \leq 0. \qquad (7.3.5)$$

One readily verifies that

$$\int_Q (g(x,t,Tu(x,t)) - g(x,t,\bar{u}(x,t)))(u(x,t) - \bar{u}(x,t))^+ \, dxdt = 0,$$

and thus by (7.3.5) and due to $(u - \bar{u})^+ \in V_0 \cap L_+^2(Q)$ one gets $(u - \bar{u})^+ = 0$. Hence, $u \leq \bar{u}$ in Q. Similarly one can prove that $\underline{u} \leq u$ holds, so that any solution of the auxiliary problem (7.3.3) belongs to the interval $[\underline{u}, \bar{u}]$, which implies $Tu = u$ such that $\Lambda_T u = \Lambda u = h$, and thus $u \in \mathcal{B}$.

Define $L := \Lambda_T$ and the operator $N : \mathcal{B} \to L^2(Q)$ by

$$Nu(x,t) := f(x,t,u(x,t),Lu(x,t)). \qquad (7.3.6)$$

The assertion of Theorem 7.3.1 is proved provided we are able to show that the equation $Lu = Nu$ has extremal solutions in $\mathcal{B}$. For this purpose let us verify the following conditions:

(i) If u, $v \in \mathcal{B}$, $u \le v$ and $Lu \le Lv$, then $h_- \le Nu \le Nv \le h_+$.

(ii) For each $h \in [h_-, h_+] \subset L^2(Q)$, there exist extremal solutions of $Lu = h$ in $\mathcal{B}$, which are increasing with respect to h.

Condition (ii) has already been shown above. Let u, $v \in \mathcal{B}$, $u \le v$, and $Lu \le Lv$; then by using hypothesis (F1) and (H1) we obtain

$$h_- = \Lambda \underline{u} = L\underline{u} \le f(\cdot, \cdot, \underline{u}, L\underline{u}) \le f(\cdot, \cdot, u, Lu) = Nu$$
$$\le f(\cdot, \cdot, v, Lv) = Nv \le f(\cdot, \cdot, \bar{u}, L\bar{u}) \le L\bar{u} = \Lambda \bar{u} = h_+,$$

which verifies condition (i). Corollary 1.1.2 can now be applied to ensure the existence of extremal solutions of $Lu = Nu$ in $\mathcal{B}$, which completes the proof.   □

7.3.2 Existence of global extremal solutions. In this subsection we are going to prove the existence of global extremal solutions of the implicit IBVP (7.3.1) under the assumptions (G2), (F1) and the following additional growth condition imposed on f:

(F2) There exist functions $p \in L_+^2(Q)$, $k \in L_+^\infty(Q)$ and $\lambda \colon Q \to [0, 1)$ with $\frac{1}{1-\lambda} \in L^\infty(Q)$ such that for a.e. $(x, t) \in Q$ and for all r, $\zeta \in \mathbb{R}$ the following growth condition for $f \colon Q \times \mathbb{R} \times \mathbb{R} \to \mathbb{R}$ is fulfilled: $|f(x, t, r, \zeta)| \le p(x, t) + k(x, t)|r| + \lambda(x, t)\,|\zeta|$.

In the next lemma we provide an a priori bound of $|\Lambda u|$ for any solution u of (7.3.1).

Lemma 7.3.1. *Let hypotheses (G2), (F1), and (F2) be satisfied, and let $v \in W_0$ be the solution of the linear IBVP*

$$\frac{\partial v}{\partial t} + Av - (l + \frac{k}{1-\lambda})v = m + \frac{p}{1-\lambda} \quad \text{in } Q,\ v = 0 \ \text{ on } \Gamma,\ v = 0 \ \text{ in } \Omega \times \{0\}.$$
$$(7.3.7)$$

Then for any solution $u \in W_0$ of the original IBVP (7.3.1) an estimate in the form

$$|\Lambda u| \le \frac{kv + p}{1 - \lambda} \tag{7.3.8}$$

holds.

Proof. Since $m + \frac{p}{1-\lambda} \in L_+^2(Q)$, it follows from Lemma 7.2.1 that $v \ge 0$. Taking the growth condition (F2) into account we get

$$|\Lambda u| = |f(x, t, u, \Lambda u)| \le k|u| + p + \lambda|\Lambda u|,$$

which yields

$$|\Lambda u| \le \frac{k|u| + p}{1 - \lambda}. \tag{7.3.9}$$

Using $\Lambda u = \frac{\partial u}{\partial t} + Au + g(x, t, u)$, we get

$$\frac{\partial u}{\partial t} + Au \le |g(x, t, u)| + |\Lambda u|.$$

By (7.3.9) and (G2) from the last inequality we obtain

$$\frac{\partial u}{\partial t} + Au \le m + l|u| + \frac{k|u|}{1 - \lambda} + \frac{p}{1 - \lambda}.$$

Subtracting equation (7.3.7) from the last inequality we get

$$\frac{\partial (u - v)}{\partial t} + A(u - v) \le (l + \frac{k}{1 - \lambda})(|u| - v).$$

Taking $\varphi = (u - v)^+ \in \mathcal{V}_0 \cap L_+^2(Q)$ as a special test function in the last inequality, we obtain in view of $(u - v)^+(x, 0) = 0$ the inequality

$$\frac{1}{2} \|(u - v)^+(\cdot, t)\|_{L^2(\Omega)}^2 + \mu \|\nabla(u - v)^+\|_{L^2(Q_t)}^2$$
$$\le \int_{Q_t} (l + \frac{k}{1 - \lambda})(|u| - v)(u - v)^+ \, dx ds.$$

Since $v \ge 0$, we have

$$\int_{Q_t} (l + \frac{k}{1 - \lambda})(|u| - v)(u - v)^+ \, dx ds$$
$$= \int_{\{u > v\}} (l + \frac{k}{1 - \lambda})(u - v)(u - v) \, dx ds$$
$$\le \|l + \frac{k}{1 - \lambda}\|_{L^\infty(Q)} \int_{Q_t} ((u - v)^+)^2 \, dx ds,$$

and thus

$$\frac{1}{2} \|(u - v)^+(\cdot, t)\|_{L^2(\Omega)}^2 + \mu \|\nabla(u - v)^+\|_{L^2(Q_t)}^2$$
$$\le \|l + \frac{k}{1 - \lambda}\|_{L^\infty(Q)} \int_{Q_t} ((u - v)^+)^2 \, dx ds,$$

for any $t \in [0, \tau]$. Applying Gronwall's lemma in a similar way as in the proof of Lemma 7.2.1 it follows that $(u - v)^+(x, t) = 0$ for a.e. $(x, t) \in Q$, that is, $u \leq v$. By analogous reasoning one can show that $-v \leq u$, and thus $|u| \leq v$ which due to (7.3.9), implies

$$|\Lambda u| \leq \frac{kv + p}{1 - \lambda},$$

and thus the assertion (7.3.8). □

Lemma 7.3.2. *Let hypotheses of Lemma 7.3.1 be satisfied. Then for any $u \in W_0$ (not necessarily a solution) satisfying $|\Lambda u| \leq \frac{kv+p}{1-\lambda}$, an estimate in the form*

$$|f(\cdot, \cdot, u, \Lambda u)| \leq \frac{kv + p}{1 - \lambda} \tag{7.3.10}$$

holds.

Proof. By definition of Λu one immediately gets the inequality

$$\frac{\partial u}{\partial t} + Au \leq |g(x, t, u)| + |\Lambda u|.$$

The assumed estimate of $|\Lambda u|$ and the growth condition (G2) yield

$$\frac{\partial u}{\partial t} + Au - l|u| - \frac{kv}{1 - \lambda} \leq m + \frac{p}{1 - \lambda}.$$

Subtracting equation (7.3.7) for v from the last inequality we get

$$\frac{\partial(u - v)}{\partial t} + A(u - v) \leq l(|u| - v),$$

which in just the same way as in the proof of Lemma 7.3.1 yields $|u| \leq v$, so that by the growth condition of (F2) we obtain

$$|f(\cdot, \cdot, u, \Lambda u)| \leq k|u| + p + \lambda|\Lambda u| \leq kv + p + \lambda|\Lambda u|.$$

This last inequality and the assumed inequality for $|\Lambda u|$ imply

$$|f(\cdot, \cdot, u, \Lambda u)| \leq kv + p + \lambda\frac{kv + p}{1 - \lambda} = \frac{kv + p}{1 - \lambda},$$

which proves the lemma. □

On the basis of Corollary 1.1.2 of chapter 1 and by means of Lemmas 7.2.2, 7.3.1, and 7.3.2 we are now able to prove the existence of global extremal solutions to the IBVP (7.3.1).

Theorem 7.3.2. *Let* $f: Q \times \mathbb{R} \times \mathbb{R} \to \mathbb{R}$ *satisfy (F1) and (F2), and let* $g: Q \times \mathbb{R} \to \mathbb{R}$ *satisfy hypothesis (G2). Then the IBVP (7.3.1) has (global) extremal solutions in* W_0, *and they are increasing with respect to* f.

Proof. We introduce the subset $V \subset W_0$ by

$$V := \{u \in W_0 \mid h_- \le \Lambda u \le h_+\}, \quad \text{with } h_\pm = \pm \frac{kv + p}{1 - \lambda},$$

where v is the unique solution of (7.3.7), and define operators L and N on V by $Lu := \Lambda u$ and $Nu := f(\cdot, \cdot, u, \Lambda u)$. From Lemma 7.3.2 it follows that if $u \in V$, then

$$|Nu| = |f(\cdot, \cdot, u, \Lambda u)| \le \frac{kv + p}{1 - \lambda},$$

and thus $N(V) \subseteq [h_-, h_+]$. This and the monotonicity of f in its last two arguments imply

   (i) If u, $v \in V$, $u \le v$ and $Lu \le Lv$, then $h_- \le Nu \le Nv \le h_+$.

By means of Lemma 7.2.2 we get

   (ii) For each $h \in [h_-, h_+]$ there exist extremal solutions of $Lu = h$ in V and they are increasing with respect to h.

Thus Corollary 1.1.2 of chapter 1 may be applied which yields the existence of the least and greatest solution u_* and u^* in V, respectively, of the operator equation $Lu = Nu$, which is equivalent with the IBVP (7.3.1), so that the extremal solutions u_* and u^* of $Lu = Nu$ must be also extremal solutions of the IBVP (7.3.1) in V. Finally, because any solution of (7.3.1) belongs to V according to Lemma 7.3.1, then u_* and u^* are (global) extremal solutions of the IBVP (7.3.1) in the whole space W_0. Again by Corollary 1.1.2 the monotone dependence of the extremal solutions of $Lu = Nu$ with respect to N implies the monotone dependence of the extremal solutions of the IBVP (7.3.1) with respect to f which completes the proof.   □

7.3.3 Generalizations and examples. Let us consider the implicit parabolic problem

$$\Lambda u = q(x, t, u, \Lambda u) \text{ in } Q, \quad u = 0 \text{ on } \Gamma, \quad u = 0 \text{ in } \Omega \times \{0\}, \quad (7.3.11)$$

where Λ is as in previous sections, but where the nonlinearity $q: Q \times \mathbb{R} \times \mathbb{R} \to \mathbb{R}$ is not necessarily monotone increasing in its last argument. Instead we impose the following hypotheses on q:

   (Q1) The nonlinearity q is superpositionally measurable, and there is a function $\alpha \in L_+^\infty(Q)$ such that $(x, t, r, \zeta) \to q(x, t, r, \zeta) + \alpha(x, t)\,\zeta$ is increasing with respect to r and ζ for a.e. $(x, t) \in Q$.

(Q2) There exist functions $\hat{p} \in L_+^2(Q)$, $\hat{k} \in L_+^\infty(Q)$ and $\hat{\lambda}: Q \to [0,1)$ satisfying $\frac{1}{1-\hat{\lambda}} \in L^\infty(Q)$ such that for a.e. $(x,t) \in Q$ and for all $r, \zeta \in \mathbb{R}$ the following estimate holds:

$$|q(x,t,r,\zeta)| \le \hat{p}(x,t) + \hat{k}(x,t)|r| + \hat{\lambda}(x,t)\,|\zeta|.$$

We are going to show that hypotheses (Q1) and (Q2) and hypothesis (G2) are sufficient to ensure the existence of extremal solutions of the IBVP (7.3.11). Furthermore, we give examples that demonstrate the applicability of the obtained results.

Theorem 7.3.3. *Let hypotheses (Q1), (Q2), and (G2) be satisfied. Then the IBVP (7.3.11) has extremal solutions, and they are increasing with respect to q.*

Proof. Define f as follows:

$$f(x,t,r,\zeta) := \frac{1}{1+\alpha(x,t)}(q(x,t,r,\zeta) + \alpha(x,t)\zeta).$$

Then the IBVP (7.3.11) is equivalent with the following one

$$\Lambda u = f(x,t,u,\Lambda u) \text{ in } Q, \quad u = 0 \text{ on } \Gamma, \quad u = 0 \text{ in } \Omega \times \{0\}, \quad (7.3.12)$$

where the nonlinearity f satisfies the hypotheses (F1) and (F2) of the subsection 7.3.2 with $k = \frac{\hat{k}}{1+\alpha}$, $p = \frac{\hat{p}}{1+\alpha}$, and $\lambda = \frac{\hat{\lambda}+\alpha}{1+\alpha}$. Thus Theorem 7.3.2 ensures that the IBVP (7.3.12), and hence also the IBVP (7.3.11), has extremal solutions. If q increases, so does also f, whence the monotone dependence of these extremal solutions with respect to q follows from the last assertion of Theorem 7.3.2. $\square$

Example 7.3.1. Let $\tau = 1$ and $\Omega = (0,1)$, and consider the IBVP

$$\begin{cases} u_t - u_{xx} = [t+x] + \frac{[t+x+u]}{1+|[t+x+u]|} + \frac{1}{2}[u_t - u_{xx}], \\ u(x,0) = 2x(1-x), \text{ and } u(0,t) = u(1,t) = 0, \ t \in (0,1), \end{cases} \quad (7.3.13)$$

where $[\cdot]: \mathbb{R} \to \mathbb{Z}$ is the integer function which means that $[z]$ denotes the greatest integer less than or equal to z. Let $\Lambda u := u_t - u_{xx}$, and let $\hat{u}$ be the unique solution of the linear IBVP

$$\Lambda\hat{u} = 0, \quad \hat{u}(x,0) = 2x(1-x), \text{ and } \quad \hat{u}(0,t) = \hat{u}(1,t) = 0, \quad t \in (0,1).$$

Substituting u by $u = \hat{u} + v$ we get an implicit IBVP of the form (7.3.11) in v, i.e.,

$$\Lambda v = [t+x] + \frac{[t + x + \hat{u} + v]}{1 + |[t + x + \hat{u} + v]|} + \frac{1}{2}[\Lambda v], \quad v(x,0) = 0, \quad v(0,t) = v(1,t) = 0.$$

$$(7.3.14)$$

In this example the function q is given by

$$q(x, t, r, \zeta) = [t + x] + \frac{[t + x + \hat{u}(x,t) + r]}{1 + |[t + x + \hat{u}(x,t) + r]|} + \frac{1}{2}[\zeta],$$

which is obviously increasing with respect to r and ζ, and satisfies the estimate

$$|q(x, t, r, \zeta)| \leq 2 + \frac{1}{2}|\zeta|$$

for a.e. $(x,t) \in (0,1) \times (0,1)$, and for all $r, \zeta \in \mathbb{R}$. Thus the hypotheses (Q1) and (Q2) can readily be verified by taking $\alpha(x,t) \equiv 0$, $p(x,t) = 2$, $\lambda(x,t) = \frac{1}{2}$ which yields $\frac{p}{1-\lambda} \equiv 4$. Hence it follows by means of Theorem 7.3.3 or Theorem 7.3.2 that there exist global extremal solutions v_* and v^* of the IBVP (7.3.14), and all solutions belong to the order interval $[-w, w]$, where w is the solution of the following IBVP

$$\Lambda w \equiv w_t - w_{xx} = 4, \quad w(x,0) = 0, \text{ and } w(0,t) = w(1,t) = 0, \quad (7.3.15)$$

which can be calculated by Fourier's method. Since $v = u - \hat{u}$, it follows that v is a solution of (7.3.14) if and only if u is a solution of (7.3.13). Thus $u^* = v^* + \hat{u}$ and $u_* = v_* + \hat{u}$ are the greatest and least solutions, respectively, of (7.3.13) and any other solution u of (7.3.13) is contained in the order interval $[v_* + \hat{u}, v^* + \hat{u}] \subset [-w + \hat{u}, w + \hat{u}]$. This yields in particular the following bounds for all solutions of the IBVP (7.3.13):

$$-w(x,t) + \hat{u}(x,t) \leq u(x,t) \leq w(x,t) + \hat{u}(x,t), \quad (7.3.16)$$

where the upper bound $w + \hat{u}$ satisfies the IBVP

$$\Lambda(w + \hat{u}) = 4, \quad (w + \hat{u})(x,0) = 2x(1-x), \quad (w + \hat{u})(0,t) = (w + \hat{u})(1,t) = 0$$

whose unique solution is given by $(w + \hat{u})(x,t) = 2x(1-x)$. Since w is given above, the lower bound can also explicitly be calculated by

$$(-w + \hat{u})(x,t) = (w + \hat{u})(x,t) - 2w(x,t) = 2x(1-x) - 2w(x,t).$$

Example 7.3.2. Let $\tau = 1$ and $\Omega = \{x \in \mathbb{R}^N : |x| < 1\}$ be the unit sphere in $\mathbb{R}^N$, and consider the IBVP

$$\begin{cases} u_t - \Delta u = [t + |x|] + \frac{[t+|x|+u]}{1+|[t+|x|+u]|} + \frac{1}{2}[u_t - \Delta u] & \text{in } Q, \\ u = 0 \ \text{in } \Omega \times \{0\}, \quad u = 0 \ \text{on } \Gamma, \end{cases} \qquad (7.3.17)$$

where $|\cdot|$ stands for the Euclidean norm in $\mathbb{R}^N$ and $[\cdot]: \mathbb{R} \to \mathbb{Z}$ is the integer function. The IBVP (7.3.17) may be considered as the higher dimensional version of (7.3.13) under homogeneous initial values. The following result can be obtained in just the same way as in Example 7.3.1.

The IBVP (7.3.17) has extremal solutions within the order interval $[-w, w]$, where w is the unique nonnegative solution of the following IBVP

$$w_t - \Delta w = 4 \ \text{ in } \ Q, \quad u = 0 \ \text{ on } \ \Gamma, \quad u = 0 \ \text{ in } \ \Omega \times \{0\}.$$

Example 7.3.3. Let $\tau = 1$ and $\Omega = (-1, 1)$, and consider the IBVP

$$\begin{cases} u_t - u_{xx} = -x + \frac{1}{2}[u_t - u_{xx}], \\ u(x,0) = 0, \ \text{and} \quad u(-1,t) = u(1,t) = 0, \quad t \in (0,1), \end{cases} \qquad (7.3.18)$$

where $[\cdot]: \mathbb{R} \to \mathbb{Z}$ is again the integer function. Let $\Lambda u := u_t - u_{xx}$, and

$$q(x,t,r,\zeta) = -x + \frac{1}{2}[\zeta],$$

which is increasing with respect to r and ζ, and satisfies the estimate

$$|q(x,t,r,\zeta)| \leq 1 + \frac{1}{2}|\zeta|,$$

for a.e. $(x,t) \in (-1,1) \times (0,1)$, and for all $r, \ \zeta \in \mathbb{R}$. Thus the hypotheses (Q1) and (Q2) can readily be verified by taking $\alpha(x,t) \equiv 0$, $p(x,t) = 1$, $\lambda(x,t) = \frac{1}{2}$ which yields $\frac{p}{1-\lambda} \equiv 2$. Hence it follows by means of Theorem 7.3.3 that there exist extremal solutions u_* and u^* of the IBVP (7.3.18) within the order interval $[-w, w]$, where w is the solution of the following IBVP

$$\Lambda w \equiv w_t - w_{xx} = 2, \quad w(x,0) = 0, \ \text{and} \ w(-1,t) = w(1,t) = 0,$$

which can be calculated by Fourier's method.

The implicit equation (7.3.18) can also be written in the form

$$f(x,t,u,\Lambda u) = 0 \quad \text{with } f(x,t,r,\zeta) = \zeta + x - \frac{1}{2}[\zeta]. \qquad (7.3.19)$$

One can easily see that $f(x,t,r,\zeta) = 0$ cannot be solved uniquely for ζ. However, by an elementary calculation one can construct explicitly the multifunction Θ defined in section 7.1 and given by

$$\Theta(x,t,r) = \{\zeta \in \mathbb{R} \mid f(x,t,r,\zeta) = 0\},$$

and one can show that Θ has neither a lower nor an upper semicontinuous multiselection such that the results that require semicontinuity of the multifunction Θ can, in general, not be applied.

7.4 Implicit elliptic problems

Here we shortly discuss the implicit elliptic BVP

$$\Lambda u = f(x,u,\Lambda u) \text{ in } \Omega, \quad u = 0 \text{ on } \partial\Omega, \qquad (7.4.1)$$

where Λ is now a semilinear elliptic operator in the form

$$\Lambda u(x) = Au(x) + g(x,u(x)), \qquad (7.4.2)$$

and A a linear, strongly elliptic operator given by

$$Au(x) = -\sum_{i,j=1}^{N} \frac{\partial}{\partial x_i}\left(a_{ij}(x)\frac{\partial u}{\partial x_j}\right),$$

with coefficients $a_{ij} \in L^\infty(\Omega)$ satisfying for all $\xi = (\xi_1,...,\xi_N) \in \mathbb{R}^N$

$$\sum_{i,j=1}^{N} a_{ij}(x)\xi_i\xi_j \geq \mu|\xi|^2, \quad \text{for a.e. } (x) \in \Omega \quad \text{with some constant } \mu > 0.$$

BVP (7.4.1) may be considered as the elliptic counterpart to the IBVP (7.1.4). Let $\Omega \subset \mathbb{R}^N$ be a bounded domain having a Lipschitz boundary $\partial\Omega$, and define the subspaces W and W_0 by $W := \{u \in W^{1,2}(\Omega) \mid Au \in L^2(\Omega)\}$ and $W_0 := \{u \in W_0^{1,2}(\Omega) \mid Au \in L^2(\Omega)\}$, respectively.

Definition 7.4.1. A function $u \in W_0$ is called a *solution* of (7.4.1) if

$$\Lambda u(x) = f(x, u(x), \Lambda u(x)) \quad \text{for a.e. } x \in \Omega.$$

Definition 7.4.2. A function $\bar{u} \in W$ is called an *upper solution* of the BVP (7.4.1) if

(i) $\bar{u} \geq 0$ on $\partial\Omega$,
(ii) $\Lambda\bar{u}(x) \geq f(x, \bar{u}(x), \Lambda\bar{u}(x))$ for a.e. $x \in \Omega$.

A *lower solution* $\underline{u}$ is similarly defined by reversing the inequality sign in (i) and (ii).

Correspondingly to the parabolic case of section 7.3 we impose the following hypotheses.

(F1') The nonlinearity $f : \Omega \times \mathbb{R} \times \mathbb{R} \to \mathbb{R}$ is sup-measurable and the function $(x, r, \zeta) \to f(x, r, \zeta)$ is increasing in r and ζ for a.e. $x \in \Omega$.

(H1') The BVP (7.4.1) has a lower solution $\underline{u}$ and an upper solution $\bar{u}$ according to Definition 7.4.2 satisfying $\underline{u} \leq \bar{u}$ and $\Lambda\underline{u} \leq \Lambda\bar{u}$.

(H2') The function g of the operator Λ is a Carathéodory function which is L^2-bounded with respect to the order interval $[\underline{u}, \bar{u}]$; i.e., there exists $m \in L^2_+(\Omega)$ such that
$|g(x, r)| \leq m(x)$, for a.e. $x \in \Omega$ and for all $r \in [\underline{u}(x), \bar{u}(x)]$.

Note again that no continuity assumptions have been imposed on f. The following theorem is the elliptic counterpart to Theorem 7.3.1 of section 7.3.

Theorem 7.4.1. *Let hypotheses (F1'), (H1'), and (H2') be satisfied. Then the implicit BVP (7.4.1) has extremal solutions within the set $\mathcal{B}$ defined by*

$$\mathcal{B} = \{u \in W_0 \mid \underline{u} \leq u \leq \bar{u} \text{ and } \Lambda\underline{u} \leq \Lambda u \leq \Lambda\bar{u}\}.$$

Proof. Let T denote the truncation operator which truncates u between $\underline{u}$ and $\bar{u}$, and define $\Lambda_T u$ by

$$\Lambda_T u(x) = Au(x) + g(x, Tu(x)). \tag{7.4.3}$$

Set $h_- := \Lambda\underline{u}$ and $h_+ := \Lambda\bar{u}$. First we prove that for each $h \in [h_-, h_+] \subset L^2(\Omega)$ the BVP

$$\Lambda_T u = h \quad \text{in } \Omega, \quad u = 0 \quad \text{on } \partial\Omega \tag{7.4.4}$$

has extremal solutions in $\mathcal{B}$ and that these extremal solutions are increasing with respect to h. By (H2') and the definition of the truncation T we have for any u

$$|g(x, Tu(x))| \leq m(x) \quad \text{for a.e. } x \in \Omega$$

with $m \in L_+^2(\Omega)$. Let $w \in W_0$ be the unique solution of the linear BVP

$$Au = |h| + m \quad \text{in } \Omega, \quad u = 0 \quad \text{on } \partial\Omega.$$

Then $w \geq 0$ and w is an upper solution of (7.4.4), and $-w$ is a lower solution for (7.4.4). Applying the extremality result of section 5.1 there exist extremal solutions of (7.4.4) within the order interval $[-w, w]$. Due to the maximum principle any solution of (7.4.4) necessarily belongs to the interval $[-w, w]$ so that (7.4.4) has extremal solutions among all its solutions. We shall show that any solution of (7.4.4) belongs to $\mathcal{B}$. For a solution u of (7.4.4) with $h \in [h_-, h_+]$ we get

$$\Lambda \underline{u} \leq \Lambda_T u \leq \Lambda \bar{u},$$

where $\bar{u}$ and $\underline{u}$ are the given upper and lower solutions, respectively. Inequality $\Lambda_T u \leq \Lambda \bar{u}$ yields the differential inequality

$$A(u - \bar{u}) + g(\cdot, Tu) - g(\cdot, \bar{u}) \leq 0, \tag{7.4.5}$$

and thus by taking the nonnegative test function $\varphi = (u-\bar{u})^+ \in W_0 \cap L_+^2(\Omega)$ we obtain from (7.4.5)

$$\mu \|\nabla(u - \bar{u})^+\|_{L^2(\Omega)}^2 \leq \langle A(u - \bar{u}), (u - \bar{u})^+ \rangle$$

$$\leq \int_\Omega (g(\cdot, \bar{u}) - g(\cdot, Tu))(u - \bar{u})^+ \, dx = 0,$$

which implies in view of $(u - \bar{u})^+ \in W_0$ that $(u - \bar{u})^+ = 0$ and thus $u \leq \bar{u}$ in Ω. Similarly one shows that $\Lambda \underline{u} \leq \Lambda_T u$ implies $\underline{u} \leq u$, and thus any solution u of the auxiliary problem (7.4.4) satisfies $u \in [\underline{u}, \bar{u}]$. Consequently $Tu = u$ and $\Lambda_T u = \Lambda u$ which yields $\Lambda \underline{u} \leq \Lambda u \leq \Lambda \bar{u}$, and thus $u \in \mathcal{B}$. To prove the monotone dependence of these extremal solutions with respect to h, let $h_1, h_2 \in [h_-, h_+]$ be such that $h_1 \leq h_2$, and denote the corresponding greatest and least solutions of (7.4.4) by v_i^* and $v_{i,*}$, respectively, $i = 1, 2$. We shall show $v_1^* \leq v_2^*$. The proof for $v_{1,*} \leq v_{2,*}$ is similar. Since $v_1^* \in W_0$ is the greatest solution of $\Lambda_T u = h_1$ and $h_1 \leq h_2$, thus v_1^* is, in particular, a lower solution of the BVP

$$\Lambda_T u = h_2 \quad \text{in } \Omega, \quad u = 0 \quad \text{on } \partial\Omega, \tag{7.4.6}$$

and obviously $v_1^* \leq w$, where w is the unique solution of the linear BVP

$$Au = |h_2| + m \quad \text{in } \Omega, \quad u = 0 \quad \text{on } \partial\Omega.$$

Because w is an upper solution of (7.4.6), there exist solutions of (7.4.6) within the interval $[v_1^*, w]$, and hence it follows $v_1^* \leq v_2^*$, since v_2^* is the greatest solution of (7.4.6).

Let $L := \Lambda_T$, and define $N : \mathcal{B} \to L^2(\Omega)$ by $Nu(x) := f(x, u(x), Lu(x))$, then the BVP (7.4.1) has extremal solutions in $\mathcal{B}$ if and only if $Lu = Nu$ has extremal solutions in $\mathcal{B}$. The latter can be proved by means of Corollary 1.1.2. To this end it remains to verify conditions (I) and (II) of this corollary, where V corresponds with $\mathcal{B}$. Condition (II) has already been proved above. The proof of condition (I) can be done in just the same way as in the proof of Theorem 7.3.1, which completes the proof.   □

Remark 7.4.1. The situation becomes more involved when we consider the existence of global extremal solutions for the BVP (7.4.1). Unlike in Theorem 7.3.2 growth conditions of the form (F2) and (G2) on f and g, respectively, as for the parabolic case are not anymore sufficient to ensure existence of global extremal solutions of the elliptic problem (7.4.1). This is mainly due to the fact that comparison and uniqueness results for semi-linear elliptic operators with Lipschitz continuous nonlinearities, in general, require small enough Lipschitz constants or an appropriate monotonicity behavior.

Example 7.4.1. The Dirichlet problem

$$-u''(x) = -x + \frac{1}{2}[-u''(x)] \quad \text{a.e. in } (-1, 1), \quad u(-1) = u(1) = 0, \quad (7.4.7)$$

is of the form (7.4.1), where $\Omega = (-1, 1)$,

$$\Lambda u = -u'' \quad \text{and} \quad f(x, r, \zeta) = -x + \frac{1}{2}[\zeta]. \qquad (7.4.8)$$

It is easy to see that f satisfies the hypothesis (F1'). Moreover, the functions

$$u_\pm(x) = \pm 2(1 - x^2), \qquad x \in [-1, 1],$$

which are the solutions of the Dirichlet problems

$$-u''(x) = \pm 4 \quad \text{a.e. in } (-1, 1), \quad u(-1) = u(1) = 0, \qquad (7.4.9)$$

are upper and lower solutions of (7.4.7); i.e., the hypothesis (H1') is satisfied when $\underline{u} = u_-$ and $\bar{u} = u_+$. Also the hypothesis (H2') holds because in this case $g(x, r) \equiv 0$. If $u \in W_0$ is any solution of (7.4.7), one directly verifies that $\underline{u} \leq u \leq \bar{u}$ and $\Lambda \underline{u} \leq \Lambda u \leq \Lambda \bar{u}$. It then follows from Theorem 7.4.1 that the Dirichlet problem (7.4.7) has extremal solutions in the whole W_0.

Next we shall show that the extremal solutions of (7.4.7) can be obtained by the method of successive approximations introduced in Proposition 1.1.2. Notice first that $u \in W_0$ is a solution of (7.4.7) if and only if $Lu = Nu$, where

$$Lu(x) := \Lambda u(x) = -u''(x) \quad \text{and} \quad Nu(x) := -x + \frac{1}{2}[Lu(x)] \quad \text{a.e. in } \Omega.$$

It is easy to show that for each $h \in L^2(\Omega)$ the function

$$u(x) = (1-x) \int_{-1}^{x} (1+t)h(t)dt + (1+x) \int_{x}^{1} (1-t)h(t)dt \qquad (7.4.10)$$

is a unique solution of equation $Lu = h$ in W_0. In particular, this solution is increasing with respect to h, whence the hypotheses (I) and (II) of Proposition 1.1.2 hold when $h_{\pm}(x) \equiv \pm 4$, $V = \{u \in W_0 \mid h_- \leq Lu \leq h_+\}$ and $X = L^2(\Omega)$. In this case the successive approximations $Lu_{n+1} = Nu_n$ can be rewritten as

$$\begin{cases} u_{n+1}(x) = (1-x) \int_{-1}^{x}(1+t)h_n(t)dt + (1+x) \int_{x}^{1}(1-t)h_n(t)dt, \\ h_n(x) = -x + \frac{1}{2}[-u_n''(x)] \quad \text{a.e. in } (-1, 1). \end{cases}$$

$$(7.4.11)$$

Since the function $z \mapsto [z]$ is right-continuous, the greatest solution of the BVP (7.4.7) can be obtained by Proposition 1.1.2 as the uniform limit of the successive approximations (7.4.11) when $u_0 = \bar{u}$. Calculations of these approximations show that $u_5 = u_6$; i.e., $u^* = u_5$ is the greatest solution of (7.4.7), and that

$$u^*(x) = \begin{cases} \frac{x^3}{6} - \frac{x^2}{4} - \frac{29}{48}x - \frac{9}{48}, & -1 \leq x < -\frac{1}{2}, \\ \frac{x^3}{6} - \frac{17}{48}x - \frac{1}{8}, & -\frac{1}{2} \leq x < 0, \\ \frac{x^3}{6} + \frac{x^2}{4} - \frac{17}{48}x - \frac{1}{8}, & 0 \leq x < \frac{1}{2}, \\ \frac{x^3}{6} + \frac{x^2}{2} - \frac{29}{48}x - \frac{1}{16}, & \frac{1}{2} \leq x \leq 1. \end{cases}$$

Similar calculations, with $u_0 = u_-$, show that the least solution of (7.4.7) is

$$u_*(x) = \begin{cases} \frac{x^3}{6} - \frac{29}{48}x - \frac{21}{48}, & -1 \leq x < -\frac{1}{2}, \\ \frac{x^3}{6} + \frac{x^2}{4} - \frac{17}{48}x - \frac{3}{8}, & -\frac{1}{2} \leq x < 0, \\ \frac{x^3}{6} + \frac{x^2}{2} - \frac{17}{48}x - \frac{1}{8}, & 0 \leq x < \frac{1}{2}, \\ \frac{x^3}{6} + \frac{3x^2}{3} - \frac{29}{48}x - \frac{15}{48}, & \frac{1}{2} \leq x \leq 1. \end{cases}$$

The differential equation of (7.4.7) can be rewritten as an explicit differential inclusion

$$-u''(x) \in \mathcal{H}(x, u(x)) = \begin{cases} \{-x, -x + \frac{1}{2}\}, & -1 < x \leq -\frac{1}{2}, \\ \{-x - \frac{1}{2}, x\}, & -\frac{1}{2} < x \leq 0, \\ \{-x - 1, -x - \frac{1}{2}\}, & 0 < t \leq \frac{1}{2}, \\ \{-x - \frac{3}{2}, -x - 1\}, & \frac{1}{2} < x \leq 1. \end{cases} \qquad (7.4.12)$$

Thus each point of the set $\{(x, y) \mid -1 < x < 1, \ u_*(x) \leq y \leq u^*(y)\}$ is a bifurcation point for the solutions of (7.4.7), so that between u_* and u^* there is a continuum of chaotically behaving solutions of (7.4.7).

Remark 7.4.2. The least (resp., greatest) solution of (7.4.7) can be obtained also by integrating twice the first (resp., second) values of $-u''(x)$ in the inclusion equation (7.4.12). The unknown eight constants in the resulting integrals can be determined by applying boundary conditions of (7.4.7) and continuity of u and u' at points $0, \pm\frac{1}{2}$. This method does not work if, e.g., $u(x)/(1 + |u(x)|)$ is added to the right-hand side of the differential equation of (7.4.7), whereas the methods we have used are applicable.

The multifunction $\mathcal{H}$ defined in (7.4.12) possesses neither a lower or an upper semicontinuous multiselection. Thus the existence of extremal solutions of (7.4.7) does not follow from the existence results based on semicontinuity assumptions, cf., e.g., [153, 177].

7.5 Notes and comments

This chapter extends the theory on implicit and discontinuous ordinary differential equations established in previous chapters to implicit partial differential equations of elliptic and parabolic type. The material of this chapter is based on recent papers by the authors; see [77, 81]. The main tools used in our treatment are the abstract fixed point theory of operator equations $Lu = Nu$ developed in chapter 1 and the extremality results of chapter 5. The advantage of our approach compared with those papers that rely on the theory of set-valued analysis (see, e.g., [27, 176, 177]) is that our assumptions can easily be verified and the existence proof becomes constructive if certain one-sided continuity conditions are satisfied.

Appendix

The purpose of this appendix is to present proofs of some results, and introduce basic concepts and well-known results, which are needed in the text. In section A we prove first a fixed point result in partially ordered sets by using a generalized iteration method, and apply it to obtain a fixed point theorem in ordered normed spaces. Inequalities, which are frequently used in the book, are collected in section B. Sobolev spaces and some of their fundamental properties are presented in section C. Section D contains basic results on pseudomonotone and quasilinear elliptic operators, and nonlinear first order evolution equations are considered in section E. Finally, in section F we provide some basic facts from nonsmooth analysis.

A Analysis in ordered spaces

A.1 Basic concepts of partially ordered sets. Given a nonempty set P, we say that a relation $x \leq y$ between certain pairs of elements of P is a *partial ordering in* P, and that P is a *poset*, if $x \leq x$ for all $x \in P$, if $x \leq y$ and $y \leq x$ imply $x = y$, and if $x \leq y$ and $y \leq z$ imply $x \leq z$. The notation $x < y$ stands for $x \leq y$ and $x \neq y$.

An element b of P is called an *upper bound* of a subset A of P if $x \leq b$ for each $x \in A$. If $b \in A$, we say that b is the *maximum* of A, and denote $b = \max A$. A lower bound of A and the minimum, $\min A$, of A are defined similarly, replacing $x \leq b$ above by $b \leq x$. If the set of all upper bounds of A has the minimum, we call it a *least upper bound of* A and denote it by $\sup A$. The greatest lower bound, $\inf A$, of A is defined similarly.

We say that a poset P is a *lattice* if $\inf\{x, y\}$ and $\sup\{x, y\}$ exist for all $x, y \in P$. A subset C of P is said to be *upward directed* if for each pair $x, y \in C$ there is a $z \in C$ such that $x \leq z$ and $y \leq z$, and C is *downward directed* if for each pair $x, y \in C$ there is a $w \in C$ such that $w \leq x$ and $w \leq y$. If C is both upward and downward directed it is called *directed*. A subset C of a poset P is called a *chain* if $x \leq y$ or $y \leq x$ for all $x, y \in C$. We say that C is *well ordered* if each nonempty subset of C has a minimum, and *inversely well ordered* if each nonempty subset of C has a maximum. Obviously, each (inversely) well ordered set is a chain and each chain is directed.

If $x \in P$ and C is a well ordered subset of P, denote
$$C^{<x} = \{y \in C \mid y < x\}.$$
If $x \in C$, then $C^{<x}$ is called a *section* of C. For the sake of completeness we shall prove the following formulation of the *principle of transfinite induction*.

(TI). *Let A be a subset of a nonempty well ordered set C. If $\min C \in A$, and if $x \in A$ whenever $\min C < x \in C$ and $C^{<x} \subseteq A$, then $A = C$.*

Proof. If $C \setminus A$ is nonempty, then $x = \min(C \setminus A)$ exists, $\min C < x \in C$ and $C^{<x} \subseteq A$, so that $x \in A$, contradicting $x \in C \setminus A$. Thus $A = C$. □

We say that a sequence $(x_n)_{n=0}^{\infty}$ of a poset P is *increasing* (resp., *strictly increasing*) if $x_n \leq x_m$ (resp., $x_n < x_m$) whenever $n < m$. A (strictly) decreasing sequence is defined similarly, replacing $n < m$ by $m < n$. A sequence of P is called *monotone* if it is increasing or decreasing. We say that a mapping $G: P \to P$ is *increasing* if $Gx \leq Gy$ whenever $x, y \in P$ and $x \leq y$.

A.2 Fixed point results in partially ordered sets. In what follows, $\underline{x}$ is an element of a poset P, and G is a mapping from P to P.

A subset B of P is called a G-*set* if the following properties hold.
(G) B is well ordered, $\underline{x} = \min B$, and if $\underline{x} < x \in B$, then $x = \sup G[B^{<x}]$.
For instance, $\{\underline{x}\}$ is a G-set. G-sets have the following property.

Lemma A.2.1. *B is a section of C if B and C are G-sets and $C \not\subseteq B$.*

Proof. Let B and C be G-sets, and assume that $C \not\subseteq B$. By a well known comparability result (cf. [43, 95]) there is a bijection φ from B onto a section of C such that $\varphi(x) \leq \varphi(y)$ whenever $x, y \in B$ and $x \leq y$. Denoting $A = \{x \in B \mid x = \varphi(x)\}$, then $\underline{x} \in A$, because $\underline{x} = \min B = \min C$ by (G). If $\underline{x} < x \in B$ and $B^{<x} \subseteq A$, then $B^{<x} = C^{<\varphi(x)}$, whence $x = \sup G[B^{<x}] = \sup G[C^{<\varphi(x)}] = \varphi(x)$ by (G), so that $x \in A$. Thus $A = B$ by (TI), and hence $B = \varphi[B]$ is a section of C. □

Lemma A.2.2. *There is a unique well ordered subset C of P, called a w.o. chain of G-iterations of $\underline{x}$, satisfying the following condition.*
  (C) *$\underline{x} = \min C$, and if $\underline{x} < x \in P$, then $x \in C$ iff $x = \sup G[C^{<x}]$.*

Proof. Let C be the union of all G-sets. As an easy consequence of Lemma A.2.1 it can be shown that C is a G-set. Thus $\underline{x} = \min C$, and if $\underline{x} < x \in C$, then $x = \sup G[C^{<x}]$. Conversely, if $\underline{x} < x \in P$, and if $x = \sup G[C^{<x}]$, then $C^{<x} \cup \{x\}$ is a G-set, whence $x \in C$. These results imply that C has property (C). If B is another well ordered subset having the properties given for C in (C), then B is a G-set, whence $B \subseteq C$. Moreover, $\underline{x} \in B$, and if $\underline{x} < x \in C$

and $C^{<x} \subseteq B$, then $B^{<x} = C^{<x}$, whence $x = \sup G[C^{<x}] = \sup G[B^{<x}]$ by (G), so that $x \in B$. Thus $B = C$ by (TI). This proves the uniqueness. $\square$

Lemma A.2.3. *Assume that $G: P \to P$ is increasing, that $\underline{x}$ is a lower bound of $G[P]$, and that C is the w.o. chain of G-iterations of $\underline{x}$.*
a) *If $x \in C$, then $x \le Gx$, and $Gx \in C$.*
b) *If $x \in C$ and $x < Gx$, then $Gx = \min\{y \in C \mid x < y\}$.*

Proof. a) Let $x \in C$ be given. If $x = \underline{x}$, then $x \le Gx$ since $\underline{x}$ is a lower bound of $G[P]$. If $\underline{x} < x$, then $x = \sup G[C^{<x}]$. Each $y \in C^{<x}$ has property $y < x$, and hence $Gy \le Gx$ because G is increasing. Thus $x = \sup G[C^{<x}] \le Gx$. In particular,

$$\sup G[C^{<x} \cup \{x\}] = \sup(G[C^{<x}] \cup \{Gx\}) = Gx. \qquad (A.2.1)$$

To prove that $Gx \in C$, notice first that it is true if $x = Gx$. Thus we may assume that $x < Gx$. The set $B = C^{<x} \cup \{x\} \cup \{Gx\}$ is well ordered and $\underline{x} = \min B$. Applying (A.2.1) we get $\sup G[B^{<Gx}] = \sup G[C^{<x} \cup \{x\}] = Gx$. Thus $\sup G[B^{<Gx}] = Gx$, so that B is a G-set, whence $Gx \in C$.

b) Assume that $x \in C$ and $x < Gx$. Then $Gx \in C$ by a), so that $z = \min\{y \in C \mid x < y\}$ exists. Applying (C) and (A.2.1) we obtain $z = \sup G[C^{<z}] = \sup G[C^{<x} \cup \{x\}] = Gx$. $\square$

The following fixed point result is an application of Lemma A.2.2 and Lemma A.2.3.

Proposition A.2.1. *Assume that $G: P \to P$ is increasing, that $\underline{x}$ is a lower bound of $G[P]$, and that $x_* = \sup G[C]$ exists, where C is the w.o. chain of G-iterations of $\underline{x}$. Then $x_* = Gx_* = \max C = \min\{x \mid Gx \le x\}$. In particular, x_* is the least fixed point of G.*

Proof. Since $x \le Gx$ for each $x \in C$ by Lemma A.2.3, then $x_* = \sup G[C]$ is an upper bound of C. $x_* = \max C$, for otherwise, $C = C^{<x_*}$, and thus $x_* = \sup G[C^{<x_*}]$, whence $x_* \in C$ by (C), contradicting $C = C^{<x_*}$. Since $x_* \le Gx_*$ and $Gx_* \in C$ by Lemma A.2.3, then $x_* = Gx_*$.

To prove that $x_* = \min\{x \mid Gx \le x\}$, let $y \in P$ satisfy $Gy \le y$, and denote $A = \{x \in C \mid x \le y\}$. Since $\underline{x}$ is a lower bound of $G[P]$, then $\underline{x} \le Gy \le y$, whence $\underline{x} \in A$. If $\underline{x} < x \in C$ and $z \in C^{<x} \subseteq A$, then $z \le y$. Thus $Gz \le Gy$ because G is increasing, and $z \le Gz$ by Lemma A.2.3, so that $z \le Gz \le Gy \le y$. This proves that y is an upper bound of $G[C^{<x}]$, whence $x = \sup G[C^{<x}] \le Gy \le y$, i.e. $x \in A$. Thus $A = C$ by (TI). In particular, $x_* = \max C \le y$. Since $x_* = Gx_*$, the above proof verifies that $x_* = \min\{x \mid Gx \le x\}$, and that x_* is the least fixed point of G. $\square$

As a consequence of Proposition A.2.1 we get the following fixed point result.

Theorem A.2.1. *Let P be a poset and $G: P \to P$ an increasing mapping.*
a) If $G[P]$ has a lower bound and each well ordered chain of $G[P]$ has the supremum, then G has the least fixed point x_*, and $x_* = \min\{x \mid Gx \leq x\}$.
b) If $G[P]$ has an upper bound and each inversely well ordered chain of $G[P]$ has the infimum, then G has the greatest fixed point $x^* = \max\{x \mid x \leq Gx\}$.

Proof. a) Let $\underline{x}$ be a lower bound of $G[P]$, and let C be the w.o. chain of G-iterations of $\underline{x}$. Since $G[C]$ is by Lemma A.2.3.a a subset of C, it is a well ordered subset of $G[P]$. Thus the hypotheses of a) imply that $x_* = \sup G[C]$ exists, whence the conclusions of a) follow from Proposition A.2.1.

b) The given hypotheses imply that the hypotheses of a) hold when the order relation $\leq$ of P is replaced by its dual $\preceq$, defined by

$$x \preceq y \text{ iff } y \leq x. \tag{A.2.2}$$

Thus G has the least fixed point x^* in $(P, \preceq)$, and $x^* = \min\{x \mid Gx \preceq x\}$. In view of (A.2.2) this means that x^* is the greatest fixed point of G in $(P, \leq)$, and that $x^* = \max\{x \mid x \leq Gx\}$. $\qquad\square$

Remark A.2.1. If the hypotheses of Lemma A.2.3 hold, its results imply that the first elements of the w.o. chain of G-iterations of $\underline{x}$ are iterations $G^n\underline{x}$, $n \in \mathbb{N}$. If $x_\omega = \sup_{n \in \mathbb{N}} G^n\underline{x}$ exists, it is the next possible element of C.

A.3 A fixed point theorem in ordered normed spaces. In many applications P is a subset of an *ordered normed space X*; i.e., X is a normed space which is ordered by a closed cone K, i.e.,

$$x \leq y \text{ if and only if } y - x \in K, \tag{A.2.3}$$

where K is a closed subset of X satisfying $K + K \subseteq K$, $K \cap (-K) = \{0\}$ and $cK \subseteq K$ for each $c \geq 0$. For instance, if K is the set $\mathbb{R}_+$ of nonnegative real numbers, the above definition yields a natural ordering in $\mathbb{R}$. By definition, K is closed and convex, and hence also weakly closed. In particular, the space X, equipped with the above-defined partial ordering and a strong or a weak topology, is an *ordered topological space*; i.e., the order intervals

$$[y) = \{x \in X \mid y \leq x\}, \ (y] = \{x \in X \mid x \leq y\}, \ [y, z] = \{x \in X \mid y \leq x \leq z\}$$

are strongly and weakly closed subsets of X for all $y, z \in X$.

In the proof of our fixed point theorem in an ordered normed space we need the following result.

Lemma A.3.1. *Let A be a well ordered subset of an ordered normed space X, and assume that each increasing sequence of A converges weakly in X. Then A contains an increasing sequence which converges weakly to $\sup A$.*

Proof. To show that A is relatively weakly compact, let $(y_n)_{n=0}^{\infty}$ be a sequence of elements of A. If $(y_n)_{n=0}^{\infty}$ has a strictly increasing subsequence, it has a weak limit by a hypothesis. Assume next that $(y_n)_{n=0}^{\infty}$ does not have any strictly increasing subsequence. Since A is well ordered, then $(y_n)_{n=0}^{\infty}$ does not have any strictly decreasing subsequence, so that $(y_n)_{n=0}^{\infty}$ has only a finite number of different elements. Thus it has a constant subsequence which has a weak limit. The above proof shows that A is relatively weakly sequentially compact, and thus relatively weakly compact (see [216, 182]).

Since A is a chain, then $\{[y) \cap \overline{A}^w \mid y \in A\}$ is a family of weakly closed subsets of the weak closure $\overline{A}^w$ of A possessing the finite intersection property. Because $\overline{A}^w$ is weakly compact, then $\bigcap\{[y) \cap \overline{A}^w \mid y \in A\}$ contains a point x. In particular, $x \in [y)$ for each $y \in A$, whence x is an upper bound of A. If $z \in X$ is an upper bound of A, it follows from $x \in \overline{A}^w \subseteq \overline{(z]}^w = (z]$ that $x \leq z$, whence $x = \sup A$. The above construction shows that $x \in \overline{A}^w$. Thus there exists by [164, Chap. 24.1] such a sequence $(x_k)_{k=0}^{\infty}$ in A that converges weakly to x. Defining $y_n = \max\{x_k \mid 0 \leq k \leq n\}$, $n \in \mathbb{N}$, we obtain an increasing sequence (y_n) in A. By a hypothesis, (y_n) has a weak limit y in X. Since $y_n \leq x$ for each n, and $x_n \leq y_n \leq y$ for each k, and since $(x]$ and $(y]$ are weakly closed, $y \leq x$ and $x \leq y$. Thus $y = x = \sup A$ is the weak limit of (y_n). □

Now we are ready to prove the first fixed point result of section 1.1.

Theorem 1.1.1. *Let P be a subset of an ordered normed space X, and let $G\colon P \to P$ an increasing mapping.*
a) If $G[P]$ has a lower bound in P and increasing sequences of $G[P]$ converge weakly in P, then G has the least fixed point x_*, and $x_* = \min\{x \mid Gx \leq x\}$.
b) If $G[P]$ has an upper bound and decreasing sequences of $G[P]$ converge weakly in P, then G has the greatest fixed point $x^* = \max\{x \mid x \leq Gx\}$.

Proof. a) To prove that the hypotheses of Theorem A.2.1.a are valid, let A be a well ordered chain in $G[P]$. By a hypothesis each increasing sequence of A has a weak limit in X, whence there is by Lemma A.3.1 an increasing sequence in A which converges weakly in X to $x = \sup A$. Moreover, the hypotheses of a) imply that x belongs to P, and that $G[P]$ has a lower bound in P, whence the assertion of a) follows from Theorem A.2.1.a.

Conclusions of b) follow from those of a) by dual argumentation used in the proof of Theorem A.2.1.b. □

B Inequalities

The following well-known inequalities are frequently used and can be found in textbooks, see, e.g., [112, 218].

B.1 Young's inequality. Let $1 < p, q < \infty$, and $1/p + 1/q = 1$; then

$$ab \le \frac{a^p}{p} + \frac{b^q}{q} \quad (a, b > 0).$$

Proof. Since the function $x \to e^x$ is convex, it follows

$$ab = e^{\log a + \log b} = e^{\frac{1}{p} \log a^p + \frac{1}{q} \log b^q} \le \frac{1}{p} e^{\log a^p} + \frac{1}{q} e^{\log b^q} = \frac{a^p}{p} + \frac{b^q}{q}$$

$\square$

B.2 Young's inequality with epsilon. Let $1 < p, q < \infty$, and $1/p + 1/q = 1$, then

$$ab \le \varepsilon a^p + C(\varepsilon) b^q \quad (a, b > 0, \ \varepsilon > 0)$$

with $C(\varepsilon) = (\varepsilon p)^{-q/p} \frac{1}{q}$.

Proof. Set $ab = ((\varepsilon p)^{1/p} a)(\frac{b}{(\varepsilon p)^{1/p}})$ and apply Young's inequality. $\square$

B.3 Let $1 \le s < \infty$, and $\xi_i \in \mathbb{R}$, $\xi_i \ge 0$, $i = 1, ..., N$, then we have the following inequality

$$a\left(\sum_{i=1}^{N} \xi_i^s\right)^{1/s} \le \sum_{i=1}^{N} \xi_i \le b\left(\sum_{i=1}^{N} \xi_i^s\right)^{1/s},$$

where a and b are some positive constants depending only on N and s.

B.4 Hölder's inequality. Let $1 \le p, q \le \infty$, $\frac{1}{p} + \frac{1}{q} = 1$. If $u \in L^p(\Omega)$, $v \in L^q(\Omega)$, then one has

$$\int_\Omega |uv| \, dx \le \|u\|_{L^p(\Omega)} \|v\|_{L^q(\Omega)}.$$

B.5 Minkowski's inequality. Let $1 \le p \le \infty$ and $u, v \in L^p(\Omega)$, then

$$\|u + v\|_{L^p(\Omega)} \le \|u\|_{L^p(\Omega)} + \|v\|_{L^p(\Omega)}.$$

B.6 Gronwall's inequality (differential form). Let $\eta : [0, a] \to \mathbb{R}$ be a nonnegative, absolutely continuous function on the interval $[0, a]$, which satisfies for a.e. t the differential inequality

$$\eta'(t) \le p(t)\eta(t) + \psi(t),$$

where $p, \psi \in L^1_+[0, a]$. Then one has

$$\eta(t) \le e^{\int_0^t p(s)ds} \left[\eta(0) + \int_0^t \psi(s)ds\right]$$

for all $t \in [0, a]$.
   Special case: If

$$\eta'(t) \le p(t)\eta(t) \quad \text{a.e. in } [0, a] \quad \text{and} \quad \eta(0) = 0,$$

then $\eta(t) \equiv 0$ on $[0, a]$.
   The following nonlinear version of the above result is used in chapter 3.

Lemma B.6.1. *Given $J = [t_0, t_1]$, assume that $p \in L^1_+(J)$, $\phi \colon \mathbb{R}_+ \to \mathbb{R}_+$ is increasing, and $\int_{0+}^1 \frac{dx}{\phi(x)} = \infty$. Then $x(t) \equiv 0$ is the only nonnegative absolutely continuous function which has properties*

$$x'(t) \le p(t)\phi(x(t)) \quad a.e. \text{ in } J, \quad x(t_0) = 0.$$

Proof. Assume on the contrary: a nonvanishing $x \in AC_+(J)$ satisfies the above inequalities. Then there exist $t_2, t_3 \in J$, $t_2 < t_3$, such that $x(t_2) = 0$ and $x(t) > 0$ on $(t_2, t_3]$. This and the hypotheses given for ϕ imply that

$$\infty = \int_{0+}^{x(t_3)} \frac{dx}{\phi(x)} = \int_{t_2+}^{t_3} \frac{x'(t)dt}{\phi(x(t))} \le \int_{t_2}^{t_3} p(t)dt < \infty,$$

a contradiction. This concludes the proof.   □

B.7   Gronwall's inequality (integral form). Let $f \in L^1_+[0, a]$ satisfy for a.e. t the integral inequality

$$f(t) \le c_1 \int_0^t f(s)ds + c_2$$

for some constants $c_1, c_2 \ge 0$. Then we have

$$f(t) \le c_2(1 + c_1 t e^{c_1 t}) \quad \text{for a.e. } t \in [0, a].$$

Special case: If

$$f(t) \le c_1 \int_0^t f(s)ds \quad \text{for a.e. } t \in [0, a],$$

then $f(t) \equiv 0$ a.e. on $[0, a]$.

The next generalization to the above Gronwall inequality is frequently applied in chapters 2 and 4; see [141, Lemma 1.5.3].

Lemma B.7.1. *Given an increasing function $\psi \colon \mathbb{R}_+ \to (0, \infty)$ for which $\int_0^\infty \frac{dx}{\psi(x)} = \infty$, then for all fixed $p \in L^1_+(J)$ and $w_0 \in \mathbb{R}_+$ the IVP*

$$w'(t) = p(t)\,\psi(w(t)) \quad \text{for a.e. } t \in J, \quad w(t_0) = w_0$$

has a unique solution $w \in AC(J)$. If $v \in C(J)$ satisfies the inequality

$$v(t) \le w_0 + \int_{t_0}^t p(s)\,\psi(v(s))\,ds \quad a.e. \text{ in } J,$$

then $v(t) \le w(t)$ for all $t \in J$.

Proof. Existence and uniqueness assertions for w can be obtained by separating variables. To prove the last conclusion, let $\overline{x}$ be the solution of the IVP

$$\overline{x}'(t) = p(t)\,\psi(\overline{x}(t)) \quad \text{for a.e. } t \in J, \quad \overline{x}(t_0) = \max\{v(t), w(t) \mid t \in J\}.$$

Denote $X = C(J)$, ordered pointwise and normed by the sup-norm, and $P = \{x \in C(J) \mid 0 \le x(t) \le \overline{x}(t), \ t \in J\}$. It is easy to see that equation

$$Gx(t) = w_0 + \int_0^t p(s)\,\psi(x(s))\,ds, \quad t \in J,$$

defines an increasing mapping $G \colon P \to P$. Moreover, w is the unique fixed point of G, and $Gv \le v$. Thus $v \le w$ by Theorem 1.1.1.a.   □

B.8   The Abstract Gronwall Lemma. (cf. [217, 7.3]) Let X be an ordered Banach space and $T\colon X \to X$ continuous, linear, and increasing operator with spectral radius $r(T) < 1$. Then equation

$$x = z + Tx$$

has for each $z \in X$ a unique solution $x \in X$. Moreover, if $y \in X$ satisfies $y \leq z + Ty$, then $y \leq x$.

B.9   Poincaré-Friedrichs inequality. Let $\Omega \subset \mathbb{R}^N$ be a bounded domain, $1 \leq p < \infty$, and $u \in W_0^{1,p}(\Omega)$ (see (C)); then we have the estimate

$$\|u\|_{L^p(\Omega)} \leq C \, \|\nabla u\|_{L^p(\Omega)},$$

where the constant C depends only on p, N, and Ω.

C   Sobolev spaces

C.1   Definition of Sobolev spaces. Let $\alpha = (\alpha_1, ..., \alpha_N)$ with nonnegative integers $\alpha_1, ..., \alpha_N$ be a *multi-index*, and denote its order by $|\alpha| = \alpha_1 + \cdots + \alpha_N$. Set $D_i = \partial/\partial x_i$, $i = 1, ..., N$, and $D^\alpha u = D_1^{\alpha_1} \cdots D_N^{\alpha_N} u$, with $D^0 u = u$. Let Ω be a domain in $\mathbb{R}^N$ with $N \geq 1$. Then $w \in L^1_{loc}(\Omega)$ is called the αth *weak* (or *generalized*) *derivative* of the function $u \in L^1_{loc}(\Omega)$ iff

$$\int_\Omega u D^\alpha \varphi \, dx = (-1)^{|\alpha|} \int_\Omega w\varphi \, dx \quad \text{for all } \varphi \in C_0^\infty(\Omega)$$

holds, where $C_0^\infty(\Omega)$ denotes the space of infinitely differentiable functions with compact support. The generalized derivative w denoted by $w = D^\alpha u$ is unique up to a change of the values of w on a set of measure zero.

Definition C.1.1. Let $1 \leq p \leq \infty$ and $m = 0, 1, 2, ...$. The *Sobolev space* $W^{m,p}(\Omega)$ is the space of all functions $u \in L^p(\Omega)$, which have generalized derivatives up to order m such that $D^\alpha u \in L^p(\Omega)$ for all $\alpha : |\alpha| \leq m$. For $m = 0$, we set $W^{0,p}(\Omega) = L^p(\Omega)$.

With the corresponding norms given by

$$\|u\|_{W^{m,p}(\Omega)} = \Big(\sum_{|\alpha| \leq m} \|D^\alpha u\|^p_{L^p(\Omega)} \Big)^{1/p}, \quad 1 \leq p < \infty,$$

$$\|u\|_{W^{m,\infty}(\Omega)} = \max_{|\alpha| \leq m} \|D^\alpha u\|_{L^\infty(\Omega)},$$

$W^{m,p}(\Omega)$ becomes a Banach space.

Definition C.1.2. $W_0^{m,p}(\Omega)$ is the closure of $C_0^\infty(\Omega)$ in $W^{m,p}(\Omega)$.

$W_0^{m,p}(\Omega)$ is a Banach space with the norm $\|\cdot\|_{W^{m,p}(\Omega)}$.

First properties: Let $\Omega \subset \mathbb{R}^N$ be a bounded domain, $N \geq 1$.

  (i) $W^{m,p}(\Omega)$ is separable for $1 \leq p < \infty$.
  (ii) $W^{m,p}(\Omega)$ is reflexive for $1 < p < \infty$.
  (iii) Let $1 \leq p < \infty$. Then $C^\infty(\Omega) \cap W^{m,p}(\Omega)$ is dense in $W^{m,p}(\Omega)$, and if $\partial\Omega$ is a Lipschitz boundary then $C^\infty(\overline{\Omega})$ is dense in $W^{m,p}(\Omega)$, where $C^\infty(\Omega)$ and $C^\infty(\overline{\Omega})$ are the spaces of infinitely differentiable functions in Ω and $\overline{\Omega}$, respectively; cf., e.g., [117].

C.2 Sobolev embedding theorem. Let $\Omega \subset \mathbb{R}^N$, $N \geq 1$, be a bounded domain with Lipschitz boundary $\partial\Omega$. Then

  (i) if $mp < N$, the space $W^{m,p}(\Omega)$ is continuously embedded in $L^{p^*}(\Omega)$, $p^* = Np/(N - mp)$, and compactly embedded in $L^q(\Omega)$ for any $q < p^*$;
  (ii) if $0 \leq k < m - \frac{N}{p} < k + 1$, the space $W^{m,p}(\Omega)$ is continuously embedded in $C^{k,\lambda}(\overline{\Omega})$, $\lambda = m - \frac{N}{p} - k$, and compactly embedded in $C^{k,\lambda'}(\overline{\Omega})$ for any $\lambda' < \lambda$;
  (iii) let $1 \leq p < \infty$, then the embeddings $L^p(\Omega) \supseteq W^{1,p}(\Omega) \supseteq W^{2,p}(\Omega) \supseteq \cdots$ are compact;

see, e.g., [117, 218]. Here $C^{k,\lambda}(\overline{\Omega})$ denotes the *Hölder space*; cf. [117].

C.3 Trace and extension. Let $\Omega \subset \mathbb{R}^N$ be a bounded domain with Lipschitz ($C^{0,1}$) boundary $\partial\Omega$, $N \geq 1$, and $1 \leq p < \infty$. Then there exists exactly one linear continuous operator

$$\gamma : W^{1,p}(\Omega) \to L^p(\partial\Omega)$$

such that

  (i) $\gamma(u) = u|_{\partial\Omega}$ if $u \in C^1(\overline{\Omega})$;
  (ii) $\|\gamma(u)\|_{L^p(\partial\Omega)} \leq C \|u\|_{W^{1,p}(\Omega)}$ with C depending only on p and Ω;
  (iii) if $u \in W^{1,p}(\Omega)$, then $\gamma(u) = 0$ in $L^p(\partial\Omega)$ if and only if $u \in W_0^{1,p}(\Omega)$.

Definition C.3.1. We call $\gamma(u)$ the *trace* (or *generalized boundary function*) of u on $\partial\Omega$.

The following result is a useful tool in the study of unbounded domain problems and is given only for the sake of completeness.

Lemma C.3.1. *Let $\Omega_0 \subset\subset \Omega$, that is, Ω_0 is compactly contained in Ω. Assume $g \in W^{1,p}(\Omega)$, $u \in W^{1,p}(\Omega_0)$ and $u - g \in W_0^{1,p}(\Omega_0)$, $1 \leq p < \infty$. Then the function w defined by*

$$w(x) = \begin{cases} u(x) & \text{if } x \in \Omega_0, \\ g(x) & \text{if } x \in \Omega \setminus \Omega_0 \end{cases}$$

is in $W^{1,p}(\Omega)$, and its generalized derivative $D_i w = \partial w / \partial x_i$ is given by

$$D_i w(x) = \begin{cases} D_i u(x) & \text{if } x \in \Omega_0, \\ D_i g(x) & \text{if } x \in \Omega \setminus \Omega_0, \end{cases}$$

$i = 1, ..., N$.

For the proof of Lemma C.3.1, see, e.g., [152, Lemma 20.14]. Its proof is based on the density property (iii) of subsection C.1 and the characterization of the traces of $W_0^{1,p}(\Omega)$ function.

Lemma C.3.2. *(cf. [180, 38.3-4]) If $f \in L^{\infty}([a,b], \mathbb{R})$, and if the function $u \colon [c,d] \to [a,b]$ is absolutely continuous, then*

$$\int_c^d f(u(t)) u'(t)\, dt = \int_{u(c)}^{u(d)} f(v)\, dv.$$

This result holds also when $f \in L^1([a,b], \mathbb{R})$ and $u \colon [c.d] \to [a,b]$ is absolutely continuous and monotone.

C.4 Chain rule and lattice structure.

Chain rule. *Let $f \in C^1(\mathbb{R})$ and $\sup_{s \in \mathbb{R}} |f'(s)| < \infty$. Let $1 \leq p < \infty$ and $u \in W^{1,p}(\Omega)$. Then the composite function $f \circ u \in W^{1,p}(\Omega)$, and its generalized derivatives are given by*

$$D_i(f \circ u) = (f' \circ u) D_i u, \quad i = 1, ..., N.$$

The generalized derivative of the following special functions are frequently used in chapters 5, 6, and 7.

Examples. *Let $1 \leq p < \infty$ and $u \in W^{1,p}(\Omega)$. Then $u^+ := \max(u,0)$, $u^- := \min(u,0)$, and $|u|$ are in $W^{1,p}(\Omega)$, and their generalized derivatives are given by*

$$(D_i u^+)(x) = \begin{cases} D_i u(x) & \text{if} \quad u(x) > 0, \\ 0 & \text{if} \quad u(x) \leq 0; \end{cases}$$

$$(D_i u^-)(x) = \begin{cases} 0 & \text{if} \quad u(x) \geq 0, \\ D_i u(x) & \text{if} \quad u(x) < 0; \end{cases}$$

$$(D_i |u|)(x) = \begin{cases} D_i u(x) & \text{if} \quad u(x) > 0, \\ 0 & \text{if} \quad u(x) = 0, \\ -D_i u(x) & \text{if} \quad u(x) < 0. \end{cases}$$

Generalized chain rule. *Let $f : \mathbb{R} \to \mathbb{R}$ be continuous and piecewise continuously differentiable with $\sup_{s \in \mathbb{R}} |f'(s)| < \infty$, and $u \in W^{1,p}(\Omega)$, $1 \leq p < \infty$. Then $f \circ u \in W^{1,p}(\Omega)$, and its generalized derivative is given by*

$$D_i(f \circ u)(x) = \begin{cases} f'(u(x)) D_i u(x) & \text{if } f \text{ is differentiable at } u(x), \\ 0 & \text{otherwise.} \end{cases}$$

The chain rule may further be extended to Lipschitz continuous f; see, e.g., [117, 218].

Lattice structure. *Let $u,\, v \in W^{1,p}(\Omega)$, $1 \leq p < \infty$. Then $\max(u,v)$ and $\min(u,v)$ are in $W^{1,p}(\Omega)$ with generalized derivatives*

$$D_i \max(u,v)(x) = \begin{cases} D_i u(x) & \text{if} \quad u(x) \geq v(x), \\ D_i v(x) & \text{if} \quad v(x) \geq u(x); \end{cases}$$

$$D_i \min(u,v)(x) = \begin{cases} D_i u(x) & \text{if} \quad u(x) \leq v(x), \\ D_i v(x) & \text{if} \quad v(x) \leq u(x). \end{cases}$$

Proof. The assertion follows easily from the above examples and the generalized chain rule by using $\max(u,v) = (u-v)^+ + v$ and $\min(u,v) = u - (u-v)^+$; see, e.g., [147, Theorem 1.20] □

Lemma C.4.1. *If u_j, $v_j \in W^{1,p}(\Omega)$ $(1 \leq p < \infty)$ are such that $u_j \to u$ and $v_j \to v$ in $W^{1,p}(\Omega)$, then $\min(u_j, v_j) \to \min(u,v)$ and $\max(u_j, v_j) \to \max(u,v)$ in $W^{1,p}(\Omega)$ as $j \to \infty$.*

For the proof see, e.g., [147, Lemma 1.22]. By means of Lemma C.4.1 we readily obtain the following result.

Lemma C.4.2. *Let $\underline{u}$, $\bar{u} \in W^{1,p}(\Omega)$ satisfy $\underline{u} \leq \bar{u}$, and let T be the truncation operator defined by*

$$Tu(x) = \begin{cases} \bar{u}(x) & \text{if} & u(x) > \bar{u}(x), \\ u(x) & \text{if} & \underline{u}(x) \leq u(x) \leq \bar{u}(x), \\ \underline{u}(x) & \text{if} & u(x) < \underline{u}(x). \end{cases}$$

Then T is a bounded continuous mapping from $W^{1,p}(\Omega)$ (respectively, $L^p(\Omega)$) into itself.

Proof. The assertion follows from Lemma C.4.1 and the following representation

$$Tu = \max(u, \underline{u}) + \min(u, \bar{u}) - u.$$

$\square$

Lemma C.4.3. *If u, $v \in W_0^{1,p}(\Omega)$, then $\max(u,v)$ and $\min(u,v)$ are in $W_0^{1,p}(\Omega)$.*

Lemma C.4.3 implies that $W_0^{1,p}(\Omega)$ has lattice structure; see, e.g., [147]. A partial ordering of traces on $\partial\Omega$ is given as follows.

Definition C.4.1. *Let $u \in W^{1,p}(\Omega)$, $1 \leq p < \infty$. Then $u \leq 0$ on $\partial\Omega$ if $u^+ \in W_0^{1,p}(\Omega)$.*

D Pseudomonotone and quasilinear elliptic operators

D.1 Main theorem on pseudomonotone operators. Let X be a real *reflexive* Banach space, X^* its *dual space* and denote by $\langle \cdot, \cdot \rangle$ the *duality pairing* between them. The norm convergence in X and X^* is denoted by $\rightarrow$ and the weak convergence by $\rightharpoonup$.

Definition D.1.1. *Let $A : X \rightarrow X^*$; then A is called*

(i) *continuous (weakly continuous) iff $u_n \rightarrow u$ implies $Au_n \rightarrow Au$ ($u_n \rightharpoonup u$ implies $Au_n \rightharpoonup Au$;*

(ii) *demicontinuous iff $u_n \rightarrow u$ implies $Au_n \rightharpoonup Au$;*

(iii) *hemicontinuous iff the real function $t \rightarrow \langle A(u+tv), w \rangle$ is continuous on $[0,1]$ for all u, v, $w \in X$;*

(iv) *strongly continuous or completely continuous iff $u_n \rightharpoonup u$ implies $Au_n \rightarrow Au$;*

(v) *bounded iff A maps bounded sets into bounded sets;*

(vi) *coercive iff $\lim_{\|u\| \rightarrow \infty} \frac{\langle Au, u \rangle}{\|u\|} = \infty$;*

(vii) *monotone (strictly monotone)* iff $\langle Au - Av, u - v \rangle \geq (>) 0$ for all $u, v \in X$;

(viii) *strongly monotone* iff there is a constant $c > 0$ such that
$\langle Au - Av, u - v \rangle \geq c\|u - v\|^2$ for all $u, v \in X$;

(ix) *uniformly monotone* iff $\langle Au - Av, u - v \rangle \geq a(\|u - v\|)\|u - v\|$ for all $u, v \in X$ where $a : [0, \infty) \to [0, \infty)$ is strictly increasing with $a(0) = 0$ and $a(s) \to +\infty$ as $s \to \infty$;

(x) *pseudomonotone* iff $u_n \rightharpoonup u$ and $\limsup_{n \to \infty} \langle Au_n, u_n - u \rangle \leq 0$ implies $\langle Au, u - w \rangle \leq \liminf_{n \to \infty} \langle Au_n, u_n - w \rangle$ for all $w \in X$;

(xi) to satisfy *(S_+)-condition* iff $u_n \rightharpoonup u$ and $\limsup_{n \to \infty} \langle Au_n, u_n - u \rangle \leq 0$ imply $u_n \to u$.

One can show (cf., e.g., [21]) that the pseudomonotonicity according to (x) is equivalent with the following definition.

Definition D.1.2. $A : X \to X^*$ is *pseudomonotone* iff $u_n \rightharpoonup u$ and $\limsup_{n \to \infty} \langle Au_n, u_n - u \rangle \leq 0$ implies $Au_n \rightharpoonup Au$ and $\langle Au_n, u_n \rangle \to \langle Au, u \rangle$.

For the following result see, e.g., [218, Proposition 27.6].

Lemma D.1.1. *Let $A, B : X \to X^*$ be operators on the real reflexive Banach space. Then*

(i) *If A is monotone and hemicontinuous, then A is pseudomonotone.*

(ii) *If A is strongly continuous, then A is pseudomonotone.*

(iii) *If A and B are pseudomonotone, then $A + B$ is pseudomonotone.*

The main theorem on pseudomonotone operators which is due to Brézis is given by (see [218, Theorem 27.A]):

Theorem D.1.1. *Let X be a real, separable, and reflexive Banach space, and let $A : X \to X^*$ be a pseudomonotone, bounded, and coercive operator, and $b \in X^*$. Then there exists a solution of the equation $Au = b$.*

D.2 Quasilinear elliptic operators.

Lemma D.2.1. *Let $\Omega \subset \mathbb{R}^N$, $N \geq 1$, be nonempty measurable set and let $f : \Omega \times \mathbb{R}^m \to \mathbb{R}$, $m \geq 1$, be a Carathéodory function; i.e., $x \to f(x, s)$ is measurable on Ω for all $s \in \mathbb{R}^m$ and $s \to f(x, s)$ is continuous on $\mathbb{R}^m$ for a.e. $x \in \Omega$. If f satisfies a growth condition in the form*

$$|f(x, s)| \leq k(x) + c \sum_{i=1}^{m} |s_i|^{p_i/q}$$

for some positive constant c and some $k \in L^q(\Omega)$, and $1 \leq q, p_i < \infty$ for all i, then the Nemytskij operator F defined by $Fu(x) := f(x, u_1(x), ..., u_m(x))$ is continuous and bounded from $L^{p_1}(\Omega) \times \cdots \times L^{p_m}(\Omega)$ into $L^q(\Omega)$. Furthermore,

$$\|Fu\|_{L^q(\Omega)} \leq const \left(\|k\|_{L^q(\Omega)} + \sum_{i=1}^{m} \|u_i\|_{L^{p_i}(\Omega)}^{p_i/q} \right).$$

Definition D.1.3. Let $\Omega \subseteq \mathbb{R}^N$, $N \geq 1$, be a nonempty measurable set. A function $f : \Omega \times \mathbb{R}^m \to \mathbb{R}$, $m \geq 1$, is called *superpositionally measurable (sup-measurable)* if the function $x \to f(x, u_1(x), ..., u_m(x))$ is measurable in Ω whenever the functions $u_i : \Omega \to \mathbb{R}$ are measurable.

Now let $\Omega \subset \mathbb{R}^N$ be a bounded domain with Lipschitz boundary $\partial\Omega$, let A_1 be the second order quasilinear differential operator in divergence form given by

$$A_1 u(x) = -\sum_{i=1}^{N} \frac{\partial}{\partial x_i} a_i(x, u(x), \nabla u(x)),$$

and let A_0 denote the operator generated by the lower order terms

$$A_0 u(x) = a_0(x, u(x), \nabla u(x)).$$

Let $1 < p < \infty$, $1/p + 1/q = 1$, and assume for the coefficients $a_i : \Omega \times \mathbb{R} \times \mathbb{R}^N \to \mathbb{R}$, $i = 0, 1, ..., N$ the following conditions:

(H1) Each $a_i(x, s, \xi)$ satisfies Carathéodory conditions, i.e., is measurable in $x \in \Omega$ for all $(s, \xi) \in \mathbb{R} \times \mathbb{R}^N$ and continuous in (s, ξ) for a.e. $x \in \Omega$. There exist a constant $c_0 > 0$ and a function $k_0 \in L^q(\Omega)$ so that

$$|a_i(x, s, \xi)| \leq k_0(x) + c_0(|s|^{p-1} + |\xi|^{p-1})$$

for a.e. $x \in \Omega$ and for all $(s, \xi) \in \mathbb{R} \times \mathbb{R}^N$.

(H2) $\sum_{i=1}^{N}(a_i(x, s, \xi) - a_i(x, s, \xi'))(\xi_i - \xi_i') > 0$ for a.e. $x \in \Omega$, for all $s \in \mathbb{R}$, and for all $\xi, \xi' \in \mathbb{R}^N$ with $\xi \neq \xi'$.

(H3) $\sum_{i=1}^{N} a_i(x, s, \xi)\xi_i \geq \nu|\xi|^p - k(x)$ for a.e. $x \in \Omega$, for all $s \in \mathbb{R}$, and for all $\xi \in \mathbb{R}^N$ with some constant $\nu > 0$ and some function $k \in L^1(\Omega)$.

Let $V_0 = W_0^{1,p}(\Omega)$ and denote by V_0^* its dual space. Under condition (H1) the differential operators A_1 and A_0 generate mappings from V_0 to V_0^* (again denoted by A_1 and A_0, respectively) defined by

$$\langle A_1 u, \varphi \rangle = \sum_{i=1}^{N} \int_{\Omega} a_i(x, u, \nabla u) \frac{\partial \varphi}{\partial x_i} \, dx, \quad \langle A_0 u, \varphi \rangle = \int_{\Omega} a_0(x, u, \nabla u) \varphi \, dx,$$

where $\langle \cdot, \cdot \rangle$ denotes the duality pairing between V_0^* and V_0. Let $A = A_1 + A_0$, then we have the following result.

Theorem D.2.1.

(i) *If (H1) is satisfied, then the mappings* $A, A_1, A_0 : V_0 \to V_0^*$ *are continuous and bounded;*

(ii) *If (H1) and (H2) are satisfied, then* $A : V_0 \to V_0^*$ *is pseudomonotone;*

(iii) *If (H1), (H2), and (H3) are satisfied, then A has the (S_+)-property.*

Conditions (H1) and (H2) are the so-called Leray-Lions conditions which guarantee that A is pseudomonotone. In their original paper Leray and Lions [173] showed the pseudomonotonicity under conditions (H1), (H2), and the following additional condition

(H4) $\limsup_{|\xi| \to \infty, \, s \in B} \sum_{i=1}^{N} \frac{a_i(x,s,\xi)\xi_i}{|\xi| + |\xi|^{p-1}} = +\infty.$
for a.e. $x \in \Omega$ and all bounded sets B.

However, Landes and Mustonen have shown in [169] that condition (H4) is redundant for the pseudomonotonicity of A. As for the proof of the results stated in Theorem D.2.1 as well as on existence theorems involving pseudomonotone operators we refer to [20, 21] and [28, 29, 118, 174, 208, 218].

If V is a closed subspace of $W^{1,p}(\Omega)$ such that $W_0^{1,p}(\Omega) \subseteq V \subseteq W^{1,p}(\Omega)$, then the result of Theorem D.2.1 holds true for the mappings $A, A_1, A_0 : V \to V^*$.

E   Nonlinear first order evolution equations

E.1 Lebesgue spaces of vector-valued functions. Let X be a Banach space with the norm $\| \cdot \|$, X^* its dual space, and $0 < \tau < \infty$. We consider functions $u : [0, \tau] \to X$ and develop some calculus notions such as measurability and integrability. Most of the material of this subsection can be found, e.g., in [112, 208, 218].

Definition E.1.1.

(i) A function $s : [0, \tau] \to X$ is called *simple* (or *step function*) if it is of the form

$$s(t) = \sum_{i=1}^{m} \chi_{E_i}(t) u_i, \quad 0 \le t \le \tau,$$

where each E_i is a Lebesgue measurable subset of the interval $[0, \tau]$, $u_i \in X$ $(i = 1, ..., m)$ and χ_{E_i} is the characteristic function of E_i.

(ii) $u : [0, \tau] \to X$ is *strongly measurable* if there exist simple functions $s_k : [0, \tau] \to X$ such that $s_k(t) \to u(t)$ for a.e. $t \in [0, \tau]$.

(iii) $u : [0, \tau] \to X$ is *weakly measurable* if for each $u^* \in X^*$ the mapping $t \to \langle u^*, u(t) \rangle$ is Lebesgue measurable.

(iv) $u : [0, \tau] \to X$ is *almost separably valued* if there exists a subset $N \subset [0, \tau]$ of zero measure such that the set $\{u(t) \mid t \in [0, \tau] \setminus N\}$ is a separable subset of X.

Theorem E.1.1 (Pettis). *The function $u : [0, \tau] \to X$ is strongly measurable if and only if u is weakly measurable and almost separably valued.*

Definition E.1.2.

(i) The *integral of the simple function* $s(t) = \sum_{i=1}^{m} \chi_{E_i}(t) u_i$ is defined by

$$\int_0^\tau s(t) \, dt := \sum_{i=1}^{m} \text{meas}\,(E_i)\, u_i.$$

(ii) $u : [0, \tau] \to X$ is *integrable* if there exists a sequence $(s_k)_{k=1}^{\infty}$ of simple functions such that

$$\int_0^\tau \|s_k(t) - u(t)\| \, dt \to 0 \text{ as } k \to \infty.$$

(iii) If u is integrable, its *integral* is defined by

$$\int_0^\tau u(t) \, dt = \lim_{k \to \infty} \int_0^\tau s_k(t) \, dt.$$

Theorem E.1.2. *The function $u : [0, \tau] \to X$ is integrable if and only if u is strongly measurable and $t \to \|u(t)\|$ is integrable, and one has*

$$\left\| \int_0^\tau u(t) \, dt \right\| \le \int_0^\tau \|u(t)\| \, dt, \quad \text{and} \quad \left\langle u^*, \int_0^\tau u(t) \, dt \right\rangle = \int_0^\tau \langle u^*, u(t) \rangle \, dt,$$

for each $u^* \in X^*$.

Definition E.1.3. Let $1 \le p \le \infty$. We denote by $L^p(0, \tau; X)$ the space of (equivalent classes of) measurable functions $u : [0, \tau] \to X$ such that $\|u(\cdot)\|$ belongs to $L^p(0, \tau; \mathbb{R})$ with

$$\|u\|_{L^p(0,\tau;X)} := \left(\int_0^\tau \|u(t)\|^p \, dt \right)^{1/p} \quad \text{for } 1 \le p < \infty,$$

$$\|u\|_{L^\infty(0,\tau;X)} := \operatorname*{ess\ sup}_{0 \leq t \leq \tau} \|u(t)\| < \infty.$$

The space $C([0,\tau];X)$ is comprised of all continuous functions $u : [0,\tau] \to X$ with

$$\|u\|_{C([0,\tau];X)} := \max_{0 \leq t \leq \tau} \|u(t)\| < \infty.$$

Theorem E.1.3. *Let X and Y be Banach spaces. Then we have the following results:*

(i) *$L^p(0,\tau;X)$ with $1 \leq p \leq \infty$ and the norm given by Definition E.1.3 is a Banach space.*

(ii) *$C([0,\tau];X)$ is dense in $L^p(0,\tau;X)$ for $1 \leq p < \infty$, and the embedding $C([0,\tau];X) \subset L^p(0,\tau;X)$ is continuous.*

(iii) *If X is a Hilbert space with scalar product $(\cdot,\cdot)_X$, then $L^2(0,\tau;X)$ is also an Hilbert space with the scalar product*

$$(u,v) = \int_0^\tau (u(t),v(t))_X \, dt.$$

(iv) *$L^p(0,\tau;X)$ is separable in the case where X is separable and $1 \leq p < \infty$.*

(v) *If the embedding $X \subseteq Y$ is continuous, then the embedding*

$$L^r(0,\tau;X) \subseteq L^q(0,\tau;Y), \quad 1 \leq q \leq r \leq \infty,$$

is also continuous.

(vi) *Let X be a reflexive and separable Banach space and let $1 < p < \infty$, $1/p + 1/q = 1$. Then $\mathcal{V} := L^p(0,\tau;X)$ is also reflexive and separable, and its dual space $\mathcal{V}^*$ is norm-isomorphic to $L^q(0,\tau;X^*)$. Therefore $\mathcal{V}^*$ and $L^q(0,\tau;X^*)$ may be identified. The duality pairing $\langle\cdot,\cdot\rangle_{\mathcal{V}}$ between $\mathcal{V}$ and its dual $\mathcal{V}^*$ can be written as*

$$\langle v,u\rangle_{\mathcal{V}} = \int_0^\tau \langle v(t),u(t)\rangle_X \, dt \quad \text{for all } u \in \mathcal{V}, \ v \in \mathcal{V}^*.$$

E.2 Evolution triples and generalized derivatives. The material of this subsection is mainly taken from [208, 218].

Definition E.2.1. We call $V \subseteq H \subseteq V^*$ an *evolution triple* if

(i) V is a real, separable, and reflexive Banach space, H is a real, separable Hilbert space;

(ii) the embedding $V \subseteq H$ is continuous, and V is dense in H;

(iii) identifying H with its dual H^* by the Riesz map, we then have $H \subseteq V^*$ with the equation

$$\langle h, v \rangle_V = (h, v) \quad \text{for } h \in H \subseteq V^*, \ v \in V.$$

Since V is reflexive and dense in H, the space H^* is dense in V^*, and hence H is dense in V^*.

Definition E.2.2. Let Y, Z be Banach spaces, and $u \in L^1(0, \tau; Y)$ and $w \in L^1(0, \tau; Z)$. Then, the function w is called the *generalized derivative* of the function u in $(0, \tau)$ iff the following relation holds:

$$\int_0^\tau \varphi'(t) u(t) \ dt = - \int_0^\tau \varphi(t) w(t) \ dt \quad \text{for all } \varphi \in C_0^\infty(0, \tau).$$

We write $w = u'$.

Theorem E.2.1. *Let $V \subseteq H \subseteq V^*$ be an evolution triple and let $1 \leq p, q \leq \infty$, $0 < \tau < \infty$. Let $u \in L^p(0, \tau; V)$; then there exists the generalized derivative $u' \in L^q(0, \tau; V^*)$ iff there is a function $w \in L^q(0, \tau; V^*)$ such that*

$$\int_0^\tau (u(t), v)_H \varphi'(t) \ dt = - \int_0^\tau \langle w(t), v \rangle_V \varphi(t) \ dt$$

for all $v \in V$ and all $\varphi \in C_0^\infty(0, \tau)$. The generalized derivative u' is uniquely defined.

Theorem E.2.2 (Lions–Aubin). *Let B_0, B, B_1 be reflexive Banach spaces with $B_0 \subseteq B \subseteq B_1$, and assume $B_0 \subseteq B$ is compactly and $B \subseteq B_1$ is continuously embedded. Let $1 < p < \infty$, $1 < q < \infty$ and define the following space*

$$W := \{ u \in L^p(0, \tau; B_0) \mid u' \in L^q(0, \tau; B_1) \}.$$

Then $W \subseteq L^p(0, \tau; B)$ is compactly embedded.

Definition E.2.3. Let V be a real, separable, and reflexive Banach space, and let $\mathcal{V} = L^p(0, \tau; V)$, $1 < p < \infty$. The Sobolev space $\mathcal{W}$ is defined by

$$\mathcal{W} = \{ u \in \mathcal{V} \mid u' \in \mathcal{V}^* \},$$

where u' is the generalized derivative, and $\mathcal{V}^* = L^q(0, \tau; V^*)$, $1/p + 1/q = 1$.

Theorem E.2.3. *Let $V \subseteq H \subseteq V^*$ be an evolution triple, and let $1 < p < \infty$, $1/p + 1/q = 1$, $0 < \tau < \infty$. Then the following hold:*

(i) *The space $\mathcal{W}$ defined in Definition E.2.3 forms a real Banach space with the norm*

$$\|u\|_{\mathcal{W}} = \|u\|_{\mathcal{V}} + \|u'\|_{\mathcal{V}^*}.$$

(ii) *The embedding $\mathcal{W} \subseteq C([0, \tau]; H)$ is continuous.*

(iii) *For all $u, v \in \mathcal{W}$ and arbitrary t, s with $0 \le s \le t \le \tau$, the following generalized integration by parts formula holds:*

$$(u(t), v(t))_H - (u(s), v(s))_H = \int_s^t \langle u'(\zeta), v(\zeta) \rangle_V + \langle v'(\zeta), u(\zeta) \rangle_V \, d\zeta.$$

E.3 Nonlinear evolution equation. The material of this subsection is mainly based on results obtained in [15, 22, 23, 208]; see also [174, 218].

Let $V \subseteq H \subseteq V^*$ be an evolution triple, and let $1 < p < \infty$, $1/p + 1/q = 1$, $0 < \tau < \infty$, and let $\mathcal{V}, \mathcal{V}^*$, and $\mathcal{W}$ be the spaces as defined in Definition E.2.3. We provide an existence result for the evolution equation

$$u'(t) + A(t)u(t) = f(t), \quad 0 < t < \tau, \quad u(0) = 0, \qquad (\text{E.3.1})$$

where $A(t) : V \to V^*$ and $f \in \mathcal{V}^*$. Without loss of generality we may assume homogeneous initial values (otherwise they can be obtained by translation). The generalized derivative defines an operator $L := d/dt$ from the subset $D(L) = \{v \in \mathcal{V} \mid v' \in \mathcal{V}^* \text{ and } v(0) = 0\}$ to $\mathcal{V}^*$ by

$$\langle Lu, v \rangle_{\mathcal{V}} = \int_0^\tau \langle u'(t), v(t) \rangle_V, \quad \text{for all } v \in \mathcal{V}.$$

One can show that $L : D(L) \to \mathcal{V}^*$ is a densely defined, closed linear maximal monotone operator; cf. [218]. Let us introduce the following conditions on the time-dependent operators $A(t) : V \to V^*$:

(H1) $\|A(t)u\|_{V^*} \le c_0 \left(\|u\|_V^{p-1} + k_0(t) \right)$ for all $u \in V$ and $t \in [0, \tau]$ with some positive constant c_0 and $k_0 \in L^q(0, \tau)$.

(H2) $A(t) : V \to V^*$ is demicontinuous for each $t \in [0, \tau]$.

(H3) The function $t \to \langle A(t)u, v \rangle$ is measurable on $(0, \tau)$ for all $u, v \in V$.

(H4) $\langle A(t)u, u \rangle \ge c_1(\|u\|_V^p - k_1(t))$ for all $u \in V$ and $t \in [0, \tau]$ with some constant $c_1 > 0$ and some function $k_1 \in L^1(0, \tau)$.

Define an operator $\hat{A}$ related with $A(t)$ by

$$\hat{A}(u)(t) := A(t)u(t), \quad t \in [0,\tau], \tag{E.3.2}$$

which may be considered as the associated Nemytskij operator. Then by (H1) and (H3) one readily verifies that $\hat{A} : V \to V^*$ is bounded. Thus problem (E.3.1) corresponds to the following one

$$Lu + \hat{A}(u) = f, \quad u \in D(L). \tag{E.3.3}$$

Definition E.3.1. Let $D(L)$ be equipped with the graph norm; that is, $\|u\|_{D(L)} = \|u\|_V + \|Lu\|_{V^*}$. The operator $\hat{A} : V \to V^*$ is called *pseudomonotone with respect to the graph norm topology of $D(L)$*, if for any sequence $(u_n) \in D(L)$ satisfying $u_n \rightharpoonup u$, in V, $Lu_n \rightharpoonup Lu$ in V^*, and $\limsup_{n\to\infty}\langle\hat{A}(u_n), u_n - u\rangle_V \leq 0$, it follows that $\hat{A}(u_n) \rightharpoonup \hat{A}(u)$ in V^* and $\langle\hat{A}(u_n), u_n\rangle_V \to \langle\hat{A}(u), u\rangle_V$. In an obvious similar way the *(S_+)-condition with respect to $D(L)$* is defined.

For the following surjectivity result which yields the existence of problem (E.3.3) we refer to [22, 174].

Theorem E.3.1. *Let $L : D(L) \subset V \to V^*$ be as given above, and let $\hat{A} : V \to V^*$ defined by (E.3.2) be bounded demicontinuous and pseudomonotone with respect to the graph norm topology of $D(L)$. If $\hat{A}$ is coercive, then $(L + \hat{A})(D(L)) = V^*$, that is, $L + \hat{A}$ is surjective.*

The next result shows that under certain conditions the Nemytskij operator $\hat{A}$ and the operators $A(t)$ related with it are in some sense of the same type; cf. [23].

Theorem E.3.2. *Let hypotheses (H1), (H2), (H3), and (H4) be satisfied. Then we have the following:*

(i) *If $A(t) : V \to V^*$ is pseudomonotone for all $t \in [0,\tau]$, then $\hat{A} : V \to V^*$ is pseudomonotone with respect to $D(L)$ according to Definition E.3.1.*

(ii) *If $A(t) : V \to V^*$ has the (S_+)-property for all $t \in [0,\tau]$, then $\hat{A} : V \to V^*$ has the (S_+)-property with respect to $D(L)$.*

(iii) *Hypotheses (H1), (H2), and (H3) imply $\hat{A} : V \to V^*$ is demicontinuous.*

(iv) *Hypothesis (H4) implies that $\hat{A} : V \to V^*$ is coercive.*

As an immediate consequence of Theorem E.3.1 and Theorem E.3.2 we get the following corollary.

Corollary E.3.1. *Let $A(t) : V \to V^*$ be pseudomonotone for all $t \in [0, \tau]$ satisfying hypotheses (H1)–(H4); then problem (E.3.3) has a solution $u \in D(L)$ for any given $f \in V^*$.*

By means of Theorem E.3.1, Theorem E.3.2, and Corollary E.3.1 one easily can get existence results for quasilinear parabolic initial boundary value problems by employing the results obtained in section C for quasilinear elliptic operators.

F Nonsmooth analysis

In this section we provide some basic facts of nonsmooth analysis, which can be found, e.g., in the monographs [18, 96, 110, 208, 219].

F.1 Subdifferential of convex functions. Let X be a Banach space and $\varphi : X \to (-\infty, +\infty]$ an extended real-valued function. Then φ is *convex* if

$$\varphi(tu + (1-t)v) \leq t\varphi(u) + (1-t)\varphi(v), \quad u, v \in X, \quad 0 \leq t \leq 1.$$

The function φ is called *proper* if $\varphi(u) < \infty$ for some $u \in X$, and its *effective domain* is $\mathrm{dom}(\varphi) = \{u \in X \mid \varphi(u) < \infty\}$. Let Y be a subset of X. A function $\varphi : Y \to \mathbb{R}$ is said to satisfy a *Lipschitz condition* on Y provided that for some positive constant K, one has

$$|\varphi(y) - \varphi(y')| \leq K\|y - y'\| \quad \text{for all } y, y' \in Y.$$

The function φ is said to be *Lipschitz near x* if there exists an ε−neighborhood of x and a constant $k > 0$ depending on this neighborhood such that the following holds:

$$|\varphi(x'') - \varphi(x')| \leq k\|x'' - x'\| \quad \text{for all } x'', x' \in x + \varepsilon B,$$

where B denotes the open unit ball in X. We call $\varphi : Y \to \mathbb{R}$ *locally Lipschitz*, if for every $y \in Y$ the function φ is Lipschitz near y.

Lemma F.1.1. *([110]) If the convex function $\varphi : X \to (-\infty, +\infty]$ is bounded from above on a neighborhood of some point, then φ is continuous at each point of the interior of $\mathrm{dom}(\varphi)$. Moreover, φ is locally Lipschitz in the interior of $\mathrm{dom}(\varphi)$.*

Corollary F.1.1. *Let $U \subseteq X$ be an open convex subset and let $\varphi : U \to \mathbb{R}$ be convex. If φ is bounded above on a neighborhood of some point of U, then φ is locally Lipschitz in U.*

Definition F.1.1. Let $\varphi : X \to (-\infty, +\infty]$ be convex and proper. The *subdifferential* of φ at $u \in X$ is the set of all functionals $u^* \in X^*$ such that

$$\langle u^*, v - u \rangle \leq \varphi(v) - \varphi(u) \quad \text{for all } v \in X,$$

and is denoted by $\partial\varphi(u)$. Each such $u^* \in \partial\varphi(u)$ is called *subgradient* of φ at u.

The set $\partial\varphi(u)$ may be empty which happens, e.g., if $\varphi(u) = +\infty$.

The following result characterizes the subdifferential of convex integral functionals.

Lemma F.1.2. *Let* $\varphi : \mathbb{R} \to (-\infty, +\infty]$ *be proper, convex, lower semicontinuous, and either* $0 = \varphi(0) = \min(\varphi)$ *or the measurable set* $\Omega \subset \mathbb{R}^N$ *has finite measure. Define* $\Phi : L^p(\Omega) \to (-\infty, +\infty]$, $1 \leq p < \infty$, *by*

$$\Phi(u) = \int_\Omega \varphi(u(x))\, dx \quad \text{if } \varphi(u) \in L^1(\Omega), \quad +\infty \text{ otherwise.}$$

Then Φ *is proper, convex, lower semicontinuous, and* $f \in \partial\Phi(u)$ *if and only if* $f \in L^q(\Omega)$, $(1/p + 1/q = 1)$, $u \in L^p(\Omega)$, *and* $f(x) \in \partial\varphi(u(x))$ *for a.e.* $x \in \Omega$.

F.2 Maximal monotone operators.

Definition F.2.1. Let X be a real Banach space and let $A : X \to 2^{X^*}$ be a multivalued mapping; i.e., to each $u \in X$ there is assigned a subset $A(u)$ of X^*. The graph of A denoted by $G(A)$ consists of all $(u, u^*) \in X \times X^*$ such that $u^* \in A(u)$. The mapping A is called

  (i) *monotone* if and only if
     $\langle u^* - v^*, u - v \rangle \geq 0$ for all $(u, u^*), (v, v^*) \in G(A)$.
 (ii) *maximal monotone* if and only if A is monotone and there is no monotone mapping $A_1 : X \to 2^{X^*}$ such that $G(A) \subset G(A_1)$.

Special case: If $X = \mathbb{R}$, then a maximal monotone mapping $\beta : \mathbb{R} \to 2^\mathbb{R}$ is called *maximal monotone graph in* $\mathbb{R}^2$. For example, an increasing function $f : \mathbb{R} \to \mathbb{R}$ generates a maximal monotone graph β in $\mathbb{R}^2$ given by $\beta(s) := [f(s - 0), f(s + 0)]$, where $f(s \pm 0)$ are the one-sided limits of f in s.

Lemma F.2.1. *Let* $\varphi : X \to (-\infty, +\infty]$ *be proper, convex, and lower semicontinuous. Then the subdifferential* $\partial\varphi : X \to 2^{X^*}$ *is maximal monotone.*

F.3   Generalized gradient. Let X be real Banach space, X^* its dual space and let $\Phi : X \to \mathbb{R}$ be a locally Lipschitz functional.

Definition F.3.1. The *generalized directional derivative* of Φ at $u \in X$ in the direction $v \in X$, denoted $\Phi^o(u; v)$, is defined as follows:

$$\Phi^o(u; v) = \limsup_{y \to u,\ t \downarrow 0} \frac{\Phi(y + tv) - \Phi(y)}{t},$$

where t is a positive real.

It is known that the function $v \to \Phi^o(u; v)$ is finite, convex, positively homogeneous, and subadditive on X, and satisfies $|\Phi^o(u; v)| \leq k(\mathcal{U}) \, |v|$, where the positive constant $k(\mathcal{U})$ depends only on a neighborhood $\mathcal{U}$ of u. By means of the generalized directional derivative Clarke's *generalized gradient* of Φ at $u \in X$ is defined as follows:

Definition F.3.2. The *generalized gradient* of Φ at $u \in X$ denoted $\partial \Phi(u)$, is defined as the subset of X^* given by

$$\partial \Phi(u) = \{\zeta \in X^* \mid \Phi^o(u; v) \geq \langle \zeta, v \rangle \text{ for all } v \in X\}.$$

Since $v \to \Phi^o(u; v)$ is convex on X and satisfies $\Phi^o(u; 0) = 0$, the generalized gradient $\partial \Phi(u)$ is nothing but the subdifferential of the functional $v \to \Phi^o(u; v)$ at $v = 0$. The generalized gradient possesses the following properties:

(i) If $\Phi_k : X \to \mathbb{R}$, $k = 1, 2$, are locally Lipschitz then their sum is locally Lipschitz and satisfies for all $u \in X$
$\partial(\Phi_1 + \Phi_2)(u) \subseteq \partial\Phi_1(u) + \partial\Phi_2(u)$.

(ii) $\partial(t\Phi)(u) = t\partial\Phi(u)$ for any scalar $t \in \mathbb{R}$.

(iii) If $\Phi : X \to \mathbb{R}$ is convex and bounded above on a neighborhood of some point of X, then Φ is locally Lipschitz in X (cf. Corollary F.1.1) and the subdifferential $\partial\Phi(u)$ defined in F.1 coincides with the generalized gradient according to Definition F.3.2 of Φ at u. This motivates us to use the same notation $\partial\Phi$ for both the generalized gradient and the subdifferential as well.

The *one-sided directional derivative* of Φ at u in the direction v is given by

$$\Phi'(u; v) = \lim_{t \downarrow 0} \frac{\Phi(u + tv) - \Phi(u)}{t}.$$

For instance, convex and locally Lipschitz functionals are one-sided directional differentiable.

Definition F.3.3. The locally Lipschitz functional Φ is called *regular* at $u \in X$ if for all $v \in X$ the one-sided directional derivative $\Phi'(u; v)$ exists and satisfies

$$\Phi^o(u; v) = \Phi'(u; v) \quad \text{for all } v \in X.$$

Finally, the functional Φ is *strictly differentiable* at $u \in X$ if there is an element $\zeta \in X^*$ such that for each $v \in X$ one has

$$\lim_{u' \to u,\ t \downarrow 0} \frac{\Phi(u' + tv) - \Phi(u')}{t} = \langle \zeta, v \rangle.$$

We set $D_s \Phi(u) = \zeta$.

Lemma F.3.1. *If Φ is strictly differentiable at u, then Φ is Lipschitz near u and $\partial \Phi(u) = \{D_s \Phi(u)\}$. Conversely, if Φ is Lipschitz near u and $\partial \Phi(u)$ reduces to a singleton $\{\zeta\}$, then Φ is strictly differentiable at u and $D_s \Phi(u) = \zeta$.*

Concerning the regularity of locally Lipschitz functionals the following results hold:

(iv) If Φ is strictly differentiable at u then Φ is regular at u.

(v) If $\Phi : X \to \mathbb{R}$ is convex then it is regular.

(vi) A finite linear combination (with nonnegative scalars) of functionals regular at u is also regular at u.

Index

References

[1]   Adje, A., *Sur et sous-solutions généralisées et problèmes aux limites du second ordre*, Bull. Soc. Math. Bel. Sér. B **42** (1990), 347–368.

[2]   Adly, S., Buttazzo, G., and Thera, M., *Critical points for nonsmooth energy functions and applications*, Nonlinear Anal. **32** (1998), 711–718.

[3]   Afrouzi, G. A. and Brown, K. J., *Unbounded principal eigenfunctions for problems on all* $\mathbb{R}^N$, Proc. Amer. Math. Soc. (to appear).

[4]   Agarwal, R. P. and Heikkilä, S., *On solvability of first order discontinuous scalar initial and boundary value problems*, Australian J. Math. (to appear).

[5]   Agarwal, R. P. and Heikkilä, S., *Extremality results for discontinuous Sturm-Liouville boundary value problems*, Math. Nachr. (to appear).

[6]   Agarwal, R. P., Hong, H. L., and Yeh, C. C., *The existence of positive solutions of the Sturm-Liouville boundary value problems*, Comput. Math. Appl. **35**, 9 (1998), 89–96.

[7]   Agarwal, R. P., O'Regan, D., and Wong, P. J. Y., *Positive Solutions to Differential, Difference and Integral Equations*, Kluwer, Dordrecht, 1999.

[8]   Akô, K., *On the Dirichlet problem for quasilinear elliptic differential equations of second order*, J. Math. Soc. Japan **13** (1961), 45–62.

[9]   Allegretto, W., *Principal eigenvalues for indefinite-weight elliptic problems in* $\mathbb{R}^N$, Proc. Amer. Math. Soc. **116** (1992), 701–706.

[10]   Amann, H., *Existence and multiplicity theorems for semilinear and elliptic boundary value problems*, Math. Z. **150** (1976), 281–295.

[11]   Ambrosetti, A. and Turner, R. E. L., *Some discontinuous variational problems*, Differential Integral Equations **1** (1988), 341–349.

[12]   Appell, J. and Zabrejko, P. P., *Nonlinear Superposition Operators*, Cambridge University Press, New York, 1990.

[13]   Arendt, W. and Favini, A., *Integrated solutions to implicit differential equations*, Rend. Sem. Mat. Univ. Politec. Torino **51** (1993), 315–329.

[14]   Aris, R., *The Mathematical Theory of Diffusion and Reaction*, Clarendon Press, Oxford, 1975.

[15] Avgerinos, E. P. and Papageorgiou, N. S., *Solutions and periodic solutions for nonlinear evolution equations with nonmonotone perturbations*, Z. Anal. Anwendungen **17** (1998), 859–875.

[16] Bajic, V. B., *Nonlinear functions and stability of motions of implicit differential systems*, Internat. J. Control **52** (1990), 1167–1187.

[17] Barbu, V. and Favini, A., *Convergence of solutions of implicit differential equations*, Differential Integral Equations **7** (1994), 665–688.

[18] Barbu, V. and Precupanu, Th., *Convexity and Optimization in Banach Spaces*, Sijthoff and Noordhoff, Netherlands, Editura Academici, Bucuresti, Romania, 1978.

[19] Bebernes, J. W. and Schmitt, K., *On the existence of maximal and minimal solutions for parabolic partial differential equations*, Proc. Amer. Math. Soc. **73** (1979), 211–218.

[20] Berkovits, J. and Mustonen, V., *On the topological degree for mappings of monotone type*, Nonlinear Anal. **10** (1986), 1373–1383.

[21] Berkovits, J. and Mustonen, V., *Nonlinear mappings of monotone type*, Report, University of Oulu (1988), pp. 53.

[22] Berkovits, J. and Mustonen, V., *Topological degree for perturbations of linear maximal monotone mappings and applications to a class of parabolic problems*, Rend. Mat. Appl., Serie VII **12** (1992), 597–621.

[23] Berkovits, J. and Mustonen, V., *Monotonicity methods for nonlinear evolution equations*, Nonlinear Anal. **27** (1996), 1397–1405.

[24] Bernfeld, S. and Lakshmikantham, V., *An Introduction to Nonlinear Boundary Value Problems*, Academic Press, New York, 1974.

[25] Biles, D. C., *Continuous dependence of nonmonotonic discontinuous differential equations*, Trans. Amer. Math. Soc. **339**, 2 (1993), 507–524.

[26] Biles, D. C., *Existence of solutions for discontinuous differential equations*, Differential Integral Equations **8**, 6 (1995), 1525–1532.

[27] Bonafede, S. and Marano, S. A., *Implicit parabolic differential equations*, Bull. Austral. Math. Soc. **51** (1995), 501–509.

[28] Brezis, H., *Équations et inéquations non linéaires dans les espaces vectoriels en dualité*, Ann. Inst. Fourier **18** (1968), 115–175.

[29] Browder, F. E., *Fixed point theory and nonlinear problems*, Bull. Amer. Math. Soc. **9** (1983), 1–39.

[30] Brown, K. J. and Stavrakakis, N. M., *Sub- and supersolutions for semilinear elliptic equations on all of* $\mathbb{R}^n$, Differential Integral Equations **7** (1994), 1215–1225.

[31]   Brown, K. J. and Stavrakakis, N. M., *On the construction of super and subsolutions for elliptic equations on all of* $\mathbb{R}^N$, Nonlinear Anal. **32** (1998), 87–95.

[32]   Brull, L. and Pallaske, U., *On differential algebraic equations with discontinuities*, Z. AMP. **43** (1992), 319–327.

[33]   Cabada, A., *The monotone method for second order problems with linear and nonlinear boundary conditions*, Differential Equations Dynam. Systems, **2**, 1 (1994), 65–80.

[34]   Cabada, A., Habets, P., and Lois, S., *Monotone method for Neumann problem with lower and upper solutions in the reverse order*, Preprint.

[35]   Cabada, A., Habets, P., and Pouso, R. L., *Optimal existence conditions for φ-Laplacian equations with upper and lower solutions in the reversed order*, J. Differential Equations (to appear).

[36]   Cabada, A. and Lois, S., *Existence results for nonlinear problems with separated boundary conditions*, Nonlinear Anal. **35** (1997), 449–456.

[37]   Cabada, A. and Pouso, R. L., *Existence result for the problem* $(\phi(u'))' = f(t, u, u')$ *with periodic and Neumann boundary conditions*, Nonlinear Anal. **30**, 3 (1997), 1733–1742.

[38]   Cabada, A. and Pouso R. L., *Existence results for the problem* $(\phi(u'))' = f(t, u, u')$ *with nonlinear boundary conditions*, Nonlinear Anal. **35** (1999), 221–231.

[39]   Cabada, A. and Pouso, R. L., *On first order discontinuous scalar differential equations*, Nonlinear Stud. **6**, 2 (1999), 161–170.

[40]   Cabada, A. and Pouso, R. L., *Extremal solutions of strongly nonlinear discontinuous second order equations with nonlinear functional boundary conditions*, Nonlinear Anal. (to appear).

[41]   Cabada, A. and Pouso, R. L., *Basic existence theory for a general class of second order O.D.E.s that includes the p-Laplacian*, Manuscript (2000).

[42]   Campell, S. L. and Griepenhog, E., *Solvability of general differential algebraic equations*, SIAM J. Sci. Comput. **16** (1995), 257–270.

[43]   Cantor, G., *Beiträge zur Begründung der transfiniten Mengenlehre II*, Math. Ann. **49** (1897), 207–246.

[44]   Carathéodory, C., *Vorlesungen über Reelle Funktionen*, Teubner, Leipzig, 1918.

[45]   Carillo, J. and Chipot, M., *On some nonlinear elliptic equations involving derivatives of the nonlinearity*, Proc. Roy. Soc. Edinburgh Sect. A **100** (1985), 281–294.

[46]   Carl, S., *A monotone iterative scheme for nonlinear reaction-diffusion systems having nonmonotone reaction terms*, J. Math. Anal. Appl. **134** (1988), 81–93.

[47]   Carl, S., *The monotone iterative technique for a parabolic boundary value problem with discontinuous nonlinearity*, Nonlinear Anal. **13** (1989), 1399–1407.

[48]   Carl, S., *An enclosing theorem and a monotone iterative scheme for elliptic systems having nonmonotone nonlinearities*, Z. Angew. Math. Mech. **70** (1990), 309–313.

[49]   Carl, S., *A combined variational-monotone iterative method for elliptic boundary value problems with discontinuous nonlinearity*, Appl. Anal. **43** (1992), 21–45.

[50]   Carl, S., *On the existence of extremal weak solutions for a class of quasilinear parabolic problems*, Differential Integral Equations **6** (1993), 1493–1505.

[51]   Carl, S., *An existence result for a class of parabolic hemivariational inequalities*, Proc. Dynam. Systems Appl. **2** (1996), 91–98.

[52]   Carl, S., *Enclosure of solutions for quasilinear dynamic hemivariational inequalities*, Nonlinear World **3** (1996), 281–298.

[53]   Carl, S., *An envelope theorem for weak solutions of quasilinear elliptic boundary value problems,* in: World Congress of Nonlinear Analysts '92 (Edited by V. Lakshmikantham) pp. 869–877, Walter de Gruyter & Co., Berlin/New York (1996).

[54]   Carl, S., *Leray-Lions operators perturbed by state-dependent subdifferentials*, Nonlinear World **3** (1996), 505–518.

[55]   Carl, S., *Quasilinear elliptic equations with discontinuous nonlinearities in $\mathbb{R}^N$*, Proceedings WCNA-96, Nonlinear Anal. **30** (1997), 1743–1751.

[56]   Carl, S., *Existence of extremal periodic solutions for quasilinear parabolic equations*, Abstr. Appl. Anal. **2**, 3–4 (1997), 257–270.

[57]   Carl, S., *A survey of recent results on the enclosure and extremality of solutions for quasilinear hemivariational inequalities*, in: From Convexity to Nonconvexity, A special volume dedicated to the memory of Professor Gaetano Fichera, Eds. R. Gilbert, P.D. Panagiotopoulos and P.A. Pardalos, Kluwer Academic Publishers (to appear).

[58]   Carl, S., *Existence and comparison results for quasilinear parabolic inclusions with state-dependent subdifferentials*, Optimization (to appear).

[59] Carl, S., *Extremal solutions of hemivariational inequalities with d.c.-superpotentials*, Proceedings of the International Conference on Differential Equations and Nonlinear Mechanics, March 1999, Orlando, Florida, Kluwer Academic Publishers (to appear).

[60] Carl, S., *Extremal solutions for quasilinear elliptic inclusions in all of $\mathbb{R}^N$ with state-dependent subdifferentials*, J. Optim. Theory Appl. **104**, 2 (2000), 323–342.

[61] Carl, S., *Existence of extremal solutions of boundary hemivariational inequalities*, J. Differential Equations (to appear).

[62] Carl, S. and Dietrich, H., *The weak upper and lower solution method for quasilinear elliptic equations with generalized subdifferentiable perturbations*, Appl. Anal. **56** (1995), 263–278.

[63] Carl, S. and Grossmann, C., *Monotone enclosure for elliptic and parabolic systems with nonmonotone nonlinearities*, J. Math. Anal. Appl. **151** (1990), 190–202.

[64] Carl, S. and Grossmann, C., *Smoothing and monotone iterations for elliptic differential inclusions*, Appl. Math. Comput. **74** (1996), 15–35.

[65] Carl, S., Grossmann, C., and Pao, C. V., *Existence and monotone iterations for parabolic differential inclusions*, Comm. Appl. Nonlinear Anal. **3** (1996), 1–24.

[66] Carl, S. and Heikkilä, S., *On a parabolic boundary value problem with discontinuous nonlinearity*, Nonlinear Anal. **15** (1990), 1091–1095.

[67] Carl, S. and Heikkilä, S., *An existence result for elliptic differential inclusions with discontinuous nonlinearity*, Nonlinear Anal. **18**, 5 (1992), 471–479.

[68] Carl, S. and Heikkilä, S., *On extremal solutions of an elliptic boundary value problem involving discontinuous nonlinearities*, Differential Integral Equations **5**, 3 (1992), 581–589.

[69] Carl, S. and Heikkilä, S., *Extremal solutions of quasilinear parabolic boundary value problems with discontinuous nonlinearities*, Dynamic Systems Appl. **3** (1994), 251–258.

[70] Carl, S. and Heikkilä, S., *On the existence of the extremal solutions for discontinuous elliptic equations under discontinuous flux conditions*, Nonlinear Anal. **23**, 12 (1994), 1499–1506.

[71] Carl, S. and Heikkilä, S., *On a second order discontinuous implicit boundary value problem*, Nonlinear Stud., **4**, 2 (1997), 219–232.

[72] Carl, S. and Heikkilä, S., *Elliptic equations with discontinuous nonlinearities in $\mathbb{R}^N$*, Nonlinear Anal. **31**, 12 (1998), 217–227.

[73]   Carl, S. and Heikkilä, S., *On discontinuous implicit evolution equations*, J. Math. Anal. Appl. **219** (1998), 455–471.

[74]   Carl, S. and Heikkilä, S., *On a second order discontinuous implicit boundary value problems with discontinuous implicit boundary conditions*, Nonlinear Anal. **33**, 3 (1998), 261–279.

[75]   Carl, S. and Heikkilä, S., *A free boundary value problem for quasilinear elliptic equations in exterior domains*, Differential Integral Equations **11**, 3 (1998), 409–423.

[76]   Carl, S. and Heikkilä, S., *On discontinuous first order implicit boundary value problems*, J. Differential Equations **148** (1998), 100–121.

[77]   Carl, S. and Heikkilä, S., *On discontinuous implicit elliptic boundary value problems*, Differential Integral Equations **11**, 6 (1998), 823–834.

[78]   Carl, S. and Heikkilä, S., *On discontinuous implicit differential equations in ordered Banach spaces with discontinuous implicit boundary conditions*, Ann. Polon. Math. **71**, 1 (1999), 1–17.

[79]   Carl, S. and Heikkilä, S., *Existence of extremal periodic solutions for discontinuous quasilinear parabolic equations*, Dynam. Contin. Discrete Impuls. Systems **5** (1999), 485–496.

[80]   Carl, S. and Heikkilä, S., *Operator and differential equations in ordered spaces*, J. Math. Anal. Appl. **234** (1999), 31–54.

[81]   Carl, S. and Heikkilä, S., *Operator equations in ordered sets and discontinuous implicit parabolic equations*, Nonlinear Anal. (to appear).

[82]   Carl, S. and Heikkilä, S., *On discontinuous implicit and explicit abstract impulsive boundary value problems*, Nonlinear Anal. (to appear).

[83]   Carl, S. and Heikkilä, S., *Discontinuous reaction-diffusion equations under discontinuous and nonlocal flux conditions*, "Advanced Topics in Nonlinear Operator Theory," Math. Comput. Modelling (to appear).

[84]   Carl, S. and Heikkilä, S., *Extremality results for first order discontinuous functional differential equations*, Comput. Math. Appl. (to appear).

[85]   Carl, S. and Heikkilä, S., *Extremality results for discontinuous functional phi-Laplacian differential equations*, Int. J. Appl. Math. (to appear).

[86]   Carl, S. and Heikkilä, S., *Operator equations and implicit impulsive differential equations in ordered spaces*, Report, Martin-Luther-Universität, Halle-Wittenberg (1999).

[87]   Carl, S., Heikkilä, S., and Koponen, P., *On first order discontinuous implicit boundary value problems*, Dynam. Contin. Discrete Impuls. Systems **6** (1999), 587–601.

[88]   Carl, S., Heikkilä, S., and Kumpulainen, M., *On a generalized iteration method with applications to fixed point theorems and elliptic systems involving discontinuities*, Nonlinear Anal. **20**, 2 (1993), 157–167.

[89]   Carl, S., Heikkilä, S., and Kumpulainen, M., *On solvability first order discontinuous scalar differential equations*, Nonlinear Times and Digest **2**, 1 (1995), 11–24.

[90]   Carl, S., Heikkilä, S., and Lakshmikantham, V., *Nonlinear elliptic differential inclusions governed by state-dependent subdifferentials*, Nonlinear Anal. **25** (1995), 729–745.

[91]   Chang, K. C., *Free boundary problems and the set-valued mappings*, J. Differential Equations **49** (1983), 1–28.

[92]   Chang, K. C., *Variational methods for non-differentiable functionals and their applications to partial differential equations*, J. Math. Anal. Appl. **80** (1981), 102–129.

[93]   Cherpion, M., De Coster, C., and Habets, P., *Monotone iterative methods for boundary value problems*, Differential Integral Equations **12** (1999), 309–338.

[94]   Chipot, M. and Rodrigues, J. F., *Comparison and stability of solutions to a class of quasilinear parabolic problems*, Proc. Roy. Soc. Edinburgh Sect. A **110** (1988), 275–285.

[95]   Ciecelski, K., *Set Theory for the Working Mathematician*, Cambridge University Press, Cambridge, 1997.

[96]   Clarke, F. H., *Optimization and Nonsmooth Analysis*, SIAM, Philadelphia, 1990.

[97]   Coddington, E. A. and Levinson, N., *Theory of Ordinary Differential equations*, McGraw-Hill, New York, 1955.

[98]   Costa, D. D. and Goncalves, J. V. A., *Critical point theory for non-differentiable functionals and applications*, J. Math. Anal. Appl. **153** (1990), 470–485.

[99]   Dancer, E. N. and Sweers, G., *On the existence of a maximal weak solution for a semilinear elliptic equation*, Differential Integral Equations **2** (1989), 533–540.

[100] De Coster, C., *Pairs of positive solutions for the one-dimensional p-Laplacian*, Nonlinear Anal. **23** (1994), 669–681.

[101] Dem'yanov, F., Stavroulakis, G. E., Polyakova, L. N., and Pana-
       giotopoulos, P. D., *Quasidifferentiability and Nonsmooth Modelling
       in Mechanics, Engineering and Economics*, Kluwer Academic Pub-
       lishers, Dordrecht, 1996.

[102] Deuel, J. and Hess, P., *A criterion for the existence of solutions of
       non-linear elliptic boundary value problems*, Proc. Roy. Soc. Edin-
       burgh Sect. A **74** (1975), 49–54.

[103] Deuel, J. and Hess, P., *Nonlinear parabolic boundary value problems
       with upper and lower solutions*, Israel J. Math. **29** (1978), 92–104.

[104] Dhage, B. C. and Heikkilä, S., *On discontinuous second order bound-
       ary value problems with deviating arguments in ordered Banach spaces*,
       Indian J. Pure Appl. Math. **30**, 8 (1999), 787–800.

[105] Diaz, J. I. and Hernandez, J., *On the existence of a free boundary
       for a class of reaction-diffusion systems*, SIAM J. Math. Anal. **15**
       (1984), 670–685.

[106] Drábek, P., Moudan, Z., and Touzani, A., *Nonlinear homogeneous
       eigenvalue problem in $\mathbb{R}^N$: nonstandard variational approach*, Com-
       ment. Math. Univ. Carolin. **38 (3)** (1997), 421–431.

[107] Dunford, N. and Schwarz, J., *Linear Operators I*, Interscience, New
       York-London, 1958.

[108] Ekeland, I., *On the variational principle*, J. Math. Anal. Appl. **47**
       (1974), 324–353.

[109] Ekeland, I., *Nonconvex minimization problems*, Bull. Amer. Math.
       Soc. **1** (1979), 443–474.

[110] Ekeland, I and Temam, R., *Convex Analysis and Variational Prob-
       lems*, North-Holland Publishing Company, Amsterdam, 1976.

[111] Erbe, L. H., Krawcewicz, W., and Kaczynski, T., *Solvability of two-
       point boundary value problems for systems of nonlinear differential
       equations of the form $y'' = g(t, y, y', y'')$)*, Rocky Mountain J. Math.
       20 (1990), 899–907.

[112] Evans, L. C., *Partial Differential Equations*, vol. 19, AMS, Provi-
       dence, 1998.

[113] Fabry, Ch. and Habets, P., *Upper and lower solutions for second-
       order boundary value problems with nonlinear boundary conditions*,
       Nonlinear Anal. **10** (1986), 985–1007.

[114] Frigon, M. and Kaczynski, T., *Boundary value problems for systems
       of implicit differential equations*, J. Math. Anal. Appl. **179** (1993),
       317–326.

[115] Gao, W. and Wang, J., *On a nonlinear second order periodic bound-
       ary value problem with Carathéodory functions*, Ann. Polon. Math.
       62, 3 (1995), 283–291.

[116] García-Huidobro, M., Manásevich, R., and Zanolin, F., *A Fredholm-
       like result for strongly nonlinear second order ODEs*, J. Differential
       Equations **114** (1994), 132–167.

[117] Gilbarg, D. and Trudinger, N. S., *Elliptic Partial Differential Equa-
       tions of Second Order*, Springer-Verlag, Berlin, 1983.

[118] Gossez, J.-P. and Mustonen, V., *Pseudomonotonicity and the Leray-
       Lions condition*, Differential Integral Equations **6** (1993), 37–45.

[119] Goncalves, J. V. and Alves, C. O., *Existence of positive solutions
       for m-Laplacian equation in $\mathbb{R}^N$ involving critical Sobolev exponents*,
       Nonlinear Anal. **32** (1998), 53–70.

[120] Goncalves, J. V. and Miyagaki, O. H., *Multiple positive solutions for
       semilinear elliptic equations in $\mathbb{R}^N$ involving subcritical exponents*,
       Nonlinear Anal. **32** (1998), 41–51.

[121] Grenon, N., *Existence result for some quasilinear parabolic problems*,
       Ann. Mat. Pura Appl. **165** (1993), 281–313.

[122] Grenon, N., *Asymptotic behaviour for some quasilinear parabolic
       equations*, Nonlinear Anal. **20** (1993), 755–766.

[123] Grossmann, C. and Carl, S., *Boundary value problems with discon-
       tinuities and monotone discretization*, J. Comput. Appl. Math. **51**
       (1994), 293–303.

[124] Guo, Z., *Boundary value problems of a class of quasilinear ordinary
       differential equations*, Differential Integral Equations **6**, 3 (1993),
       705–719.

[125] Hassan, E. R. and Rzymowski, W., *Extremal solutions of a discon-
       tinuous scalar differential equation*, Nonlinear Anal. **37** (1999), 997–
       1017.

[126] Heikkilä, S., *On fixed points through a generalized iteration method
       with applications to differential and integral equations involving dis-
       continuities*, Nonlinear Anal. **14**, 5 (1990), 413–426.

[127] Heikkilä, S., *On first order discontinuous scalar differential equa-
       tions*, Pitman Res. Notes Math. Ser. **324** (1995), Longman, 148–
       155.

[128] Heikkilä, S., *Notes on first order discontinuous differential equations
       with nonlinear boundary conditions*, Proc. Dynam. Systems Appl. **2**
       (1996), 255–260.

[129] Heikkilä, S., *On second order discontinuous scalar boundary value problems*, Nonlinear Stud., **3**, 2 (1996), 153–162.

[130] Heikkilä, S., *On discontinuously perturbed Carathéodory type differential equations*, Nonlinear Anal. **26**, 4 (1996), 775–784.

[131] Heikkilä, S., *On first-order discontinuous differential equations with functional boundary conditions*, Adv. in Nonlinear Dynamics **5** (1997), 273–281.

[132] Heikkilä, S., *Existence results for first order discontinuous differential equations of general form*, Pitman Res. Notes Math. Ser. **374** (1997), Longman, 79–83.

[133] Heikkilä, S., *First order discontinuous implicit differential equations with discontinuous boundary conditions*, Proc. 2nd World Congress of Nonlinear Analysts, Nonlinear Anal. **30**, 3 (1997), 1753–1761.

[134] Heikkilä, S., *On chain methods used in fixed point theory*, Nonlinear Stud. **6**, 2 (1999), 171–180.

[135] Heikkilä, S., *On functional differential equations in ordered Banach spaces*, Proc. Dynam. Systems Appl. **III** (to appear).

[136] Heikkilä, S., *Uniqueness and well-posedness results for first order initial and boundary value problems*, Preprint, Mathematics, Univ. of Oulu (1999), 22 pp.

[137] Heikkilä, S. and Akca, H., *On discontinuous implicit functional differential equations with implicit boundary conditions*, Preprint, Mathematics, Univ. of Oulu (1998), 21 pp.

[138] Heikkilä, S. and Cabada, A., *On first order discontinuous differential equation with nonlinear boundary conditions*, Nonlinear World **3** (1996), 487–503.

[139] Heikkilä, S. and Lakshmikantham, V., *Extension of the method of upper and lower solutions for discontinuous differential equations*, Differential Equations Dynam. Systems **1** (1993), 73–86.

[140] Heikkilä, S. and Lakshmikantham, V., *On the method of upper and lower solutions for discontinuous boundary value problems*, Nonlinear Anal. **23**, 2 (1994), 265–273.

[141] Heikkilä, S. and Lakshmikantham, V., *Monotone Iterative Techniques for Discontinuous Nonlinear Differential Equations*, Marcel Dekker Inc., New York-Basel, 1994.

[142] Heikkilä, S. and Lakshmikantham, V., *A unified theory of first order discontinuous scalar differential equations*, Nonlinear Anal. **26**, 4 (1996), 785–797.

[143] Heikkilä, S., Lakshmikantham, V., and Leela, S., *Applications of monotone techniques to differential equations with discontinuous right hand side*, Differential Integral Equations **1**, 3 (1988), 287–297.

[144] Heikkilä, S., Kumpulainen, M., and Seikkala, S., *Existence, uniqueness and comparison results for a differential equation with discontinuous nonlinearities*, J. Math. Anal. Appl. **201** (1996), 478–488.

[145] Heikkilä, S., Kumpulainen, M., and Seikkala, S., *Uniqueness, comparison and existence results for discontinuous implicit differential equations*, Dynam. Systems Appl. **7**, 2 (1998), 237–244.

[146] Heikkilä, S. and Seikkala, S., *Maximum principles and uniqueness results for phi-Laplacian boundary value problems*, J. Inequalities and Applications (to appear).

[147] Heinonen, J., Kilpeläinen, T., and Martio, O., *Nonlinear Potential Theory of Degenerate Elliptic Equations*, Clarendon Press, Oxford, 1993.

[148] Hokkanen, V. M., *Continuous dependence for an implicit nonlinear equation*, J. Differential Equations **110** (1994), 67–85.

[149] Hokkanen, V. M., *Existence of a periodic solution for implicit nonlinear equations*, Differential Integral Equations **9** (1996), 745–760.

[150] Jankowski, T., *A numerical solution of implicit ODE*, Demonstratio Math. **25** (1992), 279–295.

[151] Jiang, D. and Wang, J., *A generalized periodic boundary value problem for the one-dimensional p-Laplacian*, Ann. Polon. Math., LXV **3** (1997), 265–270.

[152] Jost, J., *Postmodern Analysis*, Springer-Verlag, Berlin/Heidelberg, 1998.

[153] Kaczynski, T., *Implicit differential equations which are not solvable for the highest derivative*, in: Delay Differential Equations and Dynamical Systems, Proceedings, Claremont, 1990 (S. Busenberg and M. Martelli, Eds.) pp. 218–224, Lecture Notes in Math. 1475, Springer-Verlag, New York/Berlin, 1991.

[154] Kaczynski, T. and Krawcewics, W., *A local Hopf bifurcation theorem for a certain class of implicit differential equations*, Canad. Math. Bull. **36** (1993), 183–189.

[155] Kolmogorov, A. N. and Fomin, S. V., *Introductory Real Analysis*, Prentice-Hall, Englewood Cliffs, N.J., 1970.

[156] Kravvaritis, D. and Papageorgiou, N. S., *Extremal periodic solutions for nonlinear parabolic equations with discontinuities*, Rend. Ist. Mat. Univ. Trieste **27** (1995), 117–135.

[157] Kravtsov, P. A., *The Cauchy problem and implicit equations (Russian)*, Partial differential equations, Leningrad Gos. Ped. Inst. Leningrad (1990).

[158] Kravtsov, P. A., *Existence and uniqueness of the solution of the Cauchy problem for an implicit system of differential equations (Russian)*, Differential equations (qualitative theory), Ryazan Gos. Ped. Inst. Ryazan (1990).

[159] Krein, S. G. and Utochkina, E. O., *An implicit canonical equation in a Hilbert space*, Ukrainian Math. J. **42** (1990), 345–347.

[160] Kuiper, H. J., *On positive solutions of nonlinear elliptic eigenvalue problems*, Rend. Circ. Mat. Palermo **20** (1971), 113–138.

[161] Kuiper, H. J., *Eigenvalue problems for noncontinuous operators associated with quasilinear elliptic equations*, Arch. Rational. Mech. Anal. **53** (1973), 178–186.

[162] Kumpulainen, M., *On extremal and unique solutions of discontinuous ordinary differential equations and finite systems of ordinary differential equations*, Dissertation, University of Oulu, Dept. of Math. Sci. (1996).

[163] Kura, T., *The weak supersolution-subsolution method for second order quasilinear elliptic equations*, Hiroshima Math. J. **19** (1989), 1–36.

[164] Köthe, G., *Topological Vector Spaces I*, Springer Verlag, Berlin, Heidelberg, New York, 1969.

[165] Ladde, G. S., Lakshmikantham, V., and Vatsala A. S., *Monotone Iterative Techniques for Nonlinear Differential Equations*, Pitman, Boston, 1985.

[166] Lagler, M. and Volkmann, P., *Über Fixpunktsätze in geordneten Mengen*, Math. Nachr. **185** (1997), 111–114.

[167] Lakshmikantham, V. and Leela, S., *Differential and Integral Inequalities I*, Academic Press, New York-London, 1969.

[168] Lakshmikantham, V. and Leela, S., *Existence and monotone method for periodic solutions of first-order differential equations*, J. Math. Anal. Appl. **91** (1983), 237–243.

[169] Landes, R. and Mustonen, V., *On pseudo-monotone operators and nonlinear noncoercive variational problems on unbounded domains*, Math. Ann. **248** (1980), 241–246.

[170] Le, V. K., *On some equivalent properties of sub-supersolutions in second order quasilinear elliptic equations*, Hiroshima Math. J. **28** (**2**) (1998), 373–380.

[171] Le, V. K., *Subsolution-supersolution method in variational inequalities*, Nonlinear Anal. (to appear).

[172] Le, V. K. and Schmitt, K., *On boundary value problems for degenerate quasilinear elliptic equations and inequalities*, J. Differential Equations **144** (1998), 170–218.

[173] Leray, J. and Lions, J. L., *Quelques résultats de Višik sur des problèmes elliptiques non linéaires par les méthodes de Minty-Browder*, Bull. Soc. Math. France **93** (1965), 97–107.

[174] Lions, J. L., *Quelques Méthodes de Résolutions des Problèmes aux Limites Nonlinéaires*, Dunod, Paris, 1969.

[175] Liz, E., *Abstract monotone iterative techniques and applications to impulsive differential equations*, Dynam. Contin. Discrete Impuls. Systems **30** (1997), 443–452.

[176] Marano, S. A., *On a boundary value problem for the differential equation $f(t, x, x', x'') = 0$*, J. Math. Anal. Appl. **182** (1994), 309–319.

[177] Marano, S. A., *Implicit elliptic differential equations*, Set-Valued Anal. **2** (1994), 545–558.

[178] Marusyak, A. G., *Determination of the periodic solution of systems of first order ordinary differential equations that are not solved with respect to the derivative (Russian)*, Akad. Nauk. Ukrain SSR, Inst. Mat. Kiev (1984), 43–53.

[179] Mawhin, J. and Schmitt, K., *Upper and lower solutions and semilinear second order elliptic equations with nonlinear boundary conditions*, Proc. Roy. Soc. Edinburgh Sect. A **97** (1984), 199–207.

[180] McShane, E. J., *Integration*, Princeton University Press, Princeton, N. J., 1974.

[181] Naniewicz, Z. and Panagiotopoulos, P. D., *Mathematical Theory of Hemivariational Inequalities and Applications*, Marcel Dekker, New York, 1995.

[182] Narici, L. and Beckenstein, E., *Topological Vector Spaces*, Marcel Dekker, New York, 1985.

[183] Niepage, H. D., *On the numerical solution of differential-algebraic equations with discontinuities*, Teubner Stuttgart (1991), 108–116.

[184] Nieto, J. J., *An abstract monotone iterative technique*, Nonlinear Anal. **28**, 12 (1997), 1923–1933.

[185] Okochi, H., *Asymtotic behavior of solutions to certain nonlinear parabolic evolution equations*, Hiroshima Math. J. **22** (1992), 237–257.

[186] O'Regan, D., *Some general principles and results for $(\phi(y'))' = qf(t, y, y')$, $0 < t < 1$*, SIAM J. Math. Anal., **24** (1993), 648–668.

[187] Panagiotopoulos, P. D., *Inequality Problems in Mechanics and Applications. Convex and Nonconvex Energy Functions*, Birkhäuser Verlag, Boston, 1985.

[188] Panagiotopoulos, P. D., *Hemivariational Inequalities and Applications in Mechanics and Engineering*, Springer-Verlag, New York, 1993.

[189] Pao, C. V., *Nonlinear Parabolic and Elliptic Equations*, Plenum Press, New York, 1992.

[190] Pao, C. V., *Nonlinear elliptic boundary-value problems in unbounded domains*, Nonlinear Anal. **18** (1992), 759–774.

[191] Pao, C. V., *Dynamics of reaction diffusion equations with nonlocal boundary conditions*, Q. Appl. Math. **53** (1995), 173–186.

[192] Petryshyn, W. V., *Solvability of various boundary value problems for equation $x'' = f(t, x, x', x'') - y$*, Pacific J. Math. **122** (1986), 169–195.

[193] del Pino, M. D., Felmer, P. L., and Miyagaki, O. H., *Existence of positive bound states of nonlinear Schrödinger equations with saddle-like potential*, Nonlinear Anal. **34** (1998), 979–989.

[194] Pouso, R. L., *Upper and lower solutions for first order discontinuous ordinary differential equations*, Manuscript (submitted) (1999).

[195] Prosenyuk, L. G., *Representation of regular solutions of a real differential equation that is not solved with respect to the derivative (Russian)*, Izv. Vyssh. Uchebn. Zaved. Mat. (1990), 70–72.

[196] Prosenyuk, L. G., *On the existence and asymptotic properties of a real system of differential equations that are not solved with respect to the derivative*, Ukrainian Math. J. **45**, 10 (1993), 1644–1648.

[197] Protter, M. H. and Weinberger, H. F., *Maximum Principles in Differential Equations*, Prentice-Hall, Englewood Cliffs, N.J., 1967.

[198] Puel, J. P., *Some results on quasi-linear elliptic equations*, Pitman Res. Notes Math. Ser. **208** (1989), 306–318.

[199] Rabier, P. J., *Implicit differential equations near a singular point*, J. Math. Anal. Appl. **2** (1989), 425–449.

[200] Rabier, P. J. and Rheinboldt, W. C., *A geometric treatment of implicit algebraic equations*, J. Differential Equations **109** (1994), 110–146.

[201] Rabier, P. J. and Rheinboldt, W. C., *On impasse points of quasilinear differential-algebraic equations*, J. Math. Anal. Appl. **181** (1994), 429–454.

[202] Ricceri, B., *Solutions lipschiziennes d'equations différentielles sous forme implicite*, C. R. Acad. Sci. Paris Ser. I Math. **295**, 3 (1982), 245–248.

[203] Rzymowski, W. and Walachowski, D., *One-dimensional differential equation under weak assumptions*, J. Math. Anal. Appl. **198**, 3 (1996), 657–670.

[204] Saint-Raymond, J., *Equations différentielles sous implicite*, Matematiche (Catania) **2** (1989), 237–257.

[205] Sattinger, D. H., *Monotone methods in nonlinear elliiptic and parabolic boundary value problem*, Indiana Univ. Math. J. **21** (1972), 979–1000.

[206] Seikkala, S. and Heikkilä, S., *On a classical Nicoletti boundary value problem with a discontinuous nonlinearity*, Proceedings of Dynam. Systems Appl. **2** (1996), 501–506.

[207] Seikkala, S. and Heikkilä, S., *Uniqueness, comparison and existence results for discontinuous implicit differential equations*, Proc. 2nd World Congress of Nonlinear Analysts, Nonlinear Anal. **30**, 3 (1997), 1771–1780.

[208] Showalter, R. E., *Monotone Operators in Banach Space and Nonlinear Partial Differential Equations*, vol. 49, AMS, Providence, 1997.

[209] Stanek, S., *On a class of functional boundary value problems for equation $x'' = f(t, x, x', x'', \lambda)$*, Ann. Polon. Math. **59** (1994), 225–237.

[210] Vakhidov, S., *Existence and uniqueness of the solution of a differential equation, which cannot be solved with respect to the derivative, in a normed space (Russian)*, Dokl. Akad. Nauk. Uz SSR **11** (1987), 13–15.

[211] Walter, W., *Differential and Integral Inequalities*, Springer, Berlin, 1970.

[212] Walter, W., *A new approach to minimum and comparison principles for nonlinear ordinary differential operators of second order*, Nonlinear Anal. **25**, 9 (1995), 1071–1078.

[213] Wang, M. X., Cabada, A., and Nieto, J. J., *Monotone method for nonlinear second order periodic boundary value problems with Carathéodory functions*, Ann. Polon. Math. **58** (1993), 221–235.

[214] Wang, J. and Gao, W., *Existence of solutions to boundary value problems for a nonlinear second order equation with weak Carathéodory functions*, Differential Equations Dynam. Systems **5**, 2 (1997), 175–185.

[215] Wang, J., Gao, W., and Lin, Z., *Boundary value problems for general second order equations and similarity of solutions to the Rayleigh problem*, Tôkohu Math. J. **47** (1997), 327–344.

[216] Wilansky, A., *Topology for Analysis*, Ginn, Waltham, Mass., 1970.

[217] Zeidler, E., *Nonlinear Functional Analysis and Its Applications, Vol. I: Fixed-Point Theorems*, Springer-Verlag, Berlin, 1985.

[218] Zeidler, E., *Nonlinear Functional Analysis and Its Applications*, Vols. II A/B, Springer-Verlag, Berlin, 1990.

[219] Zeidler, E., *Nonlinear Functional Analysis and Its Applications, Vol. III: Variational Methods*, Springer-Verlag, New York, 1985.

[220] Zernov, A. E., *The existence and asymptotic behavior of the solution of the Cauchy problem of implicit form (Russian)*, Izv. Vyssh. Uchebn. Zaved. Mat. **1** (1993), 78-81.